ILLUSTRATED GUIDE
TO Wildlife

ILLUSTRATED GUIDE
TO
Wildlife

FROM YOUR BACK DOOR
TO THE GREAT OUTDOORS

Mammals | Birds | Reptiles & Amphibians
Aquatic Life | Insects & Spiders

NATIONAL GEOGRAPHIC
WASHINGTON, D.C.

A Luna Moth, the most common giant silk moth in
the East, rests on the pink blossoms of an azalea.

PRECEDING PAGES: *The Barred Owl, seen here in a snowy
landscape, is one of the most vocal owls in North America.*

CONTENTS

Introduction 6

Purple Sea Stars and barnacles cling to rocks at low tide in southeast Alaska's Sitka Sound.

The Wide World of Wildlife
Honoring Our Animal Neighbors

No matter what territory you explore—urban or rural, wild or developed, temperate or subtropical, land or sea—we humans are vastly outnumbered by other life-forms. Take all the animals in the world (forget plant life—if we count those species, the numbers are astronomical), from tiny mites to giant whales, and the numbers make it clear: This world belongs to them.

The incredible online Encyclopedia of Life (eol.org) estimates nine to ten million animal species—so many we have not even counted. Almost 100,000 mollusk species are known, for instance: clams, oysters, squids, snails, slugs, and on and on. Top category in number is the insects and arachnids, or spiders: Five and a half million species are estimated to exist, and we humans have observed, named, described, and catalogued fewer than 20 percent of them.

All in all, the quest to identify has focused on the creatures with fur or feathers. Of all 5,600 mammal species presumed to exist worldwide, we have records of 5,501; of the estimated 10,500 bird species, we have recorded 10,064. These numbers indicate species, or types of creatures. Consider that a thriving species—earthworms, meadow mice, American robins—might be represented by one hundred, one thousand, or even more individuals per square mile, and you begin to get a sense of how many animals are out there!

What does this mean to everyday observers of wildlife?

First, it means you can simply open your eyes, learn a little about habitat, life cycle, and behaviors, and get to know the animals whose world you inhabit. This book, divided into five chapters—mammals, birds, reptiles and amphibians, insects (plus some arachnids), and aquatic life—gives you a starting point, either to identify those you often see or to learn how to see those you want to identify. Working with experts in each realm, we have whittled down the number of species to a manageable number, roughly 160 in each chapter, selected for their abundance, their interest, or their iconic importance in the natural history of the continental United States and Canada.

Next, it means—or so we hope—that as you notice the wildlife around you, your sharpened awareness strengthens your commitment to preserving this beautiful planet for all creatures on it. Many species overwhelm us in number, but others are dwindling fast. There are North American species of frog, turtle, woodpecker, beetle, salamander, and butterfly (to mention a few) that are facing extinction, as well as the better known mammals in danger of disappearing forever, such as the gray wolf, the wolverine, and the polar bear.

As stewards of this Earth, we are in a position to take conscious steps to protect the planet's biodiversity by respecting efforts to preserve habitats, by planning urban development carefully, and by pitching in to keep soil, air, and water clean. For our children and our grandchildren and their grandchildren, let's do all we can. The wildlife of this world is a legacy we should not squander.

Sharp claws on enormous paws belie the demure pose of a Grizzly Bear at Alaska's Katmai National Park.

1 Mammals

The Summit of Diversity

Our closest mammalian relatives, the nonhuman primates, are not native to the United States and Canada, but kinship pervades the anatomy, physiology, and behaviors of the many mammal species that are. We understand the fierce protectiveness of a mother bear toward her cubs, the driving urge of a Moose to find a mate, and a small rodent's need to "make hay while the sun shines" ahead of a long winter.

■ What Is a Mammal?

Mammals inhabit a wide range of shapes and sizes, from a 170-ton (154-tonne) whale to a bat the size of a bumblebee, but they all share certain characteristics. Mammals form a large class of warm-blooded vertebrates that have hair of some sort at some time of life (even whales and dolphins), three middle-ear bones, a neocortex in the brain, and mammary glands in the females that produce milk to nourish young. Most mammals bear young sustained by a placenta during pregnancy. Monotremes, an ancient order, are one exception: Their offspring develop in the leathery eggs they lay that, upon hatching, are nursed from pores on the mother's belly. These unique mammals—the Duck-billed Platypus and several species of echidna, or "spiny anteater"—are absent from the New World. Another exception is the marsupial, which gives birth to underdeveloped young that finish their development in a maternal pouch where nursing takes place. The Virginia Opossum is the only marsupial native to our area.

Placental mammals produce two kinds of young: precocial and altricial. Precocial young are basically miniatures of their parents—furred and functional and ready to move with adults on only several hours' or days' notice. Altricial young enter the world naked, often pink and wrinkled, with eyes closed shut. They need time to reach adulthood; that can be a matter of weeks in some rodents, or a matter of many years in primates.

■ Mammals in Your World

North America north of Mexico, the scope of this chapter, enjoys a wealth of mammalian species, some 474 representing 12 different orders, or related groups of mammals. They are scattered across the continent, inhabiting every ecological niche from Arctic tundra to subtropical swamp. Depending on where you live, you might see a few or a lot of mammals on a daily basis. Most people probably see a lot of only a few species. Ubiquitous mammals, such as tree and ground squirrels, are part of the background of urban and rural life in many parts of the United States and Canada. Other species are seen on occasion—the Raccoon we startle when we take out the trash, or the fox glimpsed in an arcing pounce in a grassy field. But looking for and recognizing the mammals that we don't normally see requires guidance.

Species at the larger end of the mammalian scale are easier to identify: A Grizzly Bear, a Mountain Goat, and a bull Moose are virtually unmistakable. At the smaller end, things become more difficult. Among the rodents, some species are so similar—in the genus *Peromyscus*, the deermice, for example—that you'd need to see their dental records (or intact skull specimens) to tell them apart. Even scientists find this very challenging.

■ Identifying Mammals

This chapter will help you learn about mammals and how to identify them, as well as provide the potential to observe them based on their range, habitats, and

habits. The species accounts in this chapter introduce 160 mammal species found in North America north of Mexico. It is a small but representative sample of the total number of species. These accounts include species from all the major mammalian groups except those that live their lives largely or exclusively in the water, such as whales, seals, manatees, and sea otters. These are found in Chapter 4: Aquatic Life.

The chapter sections present the species in groups arranged according to a current understanding of taxonomic classification, an understanding that is under constant revision due to the advances of molecular biology and that is not without its controversies. Scientific names provided generally follow Wilson and Reeder's *Mammal Species of the World*, 3rd edition (2005), and the common names that go with them are those with wide recognition. Within each genus presented, the species are

MAMMALS IN MOTION

Mammalian limbs display a wide range of adaptations for locomotion and other activities. Bat wings, seal flippers, and beaver webbed hind feet allow mastery of air or water. Strong, articulate feet aid tree climbing and object manipulation in opossums and Raccoons. Long hind feet help hares and kangaroo rats escape predators in elusive bounds. Prairie dogs and moles use limbs to excavate homes and tunnels. Bears amble on large paws equipped with lethal claws, whereas Mountain Lions keep their feline claws retracted while moving. Horses and other hoofed mammals need stable support for all-terrain travel and long hours of grazing and browsing. Mountain Goat hooves act like suction cups on rocky heights.

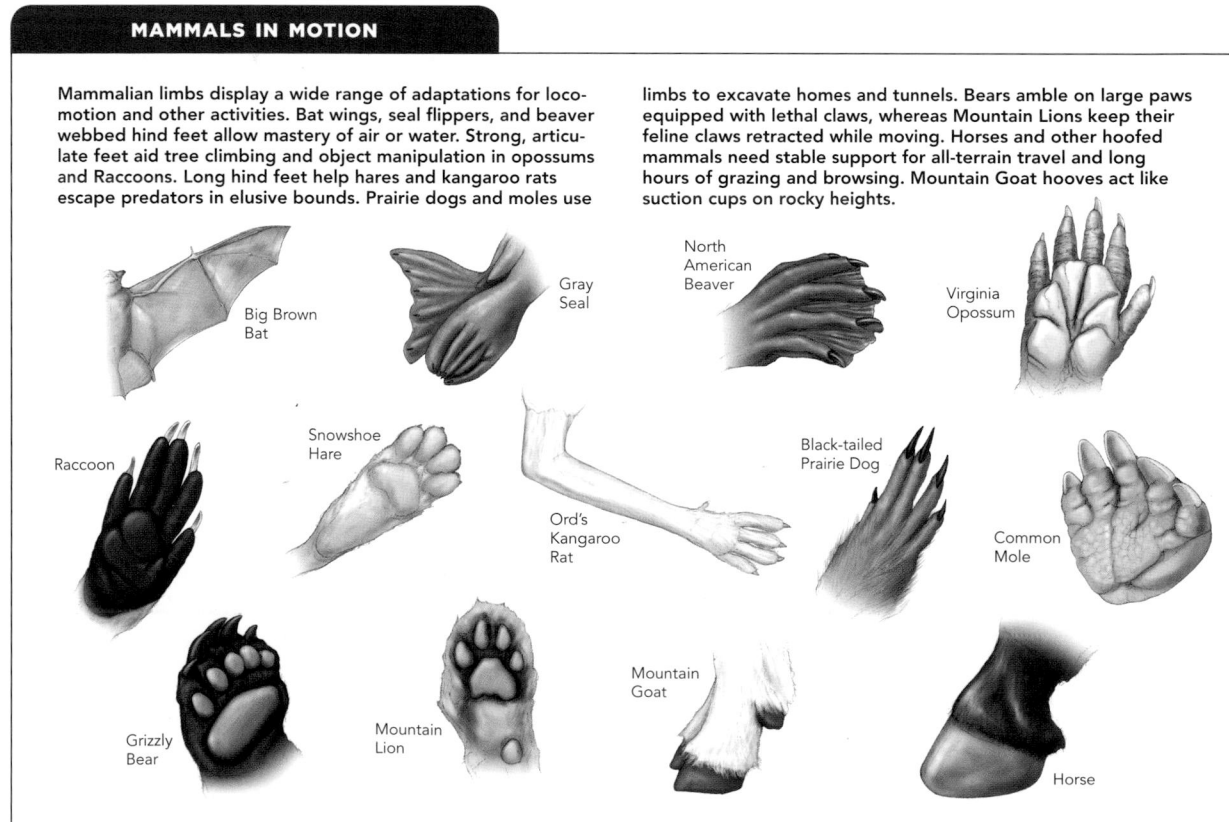

Big Brown Bat

Gray Seal

North American Beaver

Virginia Opossum

Raccoon

Snowshoe Hare

Ord's Kangaroo Rat

Black-tailed Prairie Dog

Common Mole

Grizzly Bear

Mountain Lion

Mountain Goat

Horse

Stiletto-horned male Mountain Goats posture in a duel for dominance in Utah.

listed in alphabetical order of their names, as not all species have been investigated equally and their interrelationships are imperfectly known. The measurements include the length of an animal's head, body, and tail. Height at shoulder is the standard measurement shown for the hoofed mammals. Each chapter also offers a set of characteristics that will help you identify a species and compare it with others. The species accounts also provide details about the animal's habitat, range, and diet.

■ On the Right Track

Many mammals lead lives largely hidden from our view—the vast majority are nocturnal or crepuscular (active at dawn and dusk)—but we know they are there by the signs that they leave. Signs can range from their tracks, to their droppings (known as scat), to their communication methods, to leftovers from meals, to engineering projects such as a beaver dam. All of these activities leave clues that become easy to spot if you know where and how to look.

If all the beasts were gone, men would die from a great loneliness of spirit, for whatever happens to the beasts also happens to the man.
—CHIEF SEATTLE

Tracks point us in the direction of a mammal's presence, whether they are deposited in mud, snow, or as wet marks left on a concrete sidewalk. Learning about mammalian appendages, how mammals walk, run, or hop, and the patterns they make as they move aid identification. Each account in this chapter includes the tracks made by the species; wing shape is shown for bats.

Whole books have been written on scat. Different groups of mammals have different sizes and shapes of scat with different contents. The small, regular pellets of rabbits and hares bear little resemblance to the large, lumpy plops of a Moose. Scat of the same species can show seasonal differences: the Red Fox's looser, berry-rich summer droppings look very different from the fur-matted rope of its winter scat. How scat is deposited also provides clues. Members of the cat family, for example, usually scrape the substrate before defecating, so you may find their scat in an excavated dent. Foxes often defecate in the middle of a trail or at a trail crossing. Among scat detectives, where and how is as important as what.

Many mammals leave calling cards in the form of gnawing marks and scrapes on tree trunks. These suggest feeding behaviors, construction techniques, or communication methods used to demarcate territory. Other signs to be aware of include small, flattened areas in long grass that could be the forms, or nest depressions, of rabbits, or shallow, oval pits of many sizes in dust or sand that could be evidence of wallows or dust baths of species from bison to mice. Tracks and signs form a whole category of wildlife knowledge, and many excellent books can provide a comprehensive overview.

Young Raccoons inspect the world outside their tree-cavity den.

■ Mammals in Perspective

The mammalian landscape of North America has changed greatly from the last ice age, when mammals such as mammoths, mastodons, and saber-toothed cats roamed much of the continent. Theories of the demise of these megafauna implicate climate change, habitat change, pathogens, and human impact. Closer to our time, human enterprise nearly eliminated one of the continent's most spectacular mammals, the American Bison. In the 19th century, these magnificent creatures still populated large swaths of the Great Plains, hunted sustainably for the most part by native peoples for centuries. But the buffalo got in the way of westward expansion by human newcomers, and financial gain was a strong incentive to hunt them extensively. In only 40 years, tens of millions were reduced to a thousand.

It is still possible to see great assemblages of mammals on our continent. In the Arctic, hundreds of thousands of caribou travel together for hundreds of miles to seasonal feeding grounds. Along the way, they pass under portions of oil pipeline elevated for their accommodation. At dusk in the summer months, a seemingly endless cloud of more than a million free-tailed bats emerges from under a bridge in Austin, Texas, to begin their nightly assault on swarms of flying insects.

As climate change and habitat disruption cause mammals to move into areas new to them, some species are appearing in places you might not expect, such as the Coyote in the environs of the United States capital. That icon of the Southwest may have followed burgeoning populations of deer and rabbits as it drifted closer to Washington, D.C. And while many people warm to the idea of more wildlife in the cities, its presence there indicates that we have failed to maintain natural habitats. Not all migrations are successful for the migrants or the native and long-settled species—whether animals or plants—as the perspective of time has made abundantly clear.

MAMMALS

||||||||||||||||||||||||||||||||||||||||||||||||||||||||||||||||||||

Virginia Opossum
Didelphis virginiana L 24–30 in (60–75 cm)

The only marsupial native to the United States and Canada, the Virginia Opossum lacks the cuddle factor of its Koala cousin, but it adapts well to its increasingly suburban and urban lifestyle.

KEY FACTS

Body is grayish, with long guard hairs; nose and clawed toes pink; naked prehensile tail.

+ habitat: Forests, old fields, suburban and urban areas

+ range: Central and eastern U.S. to south-eastern Canada; introduced elsewhere

+ food: Grasses, nuts, fruits, insects, snakes, and carrion

The opossum's famous play-dead routine is very convincing: A motionless body, vacant stare, and lolling tongue fool predators such as foxes, Bobcats, and Coyotes. This nocturnal species climbs agilely using the opposable thumbs of its hind feet, but waddles clumsily along the ground. It inhabits holes, crevices, and other animal dens. Twice a year an adult female may give birth to up to 20 bee-size, blind, naked newborns that crawl to her fur-lined pouch. Mortality is high. The young nurse in the pouch for about three months and ride on the mother's back when outside. Opossums do not hibernate, but become less active in winter.

Nine-banded Armadillo/Common Long-nosed Armadillo
Dasyapus novemcintus L 25–32 in (64–81 cm)

A member of an exclusive group of living mammals with bony shells, the Nine-banded Armadillo is the only armadillo species found in the United States.

KEY FACTS

Scaly body has 8 to 9 overlapping, movable bands on back; underside is furred; tail is armored.

+ habitat: Fields, woodlands, brushy areas, and roadsides

+ range: Southeastern United States; range is expanding

+ food: Insects, spiders, other invertebrates, small vertebrates, and carrion

Though it looks encumbered by its unique, scaly armor, the armadillo is flexible—and adaptable. It runs and dodges with a stiff-legged gait and can burrow quickly to escape. It can hold its breath for six minutes to cross a riverbed, weighted down by its shell, or can inflate by swallowing air to float across. Within a burrow or an aboveground nest, a female bears one litter a year of four identical offspring. Armadillo eyes don't shine in the dark, probably increasing the species' mortality rate along southern roads. Armadillos also jump straight up when startled, causing crushed backs when a car straddles a road-crossing armadillo.

Sewellel/Mountain Beaver

Aplodontia rufa L 14 in (35 cm)

Previously called the Mountain Beaver, this primitive rodent is not a beaver, but is the only member of its genus and family, the Aplodontidae. Its lineage extends 50 million years to the Eocene epoch.

KEY FACTS

Body above is dark brown or blackish; underside is lighter; white spot at base of ear; very short tail.

+ **habitat:** Thickets, meadows, and forests near water

+ **range:** Southern British Columbia to California

+ **food:** Needles and twigs of conifers and herbaceous plants, including ferns

The elusive and mainly nocturnal Sewellel looks like a Muskrat with a stubby tail. It digs tunnels that can be several hundred feet long with multiple entrances. Piles of dirt often mark the entrances, which frequently are close to water. The Sewellel also climbs trees in search of tasty twigs and needles, ascending 20 feet (6 m) up Lodgepole Pines and descending by using stubs of cut branches as ladder rungs. Sewellel females are late breeders; they have a first litter of between one and six young at about two years old. The common name likely comes from a name for fur robes that Native Americans of the Pacific Northwest wore.

Tassel-eared Squirrel

Sciurus aberti L 18 in (46 cm)

The subspecies of Tassel-eared Squirrel known as Abert's Squirrel lives south of the Grand Canyon, and the Kaibab subspecies lives north of it. Brown and black melanistic forms also occur.

KEY FACTS

Body is gray, with reddish sides and white (Abert's) or black (Kaibab) underside.

+ **habitat:** Ponderosa Pine and other coniferous or mixed forests

+ **range:** Southwestern Wyoming, Colorado, New Mexico, Arizona, and Utah

+ **food:** Pine nuts, buds, bark, berries, fungi, bones, and carrion

The Tassel-eared Squirrel is known for the long ear tufts that are especially noticeable in winter. It also sports a lush, shortish tail that is gray on top and white elsewhere in the Abert's and all white in the Kaibab subspecies. This diurnal squirrel has a close association with the Ponderosa Pine, husking pinecones for seeds and eating the inner bark of twigs that it cuts. A pile of denuded twigs under a pine is a classic sign of its presence. It builds twig nests in trees where the adult female raises a brood of two to four young. Mating is preceded by a chase involving a dominant male and his subordinates.

Eastern Gray Squirrel

Sciurus carolinensis L 18.5–26.5 in (47–67 cm)

The Eastern Gray Squirrel swishes its versatile bushy tail in courtship, flicks it in greeting, fluffs it in anger, uses it as an umbrella when it rains, and curls it around its body in the cold.

KEY FACTS

Body is gray or brownish; tail is edged in white; melanistic animals are dark brown or black.

+ **habitat:** Hardwood and mixed forests, suburban and urban areas

+ **range:** Eastern U.S. to southern Canada; introduced in West

+ **food:** Acorns, other nuts, seeds, buds, fruit, fungi, insects, and birds

Eastern Grays build leafy nests in trees, known as dreys, which are visible among bare branches in winter. Some are basic sleeping platforms; others are domed and secured for winter living. This diurnal species usually lives solitarily, but individuals may bed down together in winter while remaining active in the daytime. They breed twice a year; courtship involves spirited chasing, and a female may mate with more than one male. Litters average two to four young. Buried acorns are retrieved randomly by their odor. The town of Olney, Illinois, harbors a long-established, protected population of albino, or white, *S. carolinensis.*

Western Gray Squirrel

Sciurus griseus L 22 in (56 cm)

Western Gray Squirrels look a lot like their eastern counterparts, but are bigger and grayer. They're also shyer and more secretive, and haven't taken in a big way to urban life.

KEY FACTS

Body is gray; eyes are ringed in white; underside is white; bushy dark gray tail is edged in white.

+ **habitat:** Mainly oak, coniferous, and mixed forests

+ **range:** Southern California to northern Washington

+ **food:** Acorns, pine nuts, berries, vegetation, fungi, bark, insects, birds, and eggs

These squirrels have a preference for pine nuts. They clip cones from trees and strip them of their seeds on the ground. Western Grays may rest together in a twig nest on a tree branch, known as a drey, but otherwise adults mingle only during mating time. From two to six young are often born in a tree cavity, but heavy flea infestation can cause the mother to move them to a drey. The species sometimes raids almond and walnut orchards, with economic consequences to growers. It competes for food and habitat with the more adaptable and gregarious Eastern Gray Squirrel, which has been introduced to parts of the range.

Eastern Fox Squirrel

Sciurus niger L 20–24 in (50–60 cm)

When 18th-century taxonomist Linnaeus described the Fox Squirrel he dubbed it *niger*, believing that the black form represented the species. The most widespread form is largely orangish brown.

KEY FACTS

Colors are variable, often yellowish brown with orange cheeks, eye rings, and underside.

+ **habitat:** Open woodlands, golf courses, and suburbs

+ **range:** Eastern and central U.S., southern Canada; introduced elsewhere

+ **food:** Acorns, other nuts, seeds, buds, roots, fruits, bulbs, insects, and eggs

The largest squirrel species in the East, the Eastern Fox Squirrel weighs up to 3 pounds (1.4 kg), and it has a bushy tail that is as long as its body. Although it primarily caches and eats acorns, it is an eclectic feeder, relishing maple sap and corn on the cob. Populations have been severely reduced in the Northeast due to hunting and forest clearing. The Delmarva Fox Squirrel, a gray subspecies, is designated as endangered. Melanistic, or black, populations are found mainly in the South. The Eastern Fox Squirrel in general has extended its range and been introduced successfully in the West.

Douglas Squirrel

Tamiasciurus douglasii L 12 in (30 cm)

A resident of Pacific coast forests, the small Douglas Squirrel shares a bit of range and a number of habits, including general noisiness, with the American Red Squirrel.

KEY FACTS

Back is brown; a black stripe separates it from the orange underside. The ears have black tufts in winter.

+ **habitat:** Coniferous and mixed forests

+ **range:** Southern California to British Columbia

+ **food:** Seeds, fungi, flowers, insects, young birds, eggs, and bones

The Douglas Squirrel is known for its wide array of vocal stylings. Naturalist John Muir called it the "mockingbird of squirrels . . . barking like a dog, screaming like a hawk, chirping like a blackbird or a sparrow." The species is a cone clipper, harvesting cones from species such as fir, pine, spruce, and hemlock. Most cones are cached in a midden hoard for winter. Named for Scottish botanist David Douglas, namesake of the fir, this species nests in tree holes lined with shredded bark. A litter of four to six is born in early summer; females may have an additional litter in the fall, depending on location.

American Red Squirrel

Tamiasciurus hudsonicus L 12 in (30 cm)

The American Red Squirrel chatters loudly as a matter of routine and will scold intruders into its territory. Ojibwa Indians called the species *ajidamo*, roughly meaning "tail in air."

KEY FACTS

Body is russet above; underside is white; white eye rings; grizzled orange tail.

+ habitat: Coniferous and mixed forests, parkland, and hedgerows

+ range: Alaska and Canada into northeastern U.S., Rockies, and Appalachians

+ food: Pine nuts, berries, fungi, insects, birds, eggs, and sap

The American Red Squirrel is much less social than the Eastern Gray Squirrel and is not very tolerant of others of its own species. Its feeding habits leave a telltale sign of its presence. It clips cones from conifers and stores them to ripen underground, under rocks, or in holes until it is time to take them to a feeding station on a branch and systematically strip the seeds. The dropped scales form a midden underneath, up to 3 feet (0.9 m) high and perhaps 30 feet (9 m) long. Once a year—or twice, depending on location—males enter a breeding female's territory to mate. She will bear a litter of three to five young.

Northern Flying Squirrel

Glaucomys sabrinus L 12 in (30 cm)

Flying Squirrel is a misnomer, because it doesn't fly but glides from tree to tree by stretching out a membrane between front and hind feet. The species can cover up to 300 feet (90 m) in a glide.

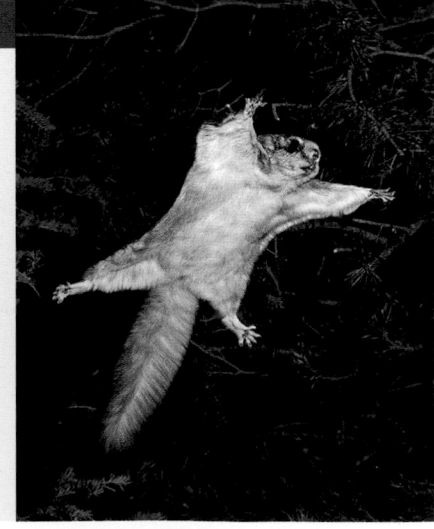

KEY FACTS

Body is brownish; membrane edge is black; tail is flat. A Pacific Northwest subspecies is darker.

+ habitat: Coniferous and mixed woodlands

+ range: Alaska through Canada into northern U.S., and southward in western and eastern mountains

+ food: Nuts, seeds, sap, fungi, lichens, birds, and eggs

Larger and heavier than its southern counterpart, the Northern Flying Squirrel shares some parts of its range with the Southern Flying Squirrel. Like the Southern, the Northern is nocturnal, coming out in the evening to partake largely of the offerings found on a tree, sometimes including birds' eggs and nestlings. This species is generally quieter in voice and less anxious in its actions than the southern one. It nests in tree cavities and builds its own nest, using twigs and leaves, lining it with finer materials, and sometimes enclosing a bird's nest. A female has one litter of two to four young in late spring.

Southern Flying Squirrel

Glaucomys volans L 9 in (23 cm)

The Southern Flying Squirrel peers from a tree hole, planning a route. It glides with membranes stretched, shifting its body to steer, then pulls up for a four-point landing on its destination tree.

KEY FACTS

Body is pale grayish brown; cheeks and underside white; membrane edge black.

+ habitat: Oak-hickory and mixed woodlands, suburban and urban areas

+ range: Eastern half of U.S. into southern Canada

+ food: Nuts, seeds, flowers, berries, bark, fungi, lichens, insects, birds, and birds' eggs

The Southern Flying Squirrel is smaller and paler than the Northern Flying Squirrel, but the two species have similar habits. The furred membrane between front and rear legs gives this species some 50 square inches (323 sq cm) with which to glide. The membrane hangs in accordion folds when not in use. Flying squirrels nest in tree cavities and use nest boxes placed in trees, which they will line with materials such as fur and feathers. Females have litters of three to five young after a 40-day gestation; those in the South may have two litters a year. These social squirrels may nest in large groups in winter.

Harris's Antelope Squirrel

Ammospermophilus harrisii L 9.5 in (24 cm)

Antelope Squirrels look somewhat like chipmunks, but lack the face stripes. The genus name of these desert species, *Ammospermophilus,* means "lover of sand and seeds."

KEY FACTS

Body is grayish brown above, with a white stripe on the sides; eyes are circled in white; tail is uniformly grizzled.

+ habitat: Desert areas with cacti and desert shrubs

+ range: Arizona and southwestern New Mexico

+ food: Cactus seeds and fruit, yucca seeds, mesquite beans, and insects

Harris's Antelope Squirrel can easily navigate prickly cacti of its desert home. It can climb to the top of a Cholla cactus without getting spines in the soft pads of its feet. It may even sit on the Cholla's summit to get a good look around. The species is diurnal, even in the heat, alternating bursts of activity with rest periods. Its burrow is frequently found at the base of a desert bush. It does not hibernate, but remains active year-round, and its body weight does not fluctuate seasonally. Breeding season may start as early as December; a litter of five to nine young is born after a one-month gestation.

White-tailed Antelope Squirrel

Ammospermophilus leucurus L 9 in (23 cm)

The White-tailed Antelope Squirrel folds its two-tone tail over its back to create shade, one of the adaptive strategies this desert species displays.

KEY FACTS

Back is grayish; side has narrow white stripe; tail is grizzled on top.

+ **habitat:** Desert shrubland, sandy and rocky areas

+ **range:** Southern Oregon and Idaho to southern California, northern Arizona, and northwestern New Mexico

+ **food:** Vegetation, seeds, insects, and other invertebrates

White-tailed Antelope Squirrels have a number of physiological adaptations to their dry desert environment, including several that conserve moisture, as they depend on drinking water. Active during the day, they sprint from one shady area to another to forage. Individual White-tails use a number of different burrows, often the abandoned burrows of kangaroo rats. This species does not hibernate, but grows a warm coat in cold areas and may refrain from activity during bad weather. A group may huddle in a burrow to stay warm. White-tails observe a dominance hierarchy rather than squabble over territories.

White-tailed Prairie Dog

Cynomys leucurus L 14 in (35 cm)

This mountain species leads a much less social life than the Black-tailed Prairie Dog. It lives in smaller colonies, and it ranges as high as 12,000 feet (3,700 m).

KEY FACTS

Back is grayish yellow; sides are yellowish; underside is buff-orange; tail has a white tip.

+ **habitat:** Grasslands and shrublands

+ **range:** Southern Montana, Wyoming, northeastern Utah, and northwestern Colorado

+ **food:** Grasses, sedges, other herbaceous plants, and woody plants

The White-tailed Prairie Dog builds complex underground chambers in its mountain habitat. This species does not require extensive clearings as the more vulnerable plains prairie dogs do for keeping a vigilant eye on the landscape. These animals spend the summer feeding voraciously, preparing for the long sleep of hibernation. By August or September many of the adults have bedded down underground; juveniles may linger above until heavy snows arrive. Four to six pups are born in May. White-tails have been targeted for extermination and fall prey to eagles, hawks, Bobcats, and badgers. They also are susceptible to plague.

Black-tailed Prairie Dog
Cynomys ludovicianus L 16 in (40 cm)

Millions of Black-tailed Prairie Dogs once excavated towns from southern Canada to the Rio Grande. Reduced in numbers, they have the most socially complex ground squirrel society.

KEY FACTS

The back is orangish brown; the sides and belly are paler; the narrow tail is tipped in black.

+ **habitat:** Short-grass prairies and other grasslands

+ **range:** Southern Saskatchewan to southwestern Texas

+ **food:** Roots and vegetation of grasses and other herbaceous plants

Prairie dog towns are vast constructions consisting of tunnels, specialized chambers, and multiple entrances, which serve as lookout points. Towns are divided into sectors containing one adult male and associated females and young; these groups hold off encroaching neighbors. Prairie dogs communicate with barks, chuckles, and a long *weee-oooo* whistle emitted during a "jump-yip," the species' signature move that may say "all's clear." Ranchers and farmers tended to despise prairie dogs for eating grasses and making town entrance holes that hobbled horses. Targeted elimination and habitat loss greatly reduced their numbers.

Hoary Marmot
Marmota caligata L 30 in (75 cm)

French-Canadian trappers called the Hoary Marmot *le siffleur*—the whistler—for the shrill warning notes that resonate in mountain meadows. Its vocal repertoire also includes barks and hisses.

KEY FACTS

Front half of body is whitish, rear half is yellowish or reddish brown; black on snout and crown.

+ **habitat:** Mountain meadows, rocky areas, and talus slopes

+ **range:** Alaska and northwestern Canada into Washington, Idaho, and Montana

+ **food:** Grasses, sedges, and other herbaceous plants

Our largest marmot species, the Hoary Marmot prefers to make its burrow on a warm, south-facing slope, often under a boulder. The chunky animal may hibernate up to nine months of the year and emerge in late spring ready to mate. Adult males take on and defend several females, and they all live in a group with their offspring. In the tolerant Hoary Marmot family, the young are not expelled as they are in many other rodent species, and may remain with the group for two years. Females breed every other year, giving birth to two to four young. The species name, *caligata*, refers to the black boots Roman soldiers wore.

Yellow-bellied Marmot/Rockchuck

Marmota flaviventris L 23 in (58 cm)

Despite the name Yellow-bellied, this marmot is anything but timid. It is noted for its raucous vocalizations, including whistles, screams, and tooth chattering.

KEY FACTS

Back is grayish brown; underside and lower legs are yellowish to bright orange; tail is reddish to dark brown; sides of neck have buff patches.

+ **habitat:** Mountain meadows in rocky areas

+ **range:** Western U.S. into southern Canada

+ **food:** Mainly flowers, grasses, other plants, and seeds

The Yellow-bellied Marmot ranges to 12,000 feet (3,700 m). Home is often a burrow in a rocky area with a nearby lookout boulder. Males often take multiple mates—one had a record of 31—and live with them and their offspring, born in litters of three to eight about a month after mating. The marmots spend the summer filling up in alpine meadows before hibernating for as long as nine months in cold climates. Those in warmer areas may estivate, or spend time in torpor, during hot summers. Some marmots at national and state parks have turned into panhandlers, sitting on their ample backsides on rock perches and playing to the tourists.

Groundhog/Woodchuck

Marmota monax L 24 in (60 cm)

Punxsutawney Phil is Pennsylvania's famous weather prognosticator. Many Groudhogs, aka Woodchucks, will hibernate in burrows—if left undisturbed—well past February 2, Groundhog Day.

KEY FACTS

Back is grizzled grayish or brownish; nose, eyes, and ears are located on the same plane.

+ **habitat:** Fields, woodland edges, stream banks, and roadsides

+ **range:** Eastern U.S. and Canada through central Canada into Alaska

+ **food:** Grasses and forbs, insects, and crops such as clover and alfalfa

The Groundhog, a kind of ground squirrel known as a marmot, is a solitary creature. Sometimes an early morning or late afternoon drive down a highway bounded by a grassy verge will reveal single Groundhogs at almost regular intervals, foraging among the many kinds of plants they enjoy. They eat heartily in summer to put on a layer of fat that will maintain them through hibernation, a remarkable physiological adaptation. The body temperature of a hibernating Woodchuck may dip to 35°F (2°C), and its heart may slow to four beats a minute. Woodchucks are champion tunnel diggers and can burrow out of sight in a minute.

Columbian Ground Squirrel

Spermophilus columbianus L 13 in (33 cm)

Both males and females of the highly social Columbian Ground Squirrel defend real estate within their home ranges. To secure their young, nesting females plug up their nest chambers at night.

KEY FACTS

Back is grizzled brownish; underside, legs, and top of snout are orange; eye rings are white.

+ **habitat:** Wet meadows and grasslands, clear-cut areas, and rangeland

+ **range:** Mountains of northwestern U.S. into British Columbia and Alberta

+ **food:** Flowers, seeds, bulbs, fruit, and vegetation

Group members in this very territorial species greet each other by "kissing" when they meet. They also scent-mark liberally, using multiple scent glands on the head and body to mark their turf. These animals cram all activity into about a hundred days before heading underground to hibernate. A great deal of care goes into preparing the hibernation den. Usually it is a carefully lined and domed chamber that incorporates a drain to help keep it dry. Young tend to hibernate with their mothers. Adult males rely on a store of food when they emerge ahead of the females, if food aboveground is scarce.

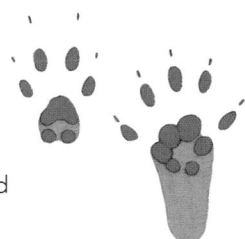

Franklin's Ground Squirrel

Spermophilus franklinii L 14 in (35 cm)

At a glance, this species resembles the Eastern Gray Squirrel, but look closely and you'll see that the ears are much smaller and the tail is thinner and less bushy.

KEY FACTS

Back is grizzled grayish brown or yellowish; underside is grayish brown or white; tail is narrow and bushy.

+ **habitat:** Grasslands, thickets, woodland and marsh edges, and embankments

+ **range:** North-central U.S. to south-central Canada

+ **food:** Vegetation, seeds, insects, birds, eggs, toads, and carrion

Franklin's Ground Squirrel keeps a low profile in the tall grasses and dense vegetation it favors. It digs extensive burrows and tends to vanish underground at the first sign of trouble. These squirrels live solitarily or in loose aggregations. Franklin's Ground Squirrel partakes of a more eclectic—and carnivorous—diet than many other ground squirrels. The diet includes duck eggs from nests in prairie potholes. Much of the species' preferred habitat of tall-grass prairie has disappeared due to agriculture and urbanization, and it is listed as threatened or endangered in a number of states within its range.

Arctic Ground Squirrel
Spermophilus parryii L 14 in (35 cm)

Orange fur is no liability for the Arctic Ground Squirrel. Hibernating in the snowy months, it needs no camouflage. Different calls signal a warning about aerial or terrestrial predators.

KEY FACTS

The back is grizzled reddish brown; the snout, underside, and limbs are a vibrant orange-brown.

+ habitat: Tundra, alpine meadows, and sandy hills

+ range: Alaska and northwestern Canada

+ food: Seeds, berries, grasses, leaves, fungi, insects, and carrion

Location is everything for the Arctic Ground Squirrel in search of a construction site. It can build its labyrinthine burrows only where at least 3 feet (0.9 m) of unfrozen surface sit above the permafrost. It crafts specialized burrows for different purposes, whether resting, hibernating, or raising young. A nesting burrow will be lined with soft grasses and perhaps caribou fur. This ground squirrel fattens up in summer—and gnaws fallen caribou antlers for minerals—before it beds down for about six months. It awakens in April or May and may have to burrow through several feet of snow to look for a mate.

California Ground Squirrel
Otospermophilus beecheyi L 16 in (40 cm)

Some California Ground Squirrels lead solitary lives in underground burrows. Others may live in connected colonies of up to a dozen adults, accessed by separate entrances.

KEY FACTS

Back is orangish brown grizzled with cream; white mantle on shoulders; underside and legs white to orangish.

+ habitat: Fields, pastures, cropland, semiarid areas

+ range: Southern Washington and western Oregon to California

+ food: Nuts, seeds, fruit, fungi, birds, insects, and garden vegetables

The California Ground Squirrel may spend up to two-thirds of the year underground, hibernating in winter and estivating in the summer. It emerges in the spring, enlarging its burrow and refurbishing it with fresh grasses. Males and females come together briefly to mate, and after a month, litters of three to nine young are born, which may remain active through the year. The species often feeds on convenient agricultural crops and vegetable plots, earning growers' wrath and making it a target for elimination. It also avails itself of roadside garbage bins. This mammal is a vector for plague and tularemia.

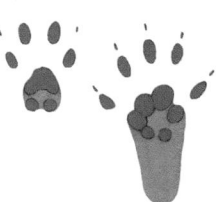

Rock Squirrel

Otospermophilus variegates L 20 in (50 cm)

A bushy tail helps to give this species the appearance of a tree squirrel, but the Rock Squirrel usually lives underground. It sometimes climbs trees, though, and occasionally nests in them.

KEY FACTS

Its colors vary, often grizzled brown; the underside is whitish; the tail is long and bushy.

+ **habitat:** Canyons, cliffs, and other rocky areas; bridges and stone walls

+ **range:** Southwestern United States

+ **food:** Nuts, seeds, fruit, vegetation, cacti, and invertebrates

Rock Squirrels generally excavate their burrows at a shallower level than those of other ground squirrels, and the burrows also tend to be less extensive. An animal may have a home burrow and another used on feeding forays. The species is social; a dominant male usually is associated with a number of adult females and their young. Mating takes place after hibernation. Females usually have one litter of three to nine young a year in colder areas, and perhaps two litters in warmer habitats with shorter winters. Burrowing Owls and other species that need underground quarters frequently use abandoned Rock Squirrel burrows.

Spotted Ground Squirrel

Xerospermophilus spilosoma L 9 in (23 cm)

Small, shy, and secretive, the Spotted Ground Squirrel is a homebody, spending much time in its burrow when the weather is hot or cold, and when there is any kind of commotion aboveground.

KEY FACTS

The back is brownish with small pale spots; underside is whitish; the long, narrow tail has a bushy tip.

+ **habitat:** Grasslands, scrublands, and deserts

+ **range:** South Dakota and eastern Wyoming to the southwestern U.S.

+ **food:** Grasses, forbs, cacti, seeds, insects, and small vertebrates

The Spotted Ground Squirrel prefers a habitat with sandy soil that sticks together. Its burrow system can be 15 feet (4.5 m) or more in length with entrances often located under a bush. The species tends to be active in the cooler parts of the day—in the morning and evening—retreating to its burrow during the hottest period. In the colder parts of the range, the species hibernates as early as late July or August. Young of the year tend to linger above into the fall. These squirrels remain underground until April or May. Young are born in the spring in litters of three to five after a gestation of about four weeks.

Thirteen-lined Ground Squirrel

Ictidomys tridecemlineatus L 10 in (25 cm)

The precise patterns of solid lines and dashes on the back of the Thirteen-lined Ground Squirrel make it unmistakable at any age, except at birth. Females produce litters of up to 14 pink babies.

KEY FACTS

Brown body has cream stripes and dashes; cheeks and sides are pale; underside is white.

+ **habitat:** Pastures, meadows, agricultural land, golf courses, and roadsides

+ **range:** Central U.S. and south-central Canada

+ **food:** Seeds, roots, nuts, insects, worms, small vertebrates, and carrion

The Thirteen-lined Ground Squirrel, also known as a gopher, prefers wide-open spaces for its shallow burrow system that may wind some 20 feet (6 m) or more. It doesn't leave mounds around entrances but scatters the dirt far away and plugs side entrances with vegetation. The species undergoes complete hibernation. Its heart rate slows from 200 beats a minute to a mere 4 or 5, and its body temperature can drop to 32°F (0°C). It must awaken before its temperature drops lower or it will freeze to death. Many predators threaten the species, including foxes, Coyotes, snakes, and the badgers that bulldoze through its burrows.

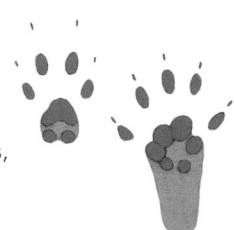

Golden-mantled Ground Squirrel

Callospermophilus lateralis L 11 in (28 cm)

The Golden-mantled Ground Squirrel has a chipmunk-like appearance, with its contrastingly striped sides and budging cheek pouches that hold hundreds of seeds.

KEY FACTS

Head is orange; back is grayish brown; black-edged white stripe on sides.

+ **habitat:** Alpine meadows, coniferous and mixed forests, brushy and rocky areas

+ **range:** Southwestern Canada to California, Arizona, and Utah

+ **food:** Seeds, nuts, vegetation, fungi, insects, birds, eggs, and carrion

The striped side of the Golden-mantled Ground Squirrel stops short of its head, a feature that distinguishes it from the Western Chipmunk; it also is heavier than the chipmunk. The Golden-mantle spends most of the summer fattening up, and sometimes caching food to prepare for the hibernation it will begin in late summer to mid-fall, depending on climate. The species is also adept at working its good looks for handouts at campsites in the mountain West. This ground squirrel lives solitarily in a burrow that often is excavated at the base of a stump, rock, or log. It curls into a tight ball, head tucked down past its chest.

Alpine Chipmunk

Tamius alpinus L 7 in (18 cm)

The Alpine Chipmunk inhabits rocky areas in a very small range in California's Sierra Nevada, occurring at elevations to almost 13,000 feet (4,000 m).

KEY FACTS

The body is pale over-all, with brown and white stripes on head and side; the sides are orangish, and the underside is whitish.

+ **habitat:** Alpine meadows and talus slopes from timberline to summit

+ **range:** Sierra Nevada in California

+ **food:** Seeds, fungi, berries, eggs, and young birds

Running close to each other at full tilt over rocky mountain slopes, a pair of Alpine Chipmunks displays the species' agility and adaptation to alpine life. This small ground squirrel is about the same size as the Least Chipmunk (*T. minimus*), but its tail is shorter and bushier, and its stripes stop short of the rump. The Alpine may raid the nest of rosy-finches and other birds to feed on eggs and young. But the bulk of its diet seems to be seeds, which it fills its cheek pouches to near bursting; then it pulls down grass stems and stuffs the seeds into them. It also enjoys what it can beg or steal at campsites.

Cliff Chipmunk

Tamias dorsalis L 9 in (23 cm)

Its sides have stripes, like other chipmunks, but on the Cliff Chipmunk they are dark and often difficult to see. A more distinct black stripe runs down the back.

KEY FACTS

Body is gray with dark mid-back stripe; sides are orangish; face has strong lines; ears are long.

+ **habitat:** Canyons and other rocky areas with brush, coniferous, and mixed forests

+ **range:** Southwest-ern U.S.

+ **food:** Seeds, nuts, berries, cacti, insects, amphibians, reptiles, birds, and eggs

The Cliff Chipmunk runs agilely across ridgelines and scales canyon walls. It forages from dawn to dusk, filling its cheek pouches with nuts, seeds, and berries to cache or to take to its cliff nest or underground burrow. This desert species faces many predators that attack by land or air, including hawks, Coyotes, badgers, and rattlesnakes. A threat can bring on a period of complete immobilization, which can last from 3 to 25 minutes. A swaying tail signals danger to others. This species has a longer breeding season than most chipmunks, but the female bears only one litter of about four or five young a year.

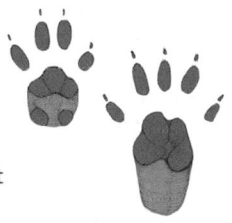

Merriam's Chipmunk

Tamias merriami L 10 in (25 cm)

A bushier-than-average chipmunk tail is the Merriam's Chipmunk's claim to fame. It may help this California species with climbing, balance, and insulation from the cold.

KEY FACTS

Body is grayish brown to reddish with dark stripes on back; lower, paler stripes are grayish or whitish.

+ **habitat:** Chaparral, brush, mixed woodlands, and rocky areas to 9,700 feet (3,000 m)

+ **range:** Central and southern California

+ **food:** Seeds, nuts, insects, lizards, birds, and eggs

More arboreal than many other chipmunk species, Merriam's Chipmunk frequently dens in stumps, hollow trees, and fallen logs, and raises its young in tree holes. It also may take over the burrows of other species, such as the woodrat. Extremely nimble, it can run up vertical cliffs on its toes and can travel through bushes and trees from branch to branch. It sometimes raids the acorn caches of Acorn Woodpeckers. At mating time, females advertise their readiness by perching in a prominent place and calling repeatedly, using the alarm *chip* sound for 10- to 30-minute periods over the course of three or four hours.

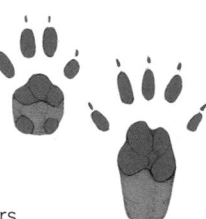

Least Chipmunk

Tamias minimus L 7.5 in (19 cm)

The Least Chipmunk is the most widespread of the western chipmunk species, ranging from alpine tundra to sagebrush flats. It may weigh no more than an ounce or two (28–56 g).

KEY FACTS

Color varies; stripes on head and back; tail as long as body.

+ **habitat:** Coniferous forests, alpine meadows and tundra, sagebrush, sand dunes, and open areas

+ **range:** Northwestern and central Canada and much of western U.S.

+ **food:** Seeds, leaves, buds, flowers, insects, eggs, and carrion

This tiny, high-strung chipmunk is "least" in name only. This highly adaptable species may lodge in an abandoned woodpecker hole, in a nest of leaves and grasses in a tree, or in burrows that it excavates underground. Its coloration varies widely according to region, ranging from very pale with pale stripes in dry areas or reddish brown with distinct stripes in the East. Its feeding habits are equally eclectic, and it stores food that it will eat in its burrow when it periodically wakes up during hibernation. Adults mate shortly after emerging from hibernation; three to six thimble-size young are born in May or June.

Eastern Chipmunk

Tamias striatus L 9.5 in (24 cm)

Eastern Chipmunks travel efficiently, making a beeline across the top of a fence rail or rock wall, before disappearing into a well-concealed burrow entrance. Danger elicits a loud, high *chip, chip.*

KEY FACTS

This species is grayish to reddish brown; dark stripe on back; dark-edged white stripe on side; eye bordered in white.

+ **habitat:** Woodlands, woodland edges, suburban and urban areas

+ **range:** Eastern half of U.S.; southeastern Canada

+ **food:** Nuts, seeds, berries, fungi, insects, worms, and garden bulbs

A rock with a messy pile of crushed acorn shells might signal the outdoor dining terrace of the Eastern Chipmunk. Its burrow has tunnels that may extend more than 100 feet (30 m) and may hold a bushel or more of stored nuts and seeds. Burrows can have a single, large multipurpose chamber or multiples. Adults live solitarily and are territorial, coming together only for mating. Females may have one or two litters a year, depending upon climate, of three to five young. Overwintering patterns vary by location and from year to year; some chipmunks awaken at intervals to eat food they cached in the fall.

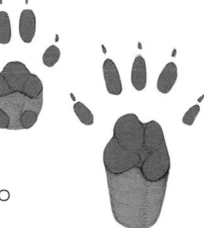

Townsend's Chipmunk

Tamius townsendii L 10 in (25 cm)

One of a number of chipmunks in the Pacific Northwest, Townsend's Chipmunk is a resident of coastal forests, where fallen logs offer den sites and new green plants take hold.

KEY FACTS

Body is tawny, grayish, or reddish brown; stripes are grayish, white, or orange and black.

+ **habitat:** Moist forests and clear-cuts with fallen logs and new growth

+ **range:** Oregon and Washington to southeastern British Columbia

+ **food:** Seeds, fruits, nuts, fungi, lichens, and insects

The Townsend's Chipmunk's day really gets going in the late morning or early afternoon when it is most active. It feeds and gathers food to store in its underground burrow; accumulation of stored food increases in late summer and fall. Some Townsend's Chipmunks will hibernate for the winter, and others in warmer coastal areas will stay active all winter. Still others may exist in an in-between state—present aboveground, but lethargic. Mating occurs in the spring, with litters of three to five born in May or June. A skillful climber, this chipmunk can scurry up a tree to hide or dive into a hole or hollow log to take cover.

American Beaver
Castor canadensis L 3–4 ft (0.9–1 m)

Nature's master engineer, the American Beaver constantly toils to create its own habitat. By damming streams and building lodges, it transforms the landscape to its needs.

KEY FACTS

Brown coat has coarse guard hairs and dense underfur; tail naked and scaly; hind feet fully webbed.

+ **habitat:** Streams, rivers, ponds, lakes, and swamps in wooded areas

+ **range:** Much of the United States and Canada

+ **food:** Bark, twigs, and leaves of trees and shrubs; herbaceous plants

The physiology of our largest rodent is geared to life in water. Paired orange incisors let it gnaw down trees for building and fell saplings for food. It uses oil from glands at the base of its tail to waterproof its fur. Webbed hind feet aid swimming, and a scaly, paddlelike tail serves as a rudder and as an alarm when slapped on water. Adult beavers form monogamous pairs and live in family groups with three or four kits of the year, as well as yearlings. Beavers along rivers usually do not dam, living in burrows in the banks. Clumsy on land, the beaver is vulnerable to predators such as foxes and Coyotes.

North American Porcupine
Erethizon dorsatum L 26 in (66 cm)

Many a curious dog has returned home whimpering with a face full of porcupine quills. The second largest North American rodent is armed with about 30,000 of them in its head, back, and tail.

KEY FACTS

Black or brown body has yellowish quills partly concealed by long guard hairs.

+ **habitat:** Brushy areas, woodlands, and desert washes

+ **range:** Alaska and Canada into western and northeastern U.S.

+ **food:** Leaves, needles, buds, nuts, fruit, and inner bark of trees; also herbaceous plants

It's an urban legend that the North American Porcupine shoots its quills. It doesn't. Instead, it presents a quill-filled back and tail to a threatening animal; shaking and stomping often causes loose quills to fly off. A lashing tail usually drives quills home. About the only predator able to kill the porcupine without mishap is the Fisher. This weasel relative bites the face until it is able to flip over the porcupine and go for the belly. Porcupines sleep by day in rock crevices, logs, or trees and feed at night on vegetation. Females give birth to a single young with soft, moist quills that quickly dry and stiffen.

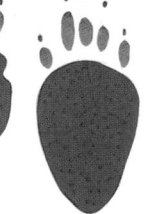

Nutria/Coypu
Myocastor coypus L 34 in (86 cm)

This South American native was introduced to the United States for fur farming; many animals escaped or were released deliberately. It is widely considered an undesirable species.

KEY FACTS

Guard hairs of yellowish or reddish brown hide gray underfur; round tail is lightly haired.

+ **habitat:** Streams, rivers, and freshwater and brackish marshes

+ **range:** Common in Gulf Coast states; scattered in southeastern and western U.S. and British Columbia

+ **food:** Aquatic plants and crops

Nearly as big as a beaver but with a narrow, rounded tail, the Nutria has worn out its welcome since its introduction in the 19th century. It decimates wetlands by eating all parts of aquatic plants, laying waste to entire stands of vegetation. It also mows down agricultural fields. The prolific Nutria—a female has up to three litters of up to 12 young a year—is particularly entrenched in Louisiana, which has a population of about 30 million. Newborn Nutrias swim and eat grass within days of birth. They can also nurse while riding on their mother's back, as her mammary glands are on her sides above the waterline.

Banner-tailed Kangaroo Rat
Dipodomys spectabilis L 13 in (33 cm)

This southwestern rodent has short front legs and elongated hind limbs, useful for jumping to evade predators. Territorial kangaroo rats fight each other by kickboxing like kangaroos.

KEY FACTS

Back is gray-brown; sides are yellowish brown; underside is white; long, striped tail has a white tip.

+ **habitat:** Short-grass plains, sandy plains, shrublands, and gravelly areas

+ **range:** Southwestern Texas, New Mexico, and southern and eastern Arizona

+ **food:** Primarily seeds

The Banner-tail is a heavy kangaroo rat, weighing about a third of a pound (150 g). Easily provoked and protective of their seed hoards, two Banner-tails will face off—jumping, twisting, kicking sand, scratching with hind claws, and biting if they get close enough. The species builds elaborate burrows that are enlarged by successive generations. They can be 3 feet (0.9 m) high and 15 feet (4.5 m) wide with multiple entrances, dead-end tunnels, and escape hatches. A storage chamber may hold several pounds of seed, and grass-lined nests shelter three litters of two or three young that may be born from January to August.

Ord's Kangaroo Rat
Dipodomys ordii L 9 in (23 cm)

Like other kangaroo rat species, the Ord's has fur-lined cheek pouches that it stuffs with grass seeds and carries back to its burrow to become part of its hoard.

KEY FACTS

Back varies from grayish to brown to orange; sides are paler; underside is white; crested tail has black tip.

+ **habitat:** Scrublands, piñon-juniper woodlands, brushlands, and sand dunes

+ **range:** Southern Alberta and Saskatchewan south to Arizona and western Texas

+ **food:** Seeds and insects

The most widespread kangaroo rat, Ord's is smaller than the Banner-tail and occupies a wider range of habitats. It will drink free water, but it can survive on water that is derived metabolically from its food. Like other kangaroo rats, it must have a nearby source of sand for dust baths to keep its fur sleek and glossy. Ord's is nocturnal, venturing out mainly on moonless nights—but braving the exposure of moonlight when it has to—to collect seeds for its hoard. It burrows underground, seldom making an aboveground mound. The species does not hibernate and breeds year-round in some parts of its range.

Plains Pocket Gopher
Geomys bursarius L 11 in (28 cm)

Pocket gophers, also known as "earth mice," are built for burrowing. They have large incisors that are always exposed, small eyes and ears, and large, curved front legs for digging.

KEY FACTS

Back is brown to reddish to black; underside is pale; tail is mostly naked; has large front claws.

+ **habitat:** Grasslands, roadsides, and croplands

+ **range:** Central U.S. from northern Texas to Manitoba

+ **food:** Roots, tubers, and vegetation

The pockets of the Plains Pocket Gopher are the animal's external, fur-lined cheek pouches that hold a haul of roots and tubers for underground storage chambers. Like other pocket gophers, this species lives its life underground. Pocket gophers are also known for their trick of causing an aboveground plant to disappear from view by pulling it into a feeding tunnel. Typically, they travel in shallow feeding tunnels and harvest the food dangling into them. Living quarters are deeper and contain a nest chamber, toilet, and food storage. Pocket gophers are solitary, except at mating time, and resent intrusion into their turf.

California Pocket Mouse

Chaetodipus californicus L 8 in (20 cm)

Pocket mice, like pocket gophers, have external cheek pouches used to transport seeds to their burrows. Some two dozen species live in the western United States; several extend into Canada.

KEY FACTS

Back is grizzled brown, sides are paler; distinct orange lines separate the sides and whitish underside; ears are long.

+ **habitat:** Chaparral, grasslands, scrublands, and woodlands

+ **range:** Central and southern California

+ **food:** Seeds, insects, and vegetation

The California Pocket Mouse lives mostly underground and alone. It remains in its burrow during the day with entrances plugged to keep up humidity below. It comes out at night to forage, then returns to the burrow to deposit its take, using paws to push out the seeds in its cheek pockets, sometimes turning the pockets inside out. The species is picky about the weather: It will become torpid and stay underground if it is too hot or too cold out, or if the weather is inclement. It does not enter true hibernation. Wildfires can kill most of an area's population, but the burned area usually is only temporarily vacated.

Woodland Jumping Mouse

Napaeozapus insignis L 8.5 in (22 cm)

The woodland counterpart to the Meadow Jumping Mouse, this species has the characteristic long tail, differentiated by a white tip. The sides of the woodland species show more color contrast.

KEY FACTS

Back is dark brown; sides bright orange or reddish brown; underside is white; white-tipped long tail.

+ **habitat:** Wet coniferous and mixed woodlands, stream banks, and swamp edges

+ **range:** Southeastern Canada and northeastern U.S. to Appalachians

+ **food:** Seeds, berries, fungi, and insects

The Woodland Jumping Mouse lives up to its name, preferring woodland and sheltered habitats to open ground. Meadow Jumping Mice are much more likely to enter the woodland mouse's habitats than the reverse, but the two species are not considered to be competitive with each other. Woodland Jumping Mice dig their own burrows or appropriate abandoned ones, preparing a nest for a long hibernation of about eight months. Like the meadow mice, they accumulate fat for hibernation, but do not store food. Three to six young, born in late spring or summer, practice jumping that usually is measured in inches as early as three weeks.

Meadow Jumping Mouse

Zapus hudsonicus L 9 in (23 cm)

The Meadow Jumping Mouse is a "bigfoot" species. Oversize rear feet on long legs give it the propulsion it needs to make jumps as high as 6 feet (1.8 m) to elude predators.

KEY FACTS

Body is dark brown with yellowish brown sides; underside is whitish; hind feet are one-third of body length; extremely long tail.

+ **habitat:** Fields, meadows, stream banks, and marshes

+ **range:** Southern Alaska and Canada into central and eastern U.S.

+ **food:** Fungi, fruit, and insects

In addition to big feet, jumping mice have long tails and tricolor coats tailored to a species' particular habitat. The nocturnal Meadow Jumping Mouse's colors of dark brown, yellow, and white blend with a meadow's colors at dusk and dawn. When a predator homes in on a mouse, it hops erratically in an evasive zigzag fashion. In summer this mouse nests on the ground or in shallow burrows. Two litters of three to seven young may be born in one year. In winter it builds a deeper burrow for hibernation; often alone or sometimes paired, it tucks up into a tight ball that cannot quite accommodate those oversize feet.

Northern Collared Lemming

Dicrostonyx groenlandicus L 5 in (13 cm)

Northern Collared Lemmings are one of the few mammals that molt to a snowy white winter coat. In Inuit legend, the white animals cascaded from the sky with snowflakes.

KEY FACTS

Back is gray in summer, orangish on throat and sides; underside is white; dark stripe extends from nose.

+ **habitat:** Rocky areas and meadows on Arctic tundra

+ **range:** Arctic from Alaska through Canada, and western edge of Greenland

+ **food:** Tundra vegetation, primarily forbs and willow

This tiny rodent is active year-round, day and night, in the challenging Arctic environment. It builds short burrows in summer; females modify these burrows to prepare for nesting. As winter nears, the forefeet of this lemming transform into tools for burrowing into snow and ice: Two middle claws lengthen, and claw pads enlarge and toughen. In spring, the feet return to their former shape. Winter nests are lined with grasses and sedges and are insulated by the snow. Lemming numbers fluctuate in four-year cycles related to the weather and the food supply. At population peaks, some undertake short migrations in search of food.

Ungava Collared Lemming

Dicrostonyx hudsonius L 5 in (13 cm)

This lemming inhabits a peninsula in northern Quebec and Labrador, which lends it the name Ungava. It undergoes the same seasonal fur changes as the Northern Collared Lemming.

KEY FACTS

Back is dark grayish brown with black stripe along spine; throat and sides orange; underside is gray.

+ **habitat:** Tundra, rocky hillsides, and meadows

+ **range:** Ungava Peninsula of northern Quebec and Labrador and islands to the west

+ **food:** Willow, birch, and aspen twigs

In their winter fur, Ungava Collared Lemmings, like the other collared lemmings that molt to white for the season, tend to look like short, stocky, white puffballs. Their forefeet also transform to digging claws in preparation for winter excavations. The range of this species abuts, but does not overlap with, that of the Northern Bog Lemming. The Ungava has the same system of short summer burrows and winter tunnels, and nests under the snow. The small range of this species is a remnant of its former presence, going back tens of thousands of years to the Pleistocene, when it inhabited the North American and Asian Arctic.

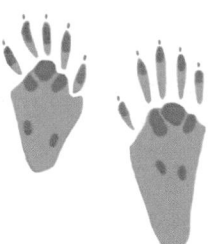

Brown Lemming

Lemmus trimucronatus L 5 in (13 cm)

Many animals of the tundra—such as foxes, weasels, and owls—rely on the Brown Lemming for sustenance. In years when the rodent is scarce, predator populations suffer.

KEY FACTS

Back and sides are yellowish brown; rump is cinnamon; has orange tufts at base of ears; tail is very short and concealed.

+ **habitat:** Low wet tundra with ample vegetation

+ **range:** Alaska through far northern Canada and into British Columbia

+ **food:** Grasses, sedges, and mosses

Predators feed on Brown Lemmings, and the lemmings feed on grasses and sedges in summer and mosses in winter. When they have depleted an area's vegetation, lemming breeding crashes. In peak lemming years, some of the rodents emigrate into towns and across bays and lakes to survive. Lemmings are typically precocious and prolific reproducers. Females can begin breeding as young as two weeks old. They breed all winter, even under the snow, and can have as many as three litters a year. In one documented case, a pair produced eight litters in 167 days—an average of a litter every 20 days—and then the male died.

Rock Vole/Yellow-nosed Vole
Microtus chrotorrhinus L 6 in (15 cm)

The elusive Rock Vole lives in scattered populations within its range. Its profile shows a more mouse-like appearance than that of other voles.

KEY FACTS

Back is brown; underside is gray; nose area is yellowish orange.

+ **habitat:** Talus slopes and other rocky areas with hardwood or mixed forests near water

+ **range:** Eastern Canada and northeastern U.S. to Appalachians

+ **food:** Bunchberries, blackberries, other vegetation, fungi, and insects

The Rock Vole is well named for its habitat, as it is for its colorful nose as an alternate common name, the Yellow-nosed Vole, and species name, *chrotorrhinus* (colored nose). The species is active year-round, day and night, spending a good part of the time in underground runs. It builds nests of plant fibers and mosses that are often tucked into rock crevices. A separate chamber is used for a latrine. The species breeds during much of the year; a female may have multiple litters of two to five young. In parts of its range, the Rock Vole shares its habitat with the Red-backed Vole, but the two do not seem competitive.

Prairie Vole
Microtus ochrogaster L 6 in (15 cm)

Runways are central to a vole's life, especially to a Prairie Vole. The rodents clip grass short and tamp down a path from repeated use. Longer grass on the sides helps to hide the passageways.

KEY FACTS

Body usually has long, grizzled, grayish brown fur above; sides are paler; underside is cream or whitish.

+ **habitat:** Prairies, grasslands, and agricultural areas

+ **range:** Central U.S. into south-central Canada

+ **food:** Seeds, roots, tubers, flowers, leaves, insects, and crops

In addition to extensive aboveground runways, Prairie Voles dig connected underground tunnels that include chambers for food caches and lined nests, which may be concealed under logs. The voles are active year-round and rely on stored roots and tubers in winter. Prairie Voles are unusual among rodents for their monogamy; once mated, they seem to mate for life, short as it is. The males also help with the care of their offspring, which can be born at any time of year, but most often in summer in litters of three to five. The mates stand together against the encroachment of other voles into their territory.

Meadow Vole
Microtus pennsylvanicus L 6.5 in (16.5 cm)

The Meadow Vole is a linchpin in the vertebrate food chain, supporting species as varied as birds, bears—even fish, which snap up swimming voles. All this makes for a short average life span.

KEY FACTS

This vole has thick fur on its dark brown back; its sides are paler; the underside is whitish; the tail is long.

+ **habitat:** Meadows, roadsides, grasslands, and orchards

+ **range:** Alaska through Canada and into northern and central U.S.

+ **food:** Green plants, roots, tubers, and bark

Meadow Voles may be the go-to lunch or dinner for many species, but when conditions are favorable, their reproduction rates can explode. The species may breed year-round, and a female usually produces five or six young in each litter; in years of population boom, an acre of meadow can house hundreds if not thousands of voles. They can strip a farm field, grazing like hoofed mammals with teeth adapted for grinding. They often sit on their haunches and hold food daintily in their forepaws. Meadow Voles construct grass-roofed runways as transit paths, and they nest in the grass or in burrows they dig with many entrances.

Woodland Vole/Pine Vole
Microtus pinetorum L 5 in (13 cm)

The burrowing evidence of the beady-eyed, short-legged, and short-tailed Woodland Vole looks like the work of a mole. This velvet-furred vole sometimes forages aboveground.

KEY FACTS

Body is auburn with grayish underside; ears and eyes are tiny; tail is very short.

+ **habitat:** Hardwood forests with thick substrate, clearings, and orchards

+ **range:** Central and eastern U.S. and slightly into southeastern Canada

+ **food:** Roots, tubers, grasses, seeds, fruits, and bark

The Woodland Vole's smooth, molelike coat helps it move easily underground and shed soil. Active day and night, the vole digs shallow tunnels, leaving a wake of ridges along the ground. It can excavate up to 15 inches (38 cm) in a minute. This vole's feeding habits frustrate gardeners and growers: It can devastate a potato crop and kill apple trees by eating roots and gnawing bark. These unpopular activities have earned it the nicknames "potato mouse" and "apple mouse." The species was first scientifically described from a specimen in the Georgia pines and named *pinetorum;* however, it is found more often in hardwood settings.

Southern Red-backed Vole

Clethrionomys gapperi L 5.5 in (14 cm)

The Southern Red-backed Vole is a resident of cool, damp forests, white cedar swamps, mountain ridges, and ferny glades. A broad chestnut stripe runs from its forehead to the base of its tail.

KEY FACTS

A chestnut stripe extends along the head and back; sides are grayish; underside is silvery or cream.

+ **habitat:** Forests, clearings, mountain meadows, and swamps

+ **range:** Canada and the northern U.S. into Rockies and Appalachians

+ **food:** Seeds, nuts, berries, fungi, lichen, insects, and carrion

Southern Red-backs travel on runways built by other rodents. The high-strung vole constructs small, spherical nests woven of grass, leaves, and moss—or at times, simple platforms—under logs or tucked into tree roots. In many areas, this species may be grayish with a sooty dorsal stripe. It is active year-round and can be nocturnal or diurnal, depending on the season. Red-backs breed about eight months of the year; females have two or three litters of four to eight young. The Northern Red-backed Vole (*C. rutilis*), a similar species that occurs in tundra and taiga habitats, has a more orange back and a hairier tail.

Round-tailed Muskrat/Florida Water Rat

Neofiber alleni L 13 in (33 cm)

The Round-tailed Muskrat makes smaller and more delicate conical homes than the Muskrat. It also has a round tail, not a laterally flattened one like the larger Muskrat.

KEY FACTS

The body is glossy brown above with buff underside, and the tail is almost hairless.

+ **habitat:** Freshwater marshes and agricultural areas

+ **range:** Throughout much of Florida and southeastern Georgia

+ **food:** Aquatic plants, including grasses and water lilies

The Round-tailed Muskrat seeks shallow, watery habitats with abundant aquatic plant life and fashions its round home on the surface with two underwater entrances. These muskrats densely populate favored areas; in central Florida, an acre can hold up to 120 adult Round-tails. The animals live individually, but may share feeding platforms—elevated pads of vegetation. The species breeds year-round; the female gives birth to one to four young in the "house," which she prepares with a lining of fine, dry grasses. Round-tails succumb to a variety of different predators, including hawks, owls, herons, snakes, and Bobcats.

Muskrat

Ondatra zibethicus L 21 in (53 cm)

This mammal exudes musk, but it is not a rat. Rather, the Muskrat is more like a large amphibious vole, or field mouse, which seldom strays from water.

KEY FACTS

Body is glossy brown above; grayish on underside; ears are tiny; hairless tail is flattened.

+ habitat: Freshwater and brackish marshes, streams, and ponds

+ range: U.S. and Canada, except arid Southwest and extreme Southeast

+ food: Aquatic plants, crayfish, fish, turtles, and mollusks

Muskrats share habitat with North American Beavers, and the two are often confused. In addition to being smaller than a beaver, the Muskrat has a naked tail and when it swims, more of its body appears above water than a beaver's. Sometimes it shows two humps in addition to the head, for a kind of Loch Ness Monster silhouette. Muskrats build their lodges out of cattails and marsh grasses, not trees. They also build tunnels, complex runways, and stream-bank burrows. They mate multiple times in a year; a female may bear up to eight litters of four to eight young. Trappers take millions of this prized fur species every year.

Western Heather Vole

Phenacomys intermedius L 5.5 in (14 cm)

This is the heather vole of higher western elevations, which does not have orange coloration on its nose and ear tufts. It shares many habits with the Eastern Heather Vole.

KEY FACTS

Back and sides are grayish to reddish brown; underside is white; tail is short.

+ habitat: Upland forests, dwarf shrubs, alpine meadows, and rocky areas

+ range: Southwestern Canada to California and Rockies to New Mexico

+ food: Shrub bark, berries, seeds, leaves, and lichens

The Western Heather Vole is usually smaller than the eastern species, and its very long whiskers extend well back beyond its ears. In the manner of the eastern species, it nests underground in the summer (caching food by the entrance) and aboveground under the snow in winter, another shared trait. Winter nests are made of grasses and may be sealed with lichens and set at the base of a shrub, rock, or stump. A family group of females and young may nest for warmth in the winter. The species otherwise is solitary and aggressively territorial. A female may bear several litters of two to eight young in spring and summer.

Eastern Heather Vole
Phenacomys ungava L 5.5 in (14 cm)

Until recently, the Eastern and Western Heather Vole were considered a single species, *P. intermedius*. Differences include the orange hairs on the Eastern's nose and in its ears.

KEY FACTS

Back and sides are brown; top of nose and tufts in front of ears are orange; tail is short.

+ **habitat:** Heather and other shrubby areas, coniferous forests, wet meadows, and woodland edges

+ **range:** Most of Canada and into northeastern Minnesota

+ **food:** Shrub bark, berries, and vegetation

Although it occupies a vast continental range, the Eastern Heather Vole, also called the Ungava Vole, seems to live in somewhat isolated populations. It is known to be very elusive, and it is noted for caching food that it gathers at the entrance to its burrow at night for next-day consumption. The species does not hibernate. It digs short burrows, mostly for use in summer, that contain nests of grasses, leaves, and lichens. In the winter it nests under snow. Females may have as many as three litters of between two to eight young each year. The species often falls prey to hawks, owls, and weasels.

Southern Bog Lemming
Synaptomys cooperi L 5 in (13 cm)

The Southern Bog Lemming is a resident of spongy bogs and woodland glades, usually near a source of grasses and sedges. It may live temporarily in scattered small colonies.

KEY FACTS

Back is grizzled brown, sides are yellowish brown; underside is silvery; tail is short.

+ **habitat:** Bogs, meadows, coniferous forests, and clear-cuts

+ **range:** Southeastern Canada and central and eastern U.S.

+ **food:** Grasses, sedges, berries, fungi, moss, and bark

Like many voles and other lemmings, Southern Bog Lemmings travel on grassy runways. They also burrow into the ground, adding chambers for feeding, caching food, and resting to the tunnels they dig. Grass clippings about an inch long may be stacked next to burrow entrances. Unlike other lemmings and their rodent cousins, bog lemmings produce bright green scat, a distinctive sign of their presence. The similar Northern Bog Lemming (*S. borealis*) tends to be larger and has orange fur at the base of its ears. It ranges from Alaska through northern Canada to the northeastern U.S. and across the borders of some western states.

Bushy-tailed Woodrat

Neotoma cinerea L 16 in (40 cm)

The Bushy-tailed Woodrat is a western species that makes itself a nuisance to cabin dwellers from the Yukon to Arizona, shredding furnishings and leaving a musky urine trail.

KEY FACTS

Back color varies from grizzled yellowish gray to black; sides are orange or yellowish; underside is whitish; tail is bushy.

+ habitat: Rocky areas, caves, canyons, and cliffs

+ range: Western Canada to California, Arizona, and the Dakotas

+ food: Vegetation, fruit, nuts, seeds, and fungi

This highly territorial large woodrat makes stick-and-vegetation nests festooned with collected materials, droppings, and food leavings. It anoints the nest with musky urine to establish ownership. Over time the urine hardens the structure, creating a midden that can last thousands of years, providing valuable scientific information. Inside the nest is a larder of seeds, nuts, and other foods to sustain the nonhibernating rodent through the winter. The Bushy-tail nests in abandoned mines, cabins, caves, and cliffs, from sea level to 14,000 feet (4,300 m). Three or four young are born each spring, and a second litter often follows.

Eastern Woodrat

Neotoma floridana L 15 in (38 cm)

Woodrats are "pack rats," those rodents whose habits are used to describe humans who hate to let any possession go. Pack rats construct houses of natural materials and collected "stuff."

KEY FACTS

Body above is grayish, dark brown, or sandy; underside and tops of feet are white; tail is long and haired.

+ habitat: Woodlands, slopes, plains, scrublands, and rocky areas

+ range: Mainly eastern and south-central U.S.

+ food: Seeds, nuts, grasses, buds, berries, and fungi

The wide-ranging Eastern Woodrat forages mainly at night, aided by its large eyes and long whiskers. It constructs a bulky house, perhaps 3 feet (0.9 m) tall and 5 feet (1.5 m) wide, out of sticks mounded haphazardly in a rock crevice, on the ground, or in a tree. At the nest, it incorporates "found" materials, ranging from jewelry to cooking utensils to small car parts. Sometimes when carrying an object it will find another that it prefers and drop its first find for the other. This has led to the moniker "trade rat," and sets up tall tales; in one tale, a woodrat stole a 50-cent piece and left two quarters behind.

Golden Mouse

Ochrotomys nuttalli L 6 in (15 cm)

If mice had beauty contests, the Golden Mouse would be a contender with its soft, burnished orangey coat. It has talent, too: It can hang like a monkey from its prehensile tail.

KEY FACTS

Body above is a golden orange-brown; sides are paler; underside is yellowish.

+ **habitat:** Woodlands with brush, vines, and briers

+ **range:** Southeastern U.S., except extreme southern Louisiana and southern Florida

+ **food:** Seeds, nuts, berries, and invertebrates

The Golden Mouse once was grouped in the genus *Peromyscus,* with species such as deermice, before it was placed in a genus of which it is the sole member: *Ochrotomys,* meaning "gold mouse." This species uses its climbing skills to evade predators and flooding. The primarily nocturnal and arboreal mouse carries seeds in its cheek pouches up vines and branches to feeding platforms. Multiple platforms are in a nest, a globular structure about 7 inches (18 cm) in diameter. Several adults may occupy a nest until breeding time, when a female with a litter of two or three young displaces all other adults, including her mate.

Northern Grasshopper Mouse

Onychomys leucogaster L 6 in (15 cm)

True to its name, the short-tailed Northern Grasshopper Mouse relishes a grasshopper meal. The Northern Grasshopper Mouse is typically bigger than the southern species.

KEY FACTS

Body color ranges from gray to brown to orangish; underside is white; individuals may have white tufts at base of their ears.

+ **habitat:** Deserts, grasslands, and shrublands

+ **range:** Western U.S. to Saskatchewan and Alberta

+ **food:** Insects and other arthropods, lizards, mice, seeds, and vegetation

Unusually for rodents, grasshopper mice are bona fide predators, stalking and killing not only insects such as grasshoppers, but other live prey such as scorpions, lizards, and even other mice. Their bodies have adapted to a carnivorous diet with its built-in infusion of water, but these species also eat seeds and vegetation when hunting is poor. The Northern Grasshopper Mouse makes nests in burrows that it digs or appropriates from other animals. The father participates in rearing two to six young in several litters a year. The species fiercely defends a territory of about 7 acres (2.8 ha) from other grasshopper mice.

Southern Grasshopper Mouse

Onychomys torridus L 5.5 in (14 cm)

Similar in many ways to the Northern Grasshopper Mouse but smaller, the southern species makes its home in southwestern arid shrublands at low elevations.

KEY FACTS

Body is gray to cinnamon to pinkish; underside is white; tail is longer than half its body length.

+ **habitat:** Desert shrublands at low elevations

+ **range:** Southwestern New Mexico and southern Arizona to southern Nevada and California

+ **food:** Insects, other arthropods, mice, and seeds

The grasshopper mouse frequently ambushes other mice from behind, dispatching them with a bite to the head, weasel-style. As it eats, this mouse may rear up on its hind feet, lift its head, and let out a high-pitched howl. The howl may be a territorial warning, which is typical of these vocal rodents. The Southern Grasshopper Mouse does not have all the physiological adaptations of other desert animals, but it is able to obtain water from its diet. Also known as the Scorpion Mouse, it has developed a method of avoiding a venomous sting while killing a scorpion by immobilizing the head before it bites the tail.

Brush Deermouse

Peromyscus boylii L 7.5 in (19 cm)

The Brush Deermouse, like the Piñon Deermouse, belongs to the genus *Peromyscus*, noted for difficulty in identification. Within it, the western species are especially challenging.

KEY FACTS

Back is grayish brown; sides are orange; underside is white; ears are large.

+ **habitat:** Piñon-juniper woodlands and chaparral with rocky areas, brush, and logs

+ **range:** Often at elevations above 2,000 feet (600 m) in the western U.S.

+ **food:** Seeds, nuts, fruit, leaves, fungi, and cacti

The Brush Deermouse gravitates to habitats with a lot of cover in the form of brush and rocks it can use for shelter and nesting. The species also infiltrates cabins in the West, where it is the most abundant *Peromyscus* in mountain regions. Genders come together only at mating time, and evidence suggests that multiple males can father the offspring in one litter, which usually numbers between two and five young. The main difference between the Brush Deermouse and the Piñon Deermouse is the latter's larger ears. Comparing ears with the size of the hind feet aids identification: The ears of these mice are usually not longer than the feet.

White-footed Deermouse
Peromyscus leucopus L 6.5 in (16.5 cm)

The White-footed Deermouse and its almost twin, the North American Deermouse, tend to confuse people, but subtle differences exist. The two species do not interbreed.

KEY FACTS

Back is dark brown; sides are orangish; underside and feet are white.

+ habitat: Woodlands, brushlands, fields, cleared areas, and riparian areas

+ range: Central and eastern U.S., except in extreme Southeast

+ food: Nuts, seeds, insects, vegetation, and fruit

A whiter underside and a slightly furred tail are two of the physical differences that distinguish the White-footed Deermouse from the North American Deermouse. Although the two species share a lot of range, the White-foot often prefers woodlands, but it also thrives in the Southwest. It swims well, a skill that extends its distribution to islands. Breeding may take place year-round, and the average litter has four young, which may be born in a tree nest, a stone wall, or an appropriated burrow. Food is often cached nearby. These abundant mammals barely live a year and are prey to many aerial and terrestrial predators.

North American Deermouse
Peromyscus maniculatus L 7 in (18 cm)

Superabundant and widespread North American Deermice make their home in nearly every kind of North American habitat, ranging from sea level to 11,500 feet (3,500 m).

KEY FACTS

Back is pale gray to reddish brown; underside is white; tail is furred.

+ habitat: Forests, grasslands, deserts, tundra, swamps, and mountains

+ range: Most of U.S. and Canada, except southeastern U.S.

+ food: Seeds, nuts, berries, fungi, insects, and other invertebrates

This nocturnal deermouse forages everywhere for the many items in its diet. Deermice do not hibernate, but groups may huddle together for warmth in cold weather. Unlike many other rodent species, females may permit their mates to stay in the nest and help with the litter of about four young. Sometimes she will leave him to care for dependent young while she moves to a new nest for her next delivery. Nests can turn up anywhere—in logs, stumps, abandoned burrows, sheds—or even an unused coat pocket. This species, along with the White-footed Deermouse, may host the Deer Tick, which can transmit Lyme disease.

Piñon Deermouse

Peromyscus truei L 7.5 in (19 cm)

The Piñon Deermouse has big feet and even bigger, almost comically large ears. It prefers a rocky habitat and an abundant supply of pine nuts and juniper seeds.

KEY FACTS

Back is gray or brown, sides are orangish; underside is white.

+ **habitat:** Arid, rocky areas with piñon, juniper, pine, and chaparral

+ **range:** Western U.S. from west Texas to California, Oregon, and southern Idaho

+ **food:** Seeds, nuts, leaves, insects, spiders, and fungi

The Piñon Deermouse sleeps by day and forages by night. It nests in hollow trees and rock crevices and often stores its surplus of seeds. Rocky slopes and cliffs are not deterrent to this species—also known as the Big-eared Cliff Mouse—and it ranges to about 10,000 feet (3,000 m). This species can be distinguished from the similar Brush Deermouse by its longer ears and usually larger size. Mortality for these mice is very high. Many factors, including predation from hawks, foxes, and Coyotes, conspire to allow only one in a litter of five or six young to reach a year of age. Females may have several litters a year.

Eastern Harvest Mouse

Reithrodontomys humulis L 4.5 in (11 cm)

The genus name for this very tiny mouse—*Reithrodontomys*—means "groove-toothed mouse"; one of its distinguishing features is a groove that runs down the front of each incisor.

KEY FACTS

Back is reddish brown; sides are paler; underside is grayish; long tail is nearly hairless.

+ **habitat:** Meadows, old fields, thickets, ditches, and marshes

+ **range:** Southeastern U.S. north to Ohio and west to eastern Texas

+ **food:** Seeds of grasses and forbs and insects

The Eastern Harvest Mouse brings home its gleanings of seeds to store. It lives snugly with a mate in a spherical nest as small as 3 inches (7 cm) in diameter, which it weaves from plant materials and attaches to a plant stem about a foot (30 cm) or more off the ground. The nest is lined, often with thistledown or cattail fuzz. Two to seven young are born in the nest from spring to fall, or year-round in the southern range. The similar Plains Harvest Mouse (*R. montanus*) is a bit larger and has a narrow, dark stripe down its back. Its south-central U.S. range slightly overlaps that of the Eastern.

Marsh Rice Rat
Oryzomys palustris L 9 in (23 cm)

The Marsh Rice Rat flocks to fields of newly planted rice to avail itself of the thoughtfully provided seeds. If the field is flooded, no problem—this rodent is an accomplished swimmer.

KEY FACTS

Body is brown to black with lighter sides; underside is whitish; tail is bicolored.

+ **habitat:** Marshes, riverbanks, wet meadows, and other wet areas

+ **range:** Atlantic coast south from New Jersey; Gulf Coast to Texas; inland to Kansas

+ **food:** Invertebrates; fish, turtles, vegetation, and carrion

The Marsh Rice Rat, one of two rice rat species that occur in the United States, makes its home in coastal and inland marshes, riverbanks, canebrakes, and wet meadows, foraging both on land and in water. In addition to seeds, these rodents enjoy a diet of shellfish, other invertebrates, fish, and turtles. This species weaves grassy nests placed on high ground, under a log, or perhaps in a tangle of weeds. In flood-prone lands the nests may hang above high water on a cattail stem or in a muskrat lodge. Marsh Rice Rats also build feeding platforms from bent grasses and connect them by trails to their nests and water sources.

Hispid Cotton Rat
Sigmodon hispidus L 10 in (25 cm)

The Hispid is the most widespread of the cotton rat species, a group of stocky vole-like rodents. "Hispid" refers to the animal's rough coat and "cotton rat" to its attraction to cotton crops.

KEY FACTS

Back is grizzled buff or gray; eastern populations are darker than western.

+ **habitat:** Grasslands, fields, agricultural lands, and scrublands

+ **range:** Southeastern and south-central U.S. to Arizona and southern California

+ **food:** Grasses and other vegetation, insects, birds, and eggs

Agricultural development in the South provided ideal circumstances for the Hispid Cotton Rat, a mainly tropical species, to extend its range and expand its population to the point that it is often the most abundant mammal wherever it is found. These cotton rats create networks of long and broad runways in the grass, which lead to nests in sheltered spots on the surface or in shallow burrows. Females may breed year-round, starting as early as six weeks old, and are very prolific. Litters average five or six young, but may be as high as 14. On the other hand, mortality is also high as a result of drought and predation.

House Mouse

Mus musculus L 6.5 in (16.5 cm)

Star of stage, screen, television, and literature, the House Mouse is a species that has been imported worldwide from Asia. It lives everywhere and eats nearly everything.

KEY FACTS

Body is brownish gray above; underside is buff; tail is long and thick.

+ habitat: Diverse, including meadows, grain fields, sand dunes, and buildings

+ range: Entire U.S. and southern Canada

+ food: Seeds, grain, insects, human food, and some nonfood household products

The House Mouse adapts to—and thrives in—most environments. Its small size and ability to squeeze through very narrow crevices gives it an "all-access pass" to anywhere it wants to go. House Mice live in colonies headed by a dominant male, several females, and their offspring. The females are potential breeding machines; they can produce up to 14 litters a year of 3 to 12 young, enduring a perpetual state of pregnancy. Mouse infestation in a building usually is announced by a musky odor, probably from urine used for territorial marking. This species has been bred into a white mouse form used in scientific experiments.

Norway Rat/Brown Rat

Rattus norvegicus L 16 in (40 cm)

In biology and behavior, the Norway Rat is consummately adaptable, leading to its unmitigated success everywhere it goes—which is everywhere. A superb swimmer, it owns the sewers.

KEY FACTS

Back is yellowish brown; underside is grayish or yellowish; tail is not as long as head and body.

+ habitat: Mainly suburban and urban areas; also fields and marshes

+ range: Throughout U.S. and populated areas of Canada

+ food: Grains, seeds, fruit, chickens, and garbage

The Norway Rat can live comfortably in the country, but it prefers a human connection, swarming to areas of human density with their endless supplies of food waste and diversity of shelter options. This species is a burrower, living on ground floors and in basements, tunnels, compost, and garbage dumps. The Norway Rat eats a wide range of foods but approaches new foods with caution and is wary of poisoned baits. It lives hierarchically, with dominant males and their females and offspring afforded closest access to the food supply. Successful breeding has created many individuals with strong resistance to anticoagulant poisons.

Black Rat/Roof Rat
Rattus rattus L 14.5 in (37 cm)

Unlike the burrowing Norway Rat, the Black Rat lives in lofty places, preferring attics and rafters of buildings and even living in trees in favorable climates.

KEY FACTS

Body is dark brown or black above; underside is grayish or whitish; tail is longer than head and body.

+ **habitat:** Buildings, ships, dockside structures, coastal forests, and fields

+ **range:** Coastal and southern U.S., West Coast to British Columbia

+ **food:** Grains, seeds, fruit, and garbage

The Black Rat colonized North America so effectively in large part by disembarking daily through ports in rat-infested ships, taking up residence in wharves and warehouses, and then dispersing to both urban and rural areas. Modern shipbuilding techniques have basically blocked that immigration route. This rat can lay waste to agricultural crops and has a detrimental effect on native species. The Black Rat is smaller and less stocky than the Norway Rat, and as an adult it lives solitarily, except for females and their newborns and at times even weaned offspring. This rat can breed year-round, producing litters of 4 to 10.

American Pika/Cony
Ochotona princeps L 8 in (20 cm)

The American Pika, also called a Cony, lives above the timberline in the West. Shaped more like a guinea pig than a rabbit, it is the only lagomorph (includes rabbits and hares) that caches food.

KEY FACTS

Body is brown, orangish, or grayish with orangish sides of neck; ears are rounded; tail is very small; feet are fully furred.

+ **habitat:** Rocky areas, often above 8,000 feet (2,400 m)

+ **range:** British Columbia and Alberta to U.S. western mountain states

+ **food:** Grasses, sedges, and forbs

Many aspects, down to its mouselike scurrying and diurnal lifestyle, distinguish the American Pika from rabbits and hares. During the summer, the Pika harvests vast amounts of grasses that it spreads out to dry before stacking them in large piles for the winter. This normally fairly tolerant animal fiercely guards its hoards against potential poachers. To make the hoards less desirable, it marks them with urine and scent from a facial gland. Like other lagomorphs, the Pika reingests its fecal pellets. It communicates by barking and whistling. The similar Collared Pika (*O. collaris*) has a more pronounced neck ring.

Pygmy Rabbit

Brachylagus idahoensis L 10.5 in (26 cm)

This tiny species was once placed in the same genus with cottontails, b
Brachylagus, which means "short hare." Its tail is gray all over, lacking †

KEY FACTS

Body is dark gray in winter, paler in summer; underside is whitish; ears are heavily furred; small tail is gray.

+ habitat: Areas with dense sagebrush

+ range: Great Basin and intermountain areas of western states, with an isolated endangered population in Washington

+ food: Mainly grasses and sagebrush

The Pygmy Rabbit depends brush for both protective cover and large portion of its nutrition. It comes out mainly at dawn and dusk to feed on scrubby, arid-land shrub and uses it to help hide its warren. Unlike most North American rabbits and hares, the Pygmy Rabbit does not nest in a den fashioned in vegetation, known as a form, but in a warren that it digs or appropriates—and then renovates—from another burrowing animal. It tends to stay close to home, and if it is alarmed while away from the nest, it scurries to one of its many warren entrances that are hidden under bushes from predators.

Antelope Jackrabbit

Lepus alleni L 25 in (63 cm)

The enormous ears of the Antelope Jackrabbit can reach up to 8 inches (20 cm) long—a quarter of the animal's body surface, helping it to cool off in the intense heat of its southwestern range.

KEY FACTS

Back is grizzled buff; sides are gray; underside is white; ears have white edges and tips; tail is black above and white below.

+ habitat: Dry grassy areas, with shrubs and cacti, to 5,000 feet (1,500 m)

+ range: Southern Arizona

+ food: Grasses, leaves, including mesquite, and cacti

Only one native mammal can outrun the Antelope Jackrabbit: the Pronghorn, a hoofed species sometimes mistakenly called an antelope. The namesake hare has been clocked at more than 40 miles (64 km) an hour, covering more than 20 feet (6 m) in a single bound. It also adopts a gait in which the hind feet hit the ground several feet ahead of where the forefeet left the ground, varying its path in a way that frustrates predators. In escaping, it displays a flashing patch of white hair on its rump, which is controlled by its back muscles. Active at dawn and dusk, and at night, the species rests camouflaged during the day.

...shoe Hare
...mericanus L 17 in (43 cm)

...northern hare changes coats with the seasons, from browns and buffs in summer to white ...winter, a trait that gives it the alternate name of Varying Hare.

KEY FACTS

Body is brown or red-dish brown, white in winter; short ears are tipped in black.

+ habitat: Brushy areas, coniferous and hardwood forests

+ range: Alaska through Canada to northern U.S., west-ern mountains, and Appalachians

+ food: Vegetation, berries, twigs, and bark

Not much larger than a cottontail and the smallest member of its genus, the Snowshoe Hare is distinguished by large, broad hind feet matted on the bottom with coarse hairs. The feet make it nimble in deep snow, able to bound away quickly under threat. Its white winter coat provides camouflage in a snowdrift; only dark eyes and black-tipped ears betray its presence. Females are a bit larger than the males and may have up to five litters a year of one to nine young, known in their first year as leverets. Snowshoe Hare popu-lations fluctuate in ten-year cycles, perhaps caused in part by disease transmission due to overcrowding.

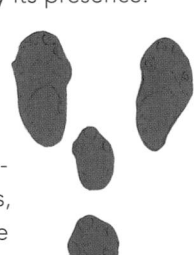

Arctic Hare
Lepus arcticus L 24 in (60 cm)

North America's heftiest hare, weighing up to 12 pounds (5.4 kg), the Arctic Hare makes its home on the frozen tundra of the Canadian high Arctic.

KEY FACTS

Body is grayish brown in summer, white in winter; always black-tipped ears are shorter than their head.

+ habitat: Treeless tundra, often in rocky areas

+ range: Canadian Arctic mainland and islands; coastal Greenland

+ food: Dwarf willow and other vegetation, berries, moss, and lichens

The Arctic Hare comes well equipped for life in a challenging habitat. Its chunky body bears thick, woolly white fur that molts to brown in the southern parts of its range in summer. Large densely furred feet offer protection against the snow and stout claws on the forefeet allow it to dig through crusty snow to reach vegetation. Pro-truding teeth help it scrape moss and lichens from rocks. This species mates in the spring, and a litter of two to eight brown young is born in the summer. The species often congregates in large groups of a hundred or more to feed or rest. A closely related species is the Alaskan Hare (*L. othus*).

Black-tailed Jackrabbit

Lepus californicus L 22 in (56 cm)

Early settlers in the Southwest named this long-eared, lean species "jackass rabbit," later shortened to jackrabbit, although it is clearly a hare in body type, development, and behavior.

KEY FACTS

Back is grizzled buff; chest orangish; underside white; long, black-tipped ears; tail black above, white below.

+ **habitat:** Brush, pastures, cropland, prairies, and desert

+ **range:** Western and central U.S., except northern Rockies and plains

+ **food:** Grasses, forbs, shrubs, cacti, and crops

The Black-tailed Jackrabbit adapts to varied environments ranging from desert scrublands to prairies. Like other hares, the Black-tail has longer legs and larger ears than rabbits, and its ears swivel independently, enhancing its hearing. Deliberate decimation of predators in the Southwest triggered population explosions of jackrabbits, which led to the devastation of crops. Fifteen jacks can eat as much vegetation as one sheep. As in other hares, young are precocial, born furred and mobile, with eyes open. The similar and more northern White-tailed Jackrabbit (*L. townsendii*) has smaller ears and an all-white tail.

European Hare

Lepus europaeus or *Lepus capensis* L 27.5 in (70 cm)

This Old World hare, with its distinctive kinked fur, is in decline and gone from parts of its former introduced range. Its antics during breeding season inspired the image of the March Hare.

KEY FACTS

Body is grizzled yellowish brown above; underside is white; large ears are tipped in black.

+ **habitat:** Fields, meadows, and pastures

+ **range:** Southeastern Ontario and adjacent areas of New York; also likely along Hudson River

+ **food:** Grasses, forbs, twigs, and bark

The European Hare is built for speed. Its long hind legs propel it downhill, uphill, or on the straightaway, where it can clock 30 miles (48 km) an hour. It often runs in a zigzag pattern, doubling back on pursuers such as foxes, Bobcats, and hawks. The largest hare in its range, it prefers open spaces where it can hunker down in vegetation, emerging to feed at dusk and dawn. It is active year-round; a female may produce her first litter when snow is still on the ground. Her leverets, or young, disperse to multiple nests for safety. Nursing them in turn, she announces her arrival with a soft grunt, which the leverets answer.

Alaskan Hare/Tundra Hare

Lepus othus L 25.5 in (65 cm)

The Alaskan Hare is similar to the Arctic Hare in appearance, habitat, and adaptation to a cold climate, but the two species' ranges are distantly separated from each other.

KEY FACTS

Body in summer is grayish or reddish brown, white in winter; underside is white; ears are black-tipped.

+ **habitat:** Tundra, especially near alder thickets and wet meadows

+ **range:** Coastal and peninsular western Alaska

+ **food:** Shrubs, especially willow; leaves, berries, and bark

The Alaskan Hare deals with the extreme conditions of its demanding habitat by often ignoring the elements and not taking cover in rain or snow. Reproduction usually is limited to one litter a year, and the precocial leverets are nursed for several months to maximize their size and fortify them for the rigors of winter. But nests are often mere depressions without any lining, and the leverets stay in place and endure the elements as well. The species begins to change to winter white in mid-September. Seasonal camouflage hinders but does not prevent predation by such species as hawks, foxes, weasels, and wolves.

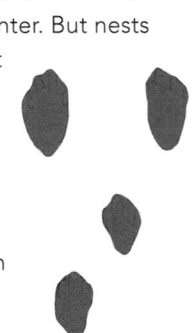

Swamp Rabbit

Sylvilagus aquaticus L 21 in (53 cm)

A more aquatic species than the Marsh Rabbit *(S. palustris),* the Swamp Rabbit favors a variety of watery environments across its widespread southern range.

KEY FACTS

Body is grizzled orange-brown above, with paler sides; underside is white; eye ring is orange; tail is small.

+ **habitat:** Bottomlands, swamps, and marshes

+ **range:** South-central U.S. from Oklahoma and Texas to northwestern South Carolina

+ **food:** Grasses, forbs, buds, and bark

The Swamp Rabbit is larger and longer eared than the Marsh Rabbit, although some authorities consider them a single species. Swamp Rabbits fashion surface nests that often are situated at the base of stumps or posts or in holes in logs and stumps. Lined with vegetation and fur, they shelter a litter of two to three young. Unlike most other cottontails, they are born with fur and with closed eyes that open in two or three days. Adult males are territorial and mark their turf with a substance from glands on their chin. The species shares some parts of its range and habitats with smaller Eastern Cottontails *(S. floridanus).*

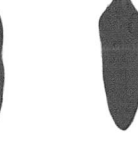

Desert Cottontail
Sylvilagus audubonii L 14–16 in (35–40 cm)

Also known as Audubon's Cottontail, the Desert Cottontail ranges the arid plains and valleys of the western U.S. states that it calls home.

KEY FACTS

Body is grizzled yellowish brown above; underside is white; ears are long and black-tipped.

+ habitat: Deserts, grasslands, stream bank brush, and piñon-juniper woodlands

+ range: North Dakota south though Texas and west to California

+ food: Plants, fruits, berries, and acorns

Slightly smaller than an Eastern Cottontail, this species has longer and pointier ears and a tail that is grayish above and white below. The Desert Cottontail rests during the day in a shallow form, or depression, in the vegetation. It comes out at sunset to feed. Under threat, it takes shelter among rocks, in vegetation, or in a burrow dug by another species, such as a prairie dog. If it must, it will also climb a low tree. In some parts of its range, the Desert Cottontail may breed year-round, and the female may have up to five litters with two to four young each year. Otherwise, the breeding season runs spring through fall.

Brush Rabbit
Sylvilagus bachmani 11–13 in (28–33 cm)

More social than a number of other rabbit species, the Brush Rabbit tolerates the company of others but maintains a sense of personal space—a zone of separation of at least a foot (0.3 m).

KEY FACTS

Back Is dark or grayish brown; sides are grayish; underside is whitish; ears are not black-tipped; tail is short.

+ habitat: Brush, brambles, woodland edges, and chaparral

+ range: West Coast from Oregon through California

+ food: Grasses, forbs, berries, and woody plants

The Brush Rabbit returns to its shelter, called a form, between foraging expeditions to groom and rest. Forms are connected to feeding areas by well-traveled runways through thick vegetation. This rabbit is never far from the cover of brush. It is cautious when entering an area, often remaining motionless while it scans for possible danger. When threatened, it thumps the ground with a hind foot and may let out a squealing distress call. It will climb a bush or low tree for safety. Females raise their young, born in litters of two to five up to four times a year in the forms, which they cover with grass before leaving to forage.

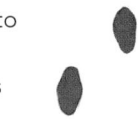

Eastern Cottontail
Sylvilagus floridanus L 15–18 in (38–45 cm)

A very common sight in its large range and varied habitats, the Eastern Cottontail displays the powder-puff rear appendage that gives it its common name.

KEY FACTS

Back is grizzled brown with orange nape; sides are paler; underside is white.

+ **habitat:** Thickets, fields, prairies, forest edges, swamps, suburban and urban areas

+ **range:** Central and eastern U.S. to southern Canada

+ **food:** Grasses, clovers, other plants; bark, twigs, and crops

The Eastern Cottontail fits the quintessential image of a rabbit, showing characteristics that separate rabbits from hares: a stockier body, shorter ears and legs, and typically naked, helpless newborns. A female may have up to seven litters a year, averaging five to seven young. If all her offspring and all of their descendants survived and reproduced similarly, this female would be responsible for more than 2.5 billion bunnies in five years—that is, if cottontails' lives were not so short. The New England Cottontail (*S. transitionalis*), which inspired the character Peter Cottontail, has struggled in its range due to habitat loss.

Mountain Cottontail
Sylvilagus nuttallii L 12–14 in (30–36 cm)

This species has taken the cottontail life to a higher level in the mountain West. Mountain Cottontails are smaller than eastern species, with proportionately shorter ears and very furry feet.

KEY FACTS

Body is grizzled buff with paler gray sides; underside is white; white-edged furred ears.

+ **habitat:** Thickets, sagebrush, conifer stands, cliffs and other rocky areas

+ **range:** Western U.S. from southern Canada to California and New Mexico

+ **food:** Grasses, sagebrush, and juniper

This usually solitary species is quite the tree climber, often climbing junipers. When feeding at dawn and dusk, primarily on grasses in summer and on sagebrush and juniper at other times, it sticks close to cover, either in thickets or rock crevices. For resting and nesting, it uses forms, or depressions in vegetation, as well as burrows likely constructed by other species. Nest forms are lined with grasses and fur and covered over with leaves and twigs. Females give birth to four to six young from late spring to early summer, producing up to five litters a year. This species is also known as Nuttall's Cottontail.

Marsh Rabbit
Sylvilagus palustris L 15 in (38 cm)

At the first sign of danger, a Marsh Rabbit may take to the water, using strong swimming skills to elude predators. It also floats motionless among vegetation, with only eyes and nose above water.

KEY FACTS

Body is grizzled red-dish brown; ears are naked inside; tail is very small and grayish; feet have long claws.

+ **habitat:** Marshes, lakeshores, and other wet areas

+ **range:** Coastal southeastern U.S. from Virginia to Alabama

+ **food:** Roots, tubers, and marsh vegetation

Most rabbits swim when pressed, but the Marsh Rabbit does so regularly; it has been seen going at a good clip more than 700 yards (640 m) from shore. The species has shorter legs than most other rabbits and frequently walks. It also stands and walks comfortably on its hind legs. This species lives and hides amid dense marsh and swamp vegetation, following trampled paths. It builds nests, known as forms, among the undergrowth that are covered over with vegetation to protect the young, which are born with fur, unusually for rabbits, but with closed eyes. Up to seven litters of two to four young may be born in a year.

Northern Short-tailed Shrew
Blarina brevicauda L 5 in (13 cm)

The ability to deliver a poisonous bite distinguishes the Northern Short-tailed Shrew, one of the continent's most common shrew species, from other North American mammals.

KEY FACTS

Body is silvery to dark gray; underside is sil-very; ears are hidden by fur.

+ **habitat:** Wood-lands, fields, bogs, marshes, and pond and stream banks

+ **range:** North-central and northeast-ern U.S. and southern Canada

+ **food:** Insects, worms, snails, other invertebrates, and small mammals

Despite a tail that measures only an inch (2.5 cm), the Short-tailed Shrew is the larg-est shrew in North America in body length and weight—weighing up to a hefty 1 ounce (28 g). The species inflicts its poisonous bite on prey; it is often fatal to insects and paralyzing to larger animals such as moles and voles. Caches of immo-bilized prey appear to be larders in the shrew's underground burrow. Burrows also are fashioned under leaf litter or snow. The female bears litters of three to seven young in burrow nests lined with grass or fur. The offspring can mate in as few as seven weeks and often live less than a year.

Least Shrew

Cryptotis parva L 3 in (8 cm)

Unlike the solitary individuals of many shrew species, the Least Shrew is gregarious, often living communally and sharing a nest with other adults.

KEY FACTS

Back is brown to black, grayer in winter; underside grayish; short, bicolor tail has tufted tip; ears are short.

+ habitat: Fields, brush, woods, marshes, and dry areas

+ range: Central and eastern U.S. to southern Ontario

+ food: Insects, small lizards and frogs, and carrion

One of the smallest North American shrews, the Least Shrew can weigh as little as a nickel. A hyperactive nature and cooperative behavior help it accomplish large tasks, such as burrow construction. Sometimes one adult will tunnel to create a new space while another shifts and packs in loosened soil. Least Shrews seem to communicate with each other by a variety of sounds, including clicking noises. The species is active day and night throughout the year and may store insects it has killed in a tunnel for future use. In southern parts of the range, females may breed year-round; a typical litter is two to seven pups.

Arctic Shrew

Sorex arcticus L 5 in (13 cm)

A tricolor body distinguishes the long-tailed Arctic Shrew. This species usually is found in open habitats within the boreal forests of its northern range.

KEY FACTS

Back is dark brown; sides are paler; underside is whitish; bicolor tail is dark above, lighter below.

+ habitat: Edges of marshes and swamps and forest clearings

+ range: Much of Canada and into northern plains of U.S.

+ food: Insects and insect larvae

Shrew species in general are noted for the frantic feeding required to meet high metabolic needs. Among the group, the Arctic Shrew may feed even more voraciously, slowing down for only short periods in its quest for food by day or night. Good eyesight seems to play a part in its daytime hunting. It has been observed in the chill air of early morning ambushing torpid grasshoppers from above, securing them with jaws and feet, and then devouring all but the legs and wings. This species suffers high mortality, and young that are born in litters of five to nine do not make it to maturity about 80 percent of the time.

Marsh Shrew/Pacific Water Shrew

Sorex bendirii L 6.5 in (16.5 cm)

North America's largest long-tailed shrew, the Marsh Shrew shares a similar semiaquatic lifestyle with the Northern Water Shrew, although its range is much more limited.

KEY FACTS

Body is blackish brown above; usually dark on underside; tail is usually all dark.

+ **habitat:** Streams, marshes, beaches, and moist woodlands

+ **range:** Northern California to southern British Columbia

+ **food:** Larvae of aquatic insects, earthworms, sow bugs, and other invertebrates

The Marsh Shrew appears entirely at home in watery habitats, having the ability to swim, dive, scull on the surface, and even dash across the water for three to five seconds. Its dark fur appears silvery when submerged due to air trapped within it. The animal uses its long, sensitive, whiskered snout to hunt for aquatic prey. In winter, the Marsh Shrew may be found farther away from a source of water and it eats a wide range of terrestrial invertebrates as well. This species is larger and lacks the more distinctly hair-fringed feet of the similar Northern Water Shrew (*S. palustris*). It is also known as Bendire's Shrew.

Masked Shrew

Sorex cinereus L 3.5 in (9 cm)

Who was that Masked Shrew? Actually, the "mask" is not readily apparent on the face of this common and widely distributed long-tailed shrew.

KEY FACTS

Back is brown, sometimes with a dark central line and paler sides; underside is paler; bicolor tail has black tip.

+ **habitat:** Woods, fields, marshes, and swamps

+ **range:** Alaska, most of Canada, and northern U.S.

+ **food:** Insects, young mice, amphibians, and carrion

Despite its abundance, the Masked Shrew remains elusive as it often travels through leaf litter or subsurface tunnels it constructs or appropriates from mice and moles. Like other long-tailed shrews, the species prefers moist environments. It makes a nest of leaves and grasses under logs or in cavities, in which the female may raise up to three litters of four to ten young each year. Life is short for this species, as it is for most shrews, lasting only about 15 months. As are many shrews, the Masked Shrew is a highly nervous animal, frequently dying of fright when threatened or startled by thunder or other loud noises.

Rock Shrew/Long-tailed Shrew
Sorex dispar L 5 in (13 cm)

Among the long-tailed shrews, the Rock Shrew has a very long tail, which it likely uses for balance while scurrying among the rocks and boulders of its habitat.

KEY FACTS

Back is dark gray, underside paler; tail is more than 90 percent total length.

+ **habitat:** Rocky regions, especially at base of forested slopes; also in artificial rock debris

+ **range:** Eastern U.S. from Tennessee and North Carolina to maritime Canada

+ **food:** Insects, spiders, and centipedes

True to its common name, the Rock Shrew is a denizen of rocky terrain in the mountainous regions of eastern North America. It lives mostly several feet underground and was long thought to be rare, as it didn't show up in collection traps near the surface. The Rock Shrew's head is adapted to hunting in rocky terrain; a narrow skull and long, narrow teeth allow the animal to pry insects and other invertebrates from cracks and crevices. Contrary to a usual correlation between larger size and higher latitude within a species, Rock Shrews in northern areas tend to be smaller than those in the southern part of the range.

Smoky Shrew
Sorex fumeus L 4.5 in (11 cm)

This medium-size shrew is mainly nocturnal and is active year-round. Like many other shrews, it echolocates by constantly emitting noises as it hunts for food.

KEY FACTS

Back is grayish brown; underside is paler; fur is grayer overall in winter.

+ **habitat:** Leaf litter, mossy rocks, and decaying logs in deciduous and coniferous woodlands

+ **range:** Southeastern U.S. into southeastern Canada

+ **food:** Insects, centipedes, sow bugs, earthworms, and salamanders

The Smoky Shrew often inhabits areas populated by Short-tailed and Masked Shrews, but it is larger than the Masked Shrew, which it resembles, and has paler feet. In addition, the Smoky Shrew has fairly prominent ears. The three species tend to minimize competition by eating different sizes of insects and other invertebrates. The Smoky Shrew does not reproduce in its first year; instead, it overwinters as a subadult. The adult female may have up to three litters of two to eight pups between March and October of its second year, but has a life span of only about 14 months and probably does not survive into another mating season.

Merriam's Shrew

Sorex merriami L 4 in (10 cm)

One of the few long-tailed shrew species that prefer dry habitats, Merriam's Shrew uses the runways and burrows of small rodents, especially those of vole species.

KEY FACTS

Back is grayish brown; sides are paler; underside is whitish; tail is dark on top, light on bottom.

+ **habitat:** Sagebrush, arid grasslands, woodlands, and shrublands

+ **range:** Western U.S. to southern British Columbia

+ **food:** Insects, spiders, and other invertebrates

The elusive Merriam's Shrew appears in isolated western colonies, where it is found at elevations ranging from 500 feet to about 9,000 feet (150–2,700 m). This small shrew has a distinctively domed head and a bicolor tail, and like many other shrew species, it molts twice a year. While Merriam's Shrew eats a varied diet of adult and larval insects as well as spiders, in the summer months it seems to have a marked preference for caterpillars. Owls often prey on the Merriam's Shrew, which other potential predators tend to avoid, perhaps because of the pungent odor produced by reproductive glands on males' flanks.

Pacific Shrew

Sorex pacificus L 6 in (15 cm)

Occupying a limited range and somewhat specialized habitats, the long-tailed Pacific Shrew manages to exploit a wide range of animal and vegetable food sources.

KEY FACTS

Body is reddish brown to dark brown above, paler orange-brown below; legs are long.

+ **habitat:** Moist thickets, brush, stream banks, and mossy fallen logs

+ **range:** Coastal northern California into Oregon

+ **food:** Insects, other invertebrates, amphibians, moss, fungi, and seeds

Its large size and reddish color distinguish the Pacific Shrew, the largest of the *Sorex*, or long-tailed shrew species. This species often deals with an abundance of food procuring opportunities in its habitats by caching the excess it has caught but cannot immediately consume under logs and in similar spots. It first immobilizes insects by sharp, swift bites to the head or thorax. The mostly nocturnal species uses both smell and hearing in hunting and is capable of pulling flying prey out of the air by using sound to locate it. Female Pacific Shrews bear litters containing from two to seven young in the summer months.

Northern Water Shrew/American Water Shrew

Sorex palustris L 6 in (15 cm)

The largest of the long-tailed shrews in eastern North America, the Northern Water Shrew usually lives near running water and is well adapted to water activities.

KEY FACTS

Back is blackish; underside is usually paler; tail has a tufted tip.

+ **habitat:** Streams, boggy areas in forests, marshes, and pond and lake edges

+ **range:** Much of Canada, northern U.S. into Sierras, Rockies, and Appalachians

+ **food:** Insects, spiders, mollusks, and small fish

The Northern Water Shrew feeds in cold, fast-moving streams where its fur traps a layer of air and keeps it buoyant while it swims to capture prey. Its partially webbed hind feet have fringes of stiff hairs that aid swimming and diving and allow it to dash briefly across the water's surface. The shrew's long, slender snout extends from a pair of tiny, beady eyes and is covered with long sensory hairs, or vibrissae, that rotate constantly. The species is active all day year-round, with peaks of activity before sunrise and after sunset. The female may bear three litters a year of two to ten young in a nest of vegetation.

Star-nosed Mole

Condylura cristata L 7 in (18 cm)

A pink, fleshy appendage of 22 radiating tentacles adorns the end of the snout of the Star-nosed Mole—a signature look unique in the mammalian world.

KEY FACTS

Body is dark brown to black; long, hairy tail is thick at base.

+ **habitat:** Moist areas in woodlands, meadows, and swamps; sometimes drier areas

+ **range:** Eastern U.S., west to Minnesota and south to Georgia; southeastern Canada

+ **food:** Aquatic insects, worms, crustaceans, and fish

Looks aren't the only distinguishing feature of the Star-nosed Mole: It leads a much more aquatic life than other moles. Its tunnels often head right through a stream bank and into the water. This mole is an excellent diver and swimmer, paddling with its wide forefeet, and takes much of its prey in muddy stream bottoms. Its nose tentacles probably aid in prey detection as well as manipulation. This species also lives more communally than other moles. Males and females remain together after mating until the young are born in a nest above the waterline. The two to seven young in a litter display nose stars at birth.

Hairy-tailed Mole

Parascalops breweri L 6.5 in (16.5 cm)

A very hairy, short tail and a preference for higher-altitude living help to distinguish Hairy-tailed Moles from other eastern mole species.

KEY FACTS

Back is grayish black; underside is silvery or spotted white; tail is short and bristled.

+ habitat: Well-drained fields, woodlands, and roadsides up to about 3,000 feet (900 m)

+ range: Northeastern U.S. to southeastern Canada

+ food: Worms, insects, other invertebrates, and plants

The Hairy-tailed Mole shows less of a presence on the surface of the ground than the Eastern Mole does. Its travel and feeding tunnels are shallower and not very noticeable when covered with vegetation. Hairy-tails forage on the surface at night. In cold weather, they build deeper burrows, throwing up short molehills on the surface. Individuals are solitary, and the sexes usually come together only for mating. Once a year, the female bears a litter of four to five whitish, wrinkled young, naked except for some hairs on the face and short sensory hairs. If avoided by predators, the species, like other moles, can live three or four years.

Eastern Mole/Common Mole

Scalopus aquaticus L 7 in (18 cm)

Bumpy tunnels that push up lawn turf signal a mole's presence, such as that of the Eastern Mole. The shallow tunnels serve as subways and places to rest and feed.

KEY FACTS

In north, back is grayish brown with paler underside; smaller and paler in south.

+ habitat: Fields, woodlands, gardens, and lawns

+ range: Eastern half of U.S., except in mountains, and to southern Canada

+ food: Earthworms, insects, other invertebrates; also plant matter

The abundant Eastern Mole demonstrates the basic mole features that make the animal a tunneling machine: strong, twisted forefeet with flattened claws, a body that tapers at both ends, and fur that moves easily backward and forward. Moles use their distinctive forefeet to excavate soil and push it out of the way. In addition to subway tunnels, moles dig deep tunnels that contain a nest chamber, where females raise litters of two to five young. The nest is advertised by the presence of a molehill of surplus soil on the surface. Despite the *aquaticus* in the scientific name, Eastern Moles are not very aquatic.

Townsend's Mole
Scapanus townsendii L 8 in (20 cm)

North America's largest mole, Townsend's Mole makes a substantial environmental impact, with its deep tunnels, high molehills, and large nest chambers.

KEY FACTS

It has a dark brown or black body; short, largely hairless tail; and very narrow snout.

+ habitat: Moist meadows, floodplains, woodlands, and alpine meadows

+ range: Northern California to southern British Columbia

+ food: Insects, worms, other invertebrates, bulbs, and other plant matter

Townsend's Mole makes elaborate excavations in the form of shallow tunnels and deeply dug burrows with secure nest chambers lined with grasses. Unlike other moles, the Townsend's shows its presence with abundant molehills, which rise higher than those of other species and are found in denser concentrations. A single Townsend's can be responsible for hundreds of hills in a small area. The soft, velvety fur of this species made it commercially valuable for purses, caps, and clothing trim. Males are somewhat larger than females, which bear two to five young in spring after a gestation period of about a month.

Southern Long-nosed Bat/Lesser Long-nosed Bat
Leptonycteris yerbabuenae L 3.5 in (9 cm)

Southern Long-nosed Bats practice chiropterophily—"bat love"—a symbiotic relationship between the flying mammals and the flowering desert plants whose blossoms they feed from and pollinate.

KEY FACTS

Body is orangish brown or grayish; face has a small noseleaf; small tail is barely visible.

+ habitat: Deserts, scrublands, and grasslands

+ range: Southern Arizona and southwestern New Mexico

+ food: Nectar, pollen, and fruit of agave, saguaro, and other desert plants; sometimes insects

It's all about the flowers for this species of the desert Southwest. At night, the Southern Long-nosed Bat leaves its roosts in caves and mines, and heads for flowering cacti. Homing in on a flower, the bat hovers hummingbird-like and extends its long tongue. Bristles on the tongue increase its surface area and allow a bat to take up large amounts of nectar and pollen. When flowering ends, the bats consume cactus fruit. Noseleafed bats tend to emit a weak sonar signal, giving them the nickname "whispering bats." Females have their young in large maternity colonies. This species, also known as *L. curasoae,* is endangered.

California Leaf-nosed Bat
Macrotus californicus L 3.5 in (9 cm)

The California Leaf-nosed Bat needs no introduction: The combination of noseleaf and enormous ears distinguish it from other North American bats.

KEY FACTS

Body is grayish; nose has distinctive, erect noseleaf; ears are very large and rounded; tail extends beyond membrane.

+ **habitat:** Desert scrublands

+ **range:** Southern California, Nevada, and Arizona

+ **food:** Insects and larvae, including caterpillars; cacti fruit

Barely able to move on the ground, the California Leaf-nosed Bat is master of the air. Short, broad wings lend it great maneuverability. Large ears—a characteristic of strong fliers—enhance hearing, allowing the bat to detect insects among thick vegetation. This species often hovers over trees and other plants, gleaning prey such as caterpillars and grasshoppers from vegetation or the ground. The leaf may indicate that echolocation takes place through the nostrils. These bats seek out roosts in mines and large caves, but they do not cluster together and seem to avoid the direct touch of other roosting individuals.

Western Mastiff Bat
Eumops perotis L 7 in (18 cm)

Large ears that join across the skull and project over the eyes give the Western Mastiff Bat the wrinkled canine appearance that suggests its common name.

KEY FACTS

Body fur is dark grayish brown or brown with white roots that often are visible.

+ **habitat:** Deserts, canyons, scrublands, and urban areas

+ **range:** Southwestern U.S. to northern California

+ **food:** Mainly flying insects, including moths, beetles, and bees

The largest U.S. bat, weighing all of 2 ounces (56 g), the Western Mastiff displays a long tail extending well past the interfemoral membrane that is characteristic of free-tailed bats. Its size and narrow wings make the bat a fast but not agile flier. These bats roost in high places such as crevices in high cliffs, allowing them to drop and launch themselves. They often climb high into the air and utter loud, piercing calls. They mate in spring, and unlike many other bat species, adult males sometimes join maternity colonies. The species, also known as the Western Bonneted Bat, leaves a telltale urine stain on cliff faces.

Brazilian Free-tailed Bat/Mexican Free-tailed Bat

Tadarida brasiliensis L 4.5 in (11 cm)

A tail that extends freely past the interfemoral membrane explains the species' common name. The species forms the largest communal roosts in the bat world.

KEY FACTS

Back is uniform gray-ish brown; underside is slightly paler; ears are wide.

+ **habitat:** Scrub-lands, deserts, and other open areas

+ **range:** Southern U.S., ranging farther north in the West than the East

+ **food:** Flying insects, including moths

Brazilian Free-tailed Bats live a life of superlatives. Narrow wings help it achieve renown as one of the fastest-flying bats, reaching speeds of up to 60 miles (96 km) an hour. It forms colonies in the millions that rank as the largest assemblages of a mammalian species anywhere in the world. Leaving their roosts in caves and buildings and under bridges at sunset, these bats spiral upward into a column and then head off for a night of hunting. Returning to the roost, they free-fall from the sky at the entrance. The species migrates to Mexico for the winter, enduring a flight of up to 1,000 miles (1,600 km).

Big Brown Bat

Eptesicus fuscus L 4.5 in (11 cm)

Familiar to many, Big Brown Bats circle above trees in evenings and enter attics, barns, and belfries to roost. This species is greatly impacted by white-nose syndrome, a fungal disease.

KEY FACTS

Body fur is dark at the roots, becoming brownish at the surface; broad face and rounded ears are blackish.

+ **habitat:** Wood-lands, fields, suburban and urban areas

+ **range:** Throughout U.S. and Canada, except far northern areas

+ **food:** Mainly fly-ing insects, especially beetles

Size and a wide-ranging presence help identify the Big Brown Bat. Its tail extends beyond the membrane between its back legs. The species emerges to hunt after sunset, using echoloca-tion to avoid obstacles and home in on insects. It mates before hibernation, although fertilization is delayed until spring. Females roost in maternity colonies that average 25 to 75 individuals. In the West they tend to have a single offspring and in the east, twins. On warm winter days these bats may awake and forage; a bat seen in midwinter may well be a Big Brown. This species plays a major role in controlling agri-cultural pests.

Eastern Red Bat

Lasiurus borealis L 4.5 in (11 cm)

Hanging by one leg from a tree branch, the solitary Eastern Red Bat resembles a dead leaf, a fact that may deter predators such as blue jays, opossums, hawks, and owls.

KEY FACTS

Male is orange to reddish brown; female less red and frosted with white; both have white patches on shoulder and thumb.

+ **habitat:** Fields, woodlands, and urban areas

+ **range:** Central and eastern U.S. and Canada (except far north), and southern Florida

+ **food:** Flying insects, especially moths and beetles

Reddish fur, long wings, and a furred tail membrane that can be wrapped around the body like a blanket distinguish the Eastern Red Bat. The males and females show a marked difference in coloring, the males being distinctly redder. These bats usually do not stray far from the roost, often in dense foliage, to hunt insects. Bats in the north may migrate to the southern part of the range in winter. Although twins are common among many bat species, the Eastern Red female may give birth to three or four young at one time. This species is sometimes found on the ground and may hibernate on the ground among leaf litter.

Hoary Bat

Lasiurus cinereus L 6 in (15 cm)

These bats are called "hoary" for the grayish white tips of their brown fur. The silvery coat helps them blend in with lichens that grow on evergreen trunks.

KEY FACTS

Body is brownish gray tipped with white; face is surrounded by yellow band; ears are short, round, and black-edged.

+ **habitat:** Deciduous and coniferous woodlands near water

+ **range:** Most of U.S. and Canada, except far north

+ **food:** Flying insects, especially moths, and also smaller bats

North America's most widespread bat, the Hoary Bat is more abundant in the western half of its range. This large bat has powerful wings built for long-distance migration, and it often chatters audibly in flight. The species has appeared far out of range and has been spotted in such distant places as Iceland and the Orkney Islands. It is the only wild land mammal to have reached and established itself in the Hawaiian Islands. This bat migrates in large waves; in spring, the pregnant female moves north ahead of the male and gives birth to two young in May or June. Like all bats, the Hoary is a fastidious groomer.

Northern Yellow Bat

Lasiurus intermedius L 5.5 in (14 cm)

Like the Seminole Bat, the Northern Yellow prefers to roost in clumps of Spanish Moss. It can be distinguished from other yellow bats by its partially furred tail membrane.

KEY FACTS

Fur is long, silky yellowish orange to yellowish brown with a grayish wash; the outer tail membrane is furred.

+ **habitat:** Forest edges with Spanish Moss and palm groves

+ **range:** Coastal southeastern U.S.

+ **food:** Mainly flying insects, including mosquitoes, flies, and bees

Northern Yellow Bats frequently hunt over open areas with water and are seen over coastal golf courses and beaches. Small groups of males and females roost together, and the species may form large feeding aggregations near the end of summer. Females often bear three or four young in summer maternity colonies; in winter, the sexes tend to segregate. The similar Southern Yellow Bat (*L. ega*) is smaller and shows a less grayish wash to its fur. In the United States it is found primarily at the southern tip of Texas. Like other *Lasiurus* species, the Northern Yellow Bat is a vesper bat, a member of the world's largest bat family.

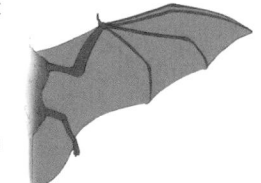

Seminole Bat/Mahogany Bat

Lasiurus seminolus L 4 in (10 cm)

Similar in size, wing shape, and color to the Eastern Red Bat, the Seminole Bat also overlaps with the southeastern portion of the Eastern Red's range.

KEY FACTS

Both male and female have reddish brown fur, white-tipped on back; shoulder and thumb have white patches.

+ **habitat:** Swamps, bogs, and forests, near water

+ **range:** Southeastern U.S.; isolated populations elsewhere

+ **food:** Flying insects, including flies and beetles; also ground insects

Unlike the similar Eastern Red Bat, male and female Seminoles share the same fur color. Eastern Red females are usually duller than Seminoles of either sex. Seminoles fly straight and swiftly, emerging after dusk to hunt on the wing or to glean insects from vegetation or the ground. They often roost individually in Spanish Moss or sometimes in pairs in the same clump of the epiphytic plant. Collection of Spanish Moss for commercial purposes has significantly reduced the habitat available to the Seminole Bat. These bats do not enter deep hibernation for the winter, but may emerge from periods of inactivity on warm winter days.

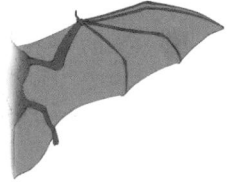

Western Pipistrelle

Pipistrellus hesperus L 3 in (8 cm)

North America's smallest bat, the Western Pipistrelle can weigh as little as a tenth of an ounce (2.8 g). Even so, adult females usually give birth to twins.

KEY FACTS

Body is blondish or yellowish; black mask and ears provide contrast.

+ **habitat:** Grasslands, scrublands, canyons, and woodlands near water

+ **range:** Arid parts of the western United States to Washington

+ **food:** Mainly small swarming insects, including mosquitoes and stoneflies

The tiny Western Pipistrelle has diurnal tendencies, sometimes beginning the nightly hunt before sunset, its slow, fluttering flight suggesting that of a butterfly. After a period of rest, it reemerges and often stays active into the morning in pursuit of swarming insects. Pale coloring echoes the species' customary habitats in arid parts of the western United States. Western Pipistrelles often roost in cracks and crevices of canyons and cliffs or in caves, always near water. Young weighing only about 0.05 ounce (1.4 g) are born in small maternity colonies; within a month or so, they are indistinguishable from adults.

Eastern Pipistrelle

Pipistrellus subflavus L 3.5 in (9 cm)

A tad larger than its western counterpart, the very abundant Eastern Pipistrelle often returns to hibernate in the same spot in the same cave each year.

KEY FACTS

Fur is dark at base, light in middle, and brown on tips; has pinkish brown face and forearms.

+ **habitat:** Farmland and woodlands, near water

+ **range:** Eastern half of U.S. and extreme southeastern Canada

+ **food:** Flying insects, including beetles and flies

Tricolor fur distinguishes this tiny bat, so small that it is sometimes mistaken for a large moth. Primarily a woodland species, the Eastern Pipistrelle often roosts in dense foliage in summer, emerging at dusk to flutter through trees in search of insects. This bat remains in the same general area year-round. It carefully chooses a location in the cave where it overwinters, seeking the right balance of chilly temperatures and moisture to ensure that it does not dry out while hibernating. Females mate in the fall and store the sperm for delayed fertilization, or mate in the spring, giving birth usually to twins.

Rafinesque's Big-eared Bat

Corynorhinus rafinesquii L 4 in (10 cm)

Elaborate, supersize ears are the hallmark of Rafinesque's Big-eared Bat. The tragus, a feature of the outer ear that stands at the entrance to the inner ear, is also prominent in this species.

KEY FACTS

Body is gray to brown on the back and lighter to whitish on the underside; bumps of flesh appear behind the nostrils.

+ habitat: Forested areas near water sources

+ range: Southeastern United States

+ food: Flying insects, especially moths

The large ears and tragus of this bat species likely serve as direction detectors. The tragus stands erect, even when the bat is resting and coils up its ears and tucks them under its wings. When disturbed by a noise, the bat moves its head and waves its ears as though tracking echoes. The elusive Rafinesque forages for insects—often hovering in flight—later in the evening than many other bats, and returns to its roost while it is still dark. This species was partial to roosting in the "twilight zone" at cave entrances, but it now roosts in buildings and other structures as human activity has breached these habitats.

Pallid Bat

Antrozous pallidus L 4.5 in (11 cm)

The Pallid Bat pursues a unique bat lifestyle: It feeds almost entirely on items from the ground, taking such prey as crickets, grasshoppers, scorpions, and lizards.

KEY FACTS

Body is sand colored; ears are enormous and point forward; nose is piglike.

+ habitat: Deserts, grasslands, canyons, and mixed forests

+ range: Southwestern U.S. to southern British Columbia

+ food: Ground and flying insects, scorpions, and nectar

The Pallid Bat is colored to match its desert surroundings. Like other big-eared bats, this species can hover in flight. But it lands often, gleaning vegetation and the ground for insects. Because of prickly prey and prickly vegetation, the Pallid Bat frequently displays holes in its wing and tail membranes. The Pallid Bat also visits cactus flowers, serving as an important pollinator. It roosts in rock crevices and buildings. During a night of foraging, it may come to rest in open, sheltered spots such as bridges and porches. This noisy bat calls to others and is known to bare its teeth and buzz when frightened or angered.

Silver-haired Bat

Lasionycteris noctivagans L 4 in (10 cm)

The Silver-haired Bat's silver-frosted fur gives it a distinctive appearance. The bottom half of the interfemoral membrane between its back legs also is furred.

KEY FACTS

Back is dark brown to black with silver frosting; face is dark; ears are rounded.

+ **habitat:** Grasslands, scrublands, and woodlands near water

+ **range:** Most of U.S. and Canada, except far northern areas

+ **food:** Mainly small flying insects including beetles, flies, and moths

Soon after sunset the Silver-haired Bat begins to hunt, often flying low and slowly in search of flying insects, especially its preferred prey: moths. It seldom is found far from water, roosting solitarily or in small groups in mines, caves, hollow trees, rock crevices, houses, and under bark. Silver-hairs mate in autumn. As with other bat species, fertilization is delayed until spring, when females form much smaller maternity groups than other species do. After a two-month gestation, they give birth, usually to twins. Some Silver-hairs appear to migrate; others stay put and may enter a state of torpor in winter.

Long-eared Myotis

Myotis evotis L 3.5 in (9 cm)

Long, dark ears—the better to hear insects rustling in the vegetation—characterize the Long-eared Myotis. This bat often appears to turn off echolocation when hunting.

KEY FACTS

Back color ranges from yellowish to dark brown; face and ears are black.

+ **habitat:** Woodlands, shrublands, grasslands, and agricultural areas

+ **range:** Temperate areas of western U.S. and Canada, to about 9,200 feet (2,800 m)

+ **food:** Flying insects, including moths and beetles

Long-eared Myotis leave their customary roosts in rocky outcroppings, dead trees, caverns, and buildings to glean insects from dense vegetation. To do so, they may hover momentarily to capture their prey. They also hunt among grass and trees along bodies of water. Active a bit longer into the night than other bats, they may hunt closer to the ground as the night air gets cooler. Females form small maternity colonies, often with males roosting nearby. They bear a single offspring in the summer months. Long-lived for a small mammal, this species can survive up to 22 years, although a typical life span is much shorter.

Little Brown Bat/Little Brown Myotis

Myotis lucifugus L 3.5 in (9 cm)

The Little Brown Bat is one of North America's most abundant bat species. As natural sites disappear, this highly adaptable bat often establishes colonies in houses and other buildings.

KEY FACTS

Back varies from tan to reddish to dark brown; underside is buff or grayish white.

+ **habitat:** Forested areas, often near water

+ **range:** U.S. and Canada, except far north and southwestern and south-central U.S.

+ **food:** Aquatic insect larvae, flies, moths, beetles, and other insects

The Little Brown Bat is a basic, no-frills species—small, glossy-furred, and lacking any facial ornamentation such as noseleafs or chin leafs. These bats mate in the fall at their hibernation sites, often in caves, but fertilization occurs with spring emergence. Females form maternity colonies with up to thousands of individuals in warm places, such as attics. When a single offspring is born after 50 to 60 days, the mother receives it in the membrane between her legs, which is cupped like a basket. Within a month it will attain adult weight: all of a quarter of an ounce (7 g). These little bats can live as long as 40 years.

Ocelot

Leopardus pardalis L 37–50 in (94–127 cm)

The beautiful, endangered Ocelot is a species that generally keeps to itself. Far less is known about its habits than about those of other wild cat species in North America.

KEY FACTS

Head, back, limbs, and tail have black spots and markings on grayish, reddish, or buff background.

+ **habitat:** Forests and brushland

+ **range:** Extreme southern Texas; former U.S. range much larger

+ **food:** Rabbits, other small mammals, birds, reptiles, fish, frogs, and some invertebrates

Fast on the ground and adept at tree climbing if pursued, the Ocelot typically rests in foliage by day and hunts by night. Both male and female Ocelots are territorial, and males try to claim females whose territories overlap with their own. The species' distinctively marked coat, which helps it blend with the dappled sun and shade of its environment, has made it a target for its fur and for the pet trade. Its endangered status makes those activities illegal in the United States and in some Latin American countries in its range. The tiny population surviving in Texas faces habitat loss and the perils of living near roads.

Canada Lynx

Lynx canadensis L 27–42 in (69–107 cm)

This stealthy cat of the boreal forest creeps up on its big cat feet, stalking undetected to within a few bounds before pouncing on its prey—usually the Snowshoe Hare.

KEY FACTS

Back is grayish buff mottled with brown; black-tipped tail is bobbed; ears have black tufts; face has sideburns.

+ **habitat:** Dense vegetation in northern boreal forests

+ **range:** Alaska and much of Canada, northwestern U.S., upper Midwest, and upper Northeast

+ **food:** Primarily Snowshoe Hares

The fate and well-being of the Canada Lynx is tied closely to that of the Snowshoe Hare. When hare populations crash, so do the lynx's, almost simultaneously. Breeding takes place in winter, even in famine, but the typical litter of one to four kittens—born in a den after 60 to 65 days—may suffer high mortality. A robust lynx litter looks and sounds much like a litter of domestic cats, but larger and a lot louder. Lynxes are larger than Bobcats, their ear tufts and sideburns are longer, and their tails are entirely black-tipped. The Canada Lynx's luxurious fur has made it a prime target of trappers for hundreds of years.

Bobcat

Lynx rufus L 24–47 in (60–119 cm)

The Bobcat's bobbed tail offers no advantage in balance or navigation, but with its white underside, it acts as a location flag for kittens trailing their mother in dense underbrush.

KEY FACTS

Back varies from brown to grayish with spots, even to solid black; underside is white with black spots; tail tip is black.

+ **habitat:** Forests, swamps, grasslands, and mountains

+ **range:** U.S. and southern Canada; absent in parts of Midwest and East

+ **food:** Rabbits, hares, rodents, birds, and deer

This solitary hunter, the most widespread wild cat in North America, is about twice the size of a domestic cat. It mainly hunts rabbits and hares but is capable of killing healthy deer. Bobcats tend to be nocturnal, although they can be spotted during the day. Adults come together at breeding time, but severe weather may find some sharing a rock shelter without much interaction. The one to six young born in spring or summer stay with their mother and disperse before the next litter arrives. The species adapts well to disturbed habitats, but heavy trapping has greatly reduced its numbers or eliminated it in some areas.

Mountain Lion/Cougar/Puma

Puma concolor L 5–8 ft (1.5–2.4 m)

The Mountain Lion, largest of the cats in the United States and Canada, answers to many regional names. In addition to Cougar and Puma, it is called Panther, Catamount, and Painter, among others.

KEY FACTS

Body is tan or russet with a whitish underside; long tail has a black tip.

+ **habitat:** Desert, forest, and mountain wilderness areas

+ **range:** Western United States into British Columbia and Alberta and southern Florida

+ **food:** Primarily large hoofed mammals

The Mountain Lion has certain traits that make it more similar to a house cat than to a lion. The species is a purring cat, not a roaring one, and it leads a solitary life except at mating time. It also employs a stalk-and-ambush hunting strategy. Cougars may mate at any time of year. A litter of one to six spotted kittens is born in a secure den after a 90-day gestation; their spots will soon give way to the single color reflected in the species name: *concolor,* or uniform color. This large cat once ranged across North America, but now is confined mainly to the West. The Florida Panther, a highly secretive subspecies, is endangered.

Coyote

Canis latrans L 3.5–4 ft (1–1.2 m)

As the Gray Wolf retreated in North America, the adaptable and resourceful Coyote fanned out from the prairies and took its place—and has since added even more territory to its range.

KEY FACTS

Back ranges from grayish to tawny brown; muzzle and legs are long; bushy tail is black-tipped.

+ **habitat:** Deserts, grasslands, woodlands; agricultural, suburban, and urban areas

+ **range:** Throughout U.S. and Canada, except some northern areas

+ **food:** Mammals, birds, snakes, insects, and carrion

The Coyote has met few habitats not to its liking. Able to adapt its living, hunting, and breeding styles to many areas, the species has even become an entrenched urbanite. Coyotes often are confused with foxes, which are smaller and have pointier faces, and with some breeds of domestic dogs, which usually have less bushy tails. They sometimes mate with dogs, producing a fertile, though not robust, offspring known as a "coydog." Coyotes live as individuals, mated pairs, or in family-based packs. Pack hunters are able to take down larger prey, such as deer. Eastern Coyotes tend to be larger than western ones.

Gray Wolf

Canis lupus L 4–6 ft (1.2–1.8 m)

The Gray Wolf once had free roam in much of North America, but centuries of trapping, shooting, and poisoning, often for a bounty, have confined it mainly to the northern U.S. and Canada.

KEY FACTS

Color from white to black, most often grayish on back; underside is usually lighter; ears are small; tail is long and bushy.

+ habitat: Forests and tundra in wilderness areas

+ range: Alaska, western and central Canada to northwestern U.S.

+ food: Hoofed mammals, hares, and rodents

The Gray Wolf most often lives in packs; lone wolves are rare. Packs comprise a half dozen or more individuals that live and hunt together to bring down large hoofed mammals. In a pack, there is a dominant, or alpha, male and an alpha female; they fill leadership roles and hold reproduction rights and other perks, such as the first go at a kill. Other adult wolves in the pack help feed and look after the pups of the alpha couple. A litter averages six young, born in spring through early summer. A proposal has been made to remove the Gray Wolf from the federal endangered species list.

Red Wolf

Canis lupus rufus or *Canis lupus* L 4–5 ft (1.2–1.5 m)

This smaller, slimmer, and redder relative of the Gray Wolf once populated much of the southeastern United States. Last-minute efforts in the 1970s saved it from extinction.

KEY FACTS

Back is gray to black to reddish brown; sides are tan; underside is white.

+ habitat: Formerly in forests, bottomlands, grasslands, other dense vegetation

+ range: Reintroduced to coastal North Carolina and South Carolina, Tennessee, and Gulf islands

+ food: Rodents, rabbits, deer, other mammals, and birds

The Red Wolf is similar in appearance to the Coyote, but the latter has more gray on its muzzle, which usually is pointier than that of the Red Wolf. The spread of the Coyote displaced the Red Wolf—an animal that lives in mated pairs or small packs—which already was reduced to low levels from targeted elimination and environmental disruption. Coyotes interbred with the wolves, leaving only a tiny Red Wolf population on the border of coastal Texas and Louisiana. Wolves there were evacuated for captive breeding and founded a population that has allowed small, but successful, reintroductions in the wild, starting in 1987.

Gray Fox
Urocyon cinereoargenteus L 2.5–3.5 ft (0.8–1 m)

The Gray Fox often uses its sharp, curved claws to scramble catlike up a tree to escape predators, to rest, or to forage. It can also jump nimbly from branch to branch.

KEY FACTS

Body is grizzled gray with reddish ears, neck, and sides; underside is white; bushy, tail tip is black.

+ **habitat:** Woodlands, old fields, brushlands, and rocky areas

+ **range:** Much of U.S., except Northwest; parts of southern Canada

+ **food:** Mammals, lizards, frogs, insects, nuts, fruit, and carrion

The Gray Fox is active at dawn, dusk, and during the night, availing itself of a wide range of animal and vegetable food sources. During the day it rests, often in dense vegetation or in the shade of a tree or rock. Dens mostly come into play during whelping season. The female gives birth to an average of four kits in a hollow log, rock crevice, abandoned building, brush pile, or burrow abandoned by other species, including the Red Fox and Groundhog. The male, or dog fox, helps care for the young, but the female fox, or vixen, meets most of their needs until they can fend for themselves at about seven months.

Arctic Fox
Vulpes lagopus L 2.5–3.5 ft (0.8–1 m)

The grayish brown coat of the Arctic Fox molts to a snow-compatible white one as winter begins. Other Arctic adaptations include a compact body, short legs, small ears, and furred feet.

KEY FACTS

Body is grayish brown above and cream on the underside in summer; coat turns white in winter.

+ **habitat:** Tundra, coastal areas, and ice floes

+ **range:** Arctic areas of Alaska and Canada

+ **food:** Rodents, hares, fish, birds, eggs, carrion, and garbage

The Arctic Fox needs camouflage because it remains active all winter, retreating to a snow bank or den only in severe weather. A rare variety of the species has a coat of bluish gray in summer that gets paler in the winter. Arctic Fox mates work together to rear the 6 to 12 pups born in a den that is reused and enlarged for decades or longer, because of the difficulty of new excavation in the permafrost. An old den may be a honeycomb of tunnels and entrances; the fox often sits watch on top. The Arctic Fox eats a wide range of seasonal foods, including the lemming, its traditional prey. It also scavenges Polar Bear kills on ice.

Swift Fox
Vulpes velox L 2.5–3 ft (0.8–0.9 m)

True to its name, the lithe Swift Fox, with its elegant, wide-set ears and dark-sided muzzle, can outrun all but the fastest predators on its home turf of flat, open terrain.

KEY FACTS

Back is grizzled gray with yellowish cast; neck, sides, and legs are yellowish orange; throat and underside are white.

+ habitat: Grasslands, shrublands, and desert

+ range: High plains from Texas into southern Canada

+ food: Rodents, rabbits, birds, lizards, insects, and vegetation

The Swift Fox is well camouflaged to blend with tawny grasses of the high plains; it sometimes rests in the sun outside its underground den. This fox appears to have strong family bonds. Two to seven young are born in the den as early as February; they emerge at three weeks, and the parents begin training them to hunt. If the female dies, the male may raise the pups alone. In the fall, they are ready to go off on their own. A hundred or so years ago, the Swift Fox population declined drastically from habitat destruction because of conversion of prairie to cropland, trapping, and poison intended for other species.

Red Fox
Vulpes vulpes L 3–3.5 ft (0.9–1 m)

Settlement and development, along with the disappearance of wolves, helped the Red Fox expand its horizons and overtake the Gray Fox in abundance in eastern woodlands.

KEY FACTS

Body is usually orangish red; legs and ears are black; tail tip is white.

+ habitat: Fields, forests, brushland, marshes, suburban and urban areas

+ range: Most of U.S. and Canada, except parts of western U.S.

+ food: Mammals, birds, eggs, frogs, insects, fruit, berries, and carrion

European settlers found the Gray Fox to be frustratingly arboreal, so they imported the Red Fox from Europe for foxhunting, only to discover that the species was a native here. The Red Fox hunts in a catlike manner, sneaking up to pounce on prey with a pronounced jump. The species seems to mate for life. The male and female dig a den in winter or find a suitable cave or crevice in which the female gives birth to about five kits in the spring. Fox families often incorporate daughters from the previous year. From a distance, foxes are sometimes confused with domestic cats; larger ears and brushy tails help identify the fox.

American Black Bear

Ursus americanus L 4–6 ft (1.2–1.8 ft)

Known for shaking down visitors to national parks and trashing cars, tents, and cabins, the American Black Bear achieved an eclectic appetite through exposure to the human world.

KEY FACTS

Color black to reddish to white; dark bears have lighter snouts.

+ **habitat:** Forests, swamps, tundra, and mountains

+ **range:** Canada and western U.S.; north-eastern U.S.; isolated populations elsewhere

+ **food:** Grasses, acorns, fruits, insects, mammals, fish, birds, and carrion

More adaptable than the other two North American bear species *(U. arctos* and *U. maritimus),* the American Black Bear also is the most wide-ranging. It has larger ears and eyes and a longer snout than the other species. Males are larger than females, weighing up to 900 pounds (400 kg). Black Bears climb trees efficiently and can sprint up to 35 miles (56 km) an hour. They den in winter in such spots as caves, under fallen trees, and in tree hollows. They do not hibernate—or undergo a marked metabolic slow-down—but are merely dor-mant. Females give birth to one to three cubs in the den and nurse them throughout the winter.

Grizzly Bear/North American Brown Bear

Ursus arctos horribilis L 5.5–8.5 ft (1.7–2.6 m)

In some areas these bears have brown fur with grayish tips that give the Grizzly Bear its common name: Grizzled means "grayish."

KEY FACTS

Body is tan, grayish, reddish, or brown; ears are small, rounded; shoulder is humped.

+ **habitat:** Forests and open area in mountains; rivers and coasts

+ **range:** Northern U.S. Rockies into west-ern Canada and Alaska

+ **food:** Vegetation, nuts, berries, insects, birds, eggs, fish, mam-mals, and carrion

A Grizzly Bear in profile with all feet on the ground shows a higher shoulder; a Black Bear shows a higher rump. The Grizzly's large forefeet armed with up to 4-inch (10 cm) curved claws can deliver a lethal blow. Adult males dominate females and young bears, especially during the feeding frenzy of the salmon run. Mothers will secure their offspring, born in the winter den, from adult males, which are known to sometimes kill and eat the young. Like Black Bears, Grizzlies take advan-tage of food sources asso-ciated with human activity, especially in national parks, although measures are taken to mitigate this situation.

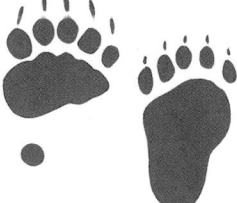

Kodiak Bear

Ursus arctos middendorffi L to 9 ft (2.7 m)

The largest North American land mammal, outweighing even the Polar Bear, the Kodiak Bear is a subspecies of North American Brown Bear. It is named for Alaska's Kodiak Island.

KEY FACTS

Body is blondish brown to dark brown; has pronounced shoulder hump.

+ **habitat:** Grasslands, meadows, wet tundra, shrublands, and spruce forest

+ **range:** Kodiak Island and other islands and coast of southern Alaska

+ **food:** Vegetation, nuts, berries, fish, small mammals, and carrion

Kodiaks of southern Alaska display the same humped shoulder, scooped face, and long claws that other Brown Bears do, including inland Grizzlies. Yet these beachcombers can weigh up to 1,600 pounds (700 kg), more than an average Grizzly. Like other Brown Bears, Kodiaks feed to maximize weight gain, utilizing a variety of foods to create the bulk that will carry them through many months in the den without eating and drinking—or urinating or defecating. Alaska protects its Kodiak Bears, allowing very regulated hunting and safeguarding traditional habitat. Aggressive encounters between Kodiaks and humans tend to be rare.

Polar Bear

Ursus maritimus L 6–8.5 ft (1.8–2.6 m)

The massive Polar Bear spends much of its time adrift on pack ice and floes, stalking the Ringed Seals that compose its main source of food.

KEY FACTS

The fur is white to yellowish white; the neck is long; the head is small; the eyes and nose are black.

+ **habitat:** Pack ice and ice floes, rocky shores, and islands

+ **range:** Canadian Arctic coasts and islands

+ **food:** Seals, walrus, small whales, fish, ducks, plants, and carrion

The Polar Bear has developed a unique Arctic lifestyle. This strong swimmer has an insulating white coat that provides effective camouflage. After mating, the adult female digs a den in a snowbank that is vented to the surface. The female enters the den in the fall and gives birth, usually to twins, within a few months; mother and cubs emerge in the spring. Like other bear species, the Polar Bear seeks out food associated with human habitation, including garbage dumps. Attempts to relocate "nuisance" bears often fail, as they are accustomed to traveling hundreds of miles during the year in search of food and then returning.

North American River Otter

Lontra canadensis L 38–49 in (96–125 cm)

Otters often build play into their daily routines. For example, family groups body-sled one at a time down snowbanks or slippery mud banks, often ending the run with a loud splash into a river.

KEY FACTS

Body is brown above, silvery below; short legs have webbed feet; tail is thick at base.

+ habitat: Streams, rivers, lakes, swamps, and coastal areas

+ range: Alaska and Canada, northwestern and eastern U.S.; reintroduced elsewhere

+ food: Fish, crayfish, frogs, small mammals, and insects

Among the North American River Otter's aquatic adaptations are a streamlined body, waterproof coat, and flaps of skin that close nose and ears when it dives. Otters prefer bodies of water replete with dens and burrows of species such as Muskrats and American Beavers. Otters mate in water, and implantation is often delayed as long as nine months. Females bear one to six young in a den or lodge; males do not participate in rearing the young. Overhunting and pollution decimated otter populations, which once occurred in much of North America. The species is now making a comeback through protection and reintroduction.

Wolverine

Gulo gulo L 32–46 in (80–116 cm)

The Wolverine's fierce reputation is not hype: The strength of this giant weasel likely exceeds that of any other mammal its size. It can take down large hoofed mammals.

KEY FACTS

Body brown to black; lighter face rim and ears; often has white chest patches.

+ habitat: Boreal forests, tundra, taiga; mountainous areas

+ range: Northern Canada to northern Rockies; some in western coastal states

+ food: Mammals, birds, berries and other plant material, and carrion

The elusive Wolverine is equipped for long, northern winters. It does not migrate or hibernate, but continues its opportunistic search for food with the aid of a thick fur coat and broad, furred feet that allow it to lope over crusty snow. Its signature weasel scent glands exude a highly odoriferous substance that marks territory and stakes a claim on its kills. Strong, smart, and often at odds with human activities, the species was targeted for extermination with traps and poisons, and suffered widespread habitat destruction. These factors nearly wiped out the Wolverine in the United States; only remnant populations survive.

American Marten/Pine Marten

Martes americana L 18–24 in (45–60 cm)

Mature evergreen forests form the prime habitat of this solitary, catlike carnivore. Adult American Martens often establish a home range of up to 15 square miles (39 sq km) when food is scarce.

KEY FACTS

Body is dark brown; head is paler; throat is whitish or orange.

+ habitat: Coniferous and mixed forests

+ range: Much of Alaska and Canada; small populations elsewhere, mainly in western U.S.

+ food: Small mammals, birds, eggs, insects, worms, nuts, berries, and carrion

A skilled hunter in the trees or on the ground, the American Marten exploits a wide range of food sources during the year. Agile and fast enough to pursue a squirrel from branch to branch—and catch it—the marten can also track a Snowshoe Hare or vole in the snow, traveling when necessary through tunnels made by other animals. This species once ranged well into the United States, but has suffered significant habitat loss; reintroduction programs have been successful. It differs from the similar American Mink by its less uniform coat color and dark appendages and from the Fisher by its smaller, more slender size and shape.

Fisher

Martes pennanti L 29–47 in (74–119 cm)

Despite the name, Fishers don't fish. They primarily target hares, rabbits, and small rodents for a meal. They also successfully take on adult porcupines, launching an attack from the front.

KEY FACTS

Body has long dark brown fur; head is grizzled; ears are small; nose is pointy.

+ habitat: Mature coniferous and mixed forests

+ range: Southeastern Alaska through Canada into eastern U.S. and western mountains

+ food: Rodents, rabbits, hares, birds, fruits, nuts, and carrion

Fishers are omnivores that often exploit the same habitats and food resources as American Martens. Where the two occur together, the larger Fisher often takes the larger prey. Foresters welcome the Fisher's willingness to go after North American Porcupines, as the quilled rodents decimate trees by their fondness for inner tree bark. The Fisher bites the porcupine in the face and neck before rolling it over to tear into the belly. A female Fisher mates about a week after giving birth. The embryo doesn't implant for another ten months or so. One to six young are born in the early spring, often in a den in a hollow tree or log.

Ermine/Stoat/Short-tailed Weasel

Mustela erminea L 7–12 in (18–30 cm)

Whether in its winter white or summer brown coat, the Ermine is a long, sleek, and very efficient hunter, able to pursue small mammals into any burrow or hole it can get its head into.

KEY FACTS

Back is brown to reddish in summer (whitish in winter); underside is white; tail tip is black.

+ **habitat:** Forests, tundra, meadows, marshes, riverbanks, and hedgerows

+ **range:** Alaska and Canada into western U.S., upper Midwest, and Northeast

+ **food:** Rodents, rabbits, frogs, and earthworms

Lightning-quick reflexes make the Ermine a successful predator, able to take down prey larger than itself and dispatch it with bites to the neck. Male Ermines are almost twice the size of females. Adults breed in summer, and after delayed implantation, four to ten young are born in spring, often in a nest in a rodent burrow. The species tends to take advantage of an abundant food supply by killing whatever and whenever it can and then storing surplus, often in a storeroom formed from a side tunnel of its den. This may lead to mass slaughter when the weasel encounters a bounty of potential prey, as in a hen house.

Long-tailed Weasel

Mustela frenata L 10–17 in (25–43 cm)

The largest weasel in North America, this species' tail measures about half its total length. Coat color shows a lot of regional variation, including a "bridled" variety with white markings on the face.

KEY FACTS

Body is brown to reddish to orange to white; underside is often whitish; tail tip is black.

+ **habitat:** Woodlands, fields, and meadows, usually near water

+ **range:** Southern Canada and most of U.S., except desert Southwest

+ **food:** Rodents, rabbits, birds, eggs, snakes, insects, and carrion

Long-tails are a generalized weasel species, occupying a wide geographic range and habitat distribution, and exploiting a wide variety of food sources. Active day and night, the species feeds voraciously to meet its tremendous energy needs, and is equally adept hunting in trees, on the ground, and underground. This weasel shares part of its range with the Ermine and Least Weasel. The Long-tail tends to go after larger prey animals to lessen competition for resources, but it sometimes includes the other weasels on its menu. Like other weasels, Long-tails in the northern parts of the range molt to white coats for the winter.

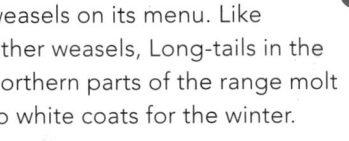

Black-footed Ferret

Mustela nigripes L 18–23 in (45–58 cm)

The survival of the endangered Black-footed Ferret is tied to the existence of prairie dogs, whose prairie ecosystem has dwindled to about 2 percent of its original expanse.

KEY FACTS

Body is tawny brown; face has black mask; feet and tail tip are black.

+ habitat: Great Plains grasslands

+ range: Formerly central U.S. to south-central Canada; now, small populations formed by reintro-duced captive-bred animals

+ food: Prairie dogs, other ground squirrels, and mice

The masked Black-footed Ferret not only dines on the prairie dog, but also appropriates and remodels the rodent's labyrinthine burrows for its own use. Active mainly at night and at dawn, the ferret hunts mostly underground; when it finds a prairie dog, the ferret seizes it by the throat, kills it, and takes it to its own burrow. The ferret was believed to be extinct until a very small population was found in Wyoming in the 1980s. That population dwindled, and the species again was believed extinct in the wild in 1987. Research suggests that captive-breds familiar with prairie dog burrows before release fare better upon reintroduction.

Least Weasel

Mustela nivalis L 5.5–8 in (14–20 cm)

The smallest carnivore, the Least Weasel packs a lot of predatory punch into a small body that may weigh less than 1 ounce (28 g) or top out around 2 ounces (56 g).

KEY FACTS

Body is brown in sum-mer; throat and under-side are white; body in winter is white with black-tipped tail.

+ habitat: Meadows, marshes, brushlands, coniferous and mixed forests

+ range: Alaska through Canada and north-central U.S.

+ food: Mice, voles, other rodents, birds, eggs, and insects

A voracious appetite keeps the Least Weasel on overdrive, alternating periods of intense activity with rest around the clock. Its main quarry are voles and mice and, like other weasels, it may store a surplus in a chamber of its den, which—again, weasel-like—it may have taken over from a burrowing rodent. Unlike other weasels, the female Least Weasel does not experience delayed implantation after mating. She can produce two litters of three to ten young each year, which she nurtures in a den nursery that she sometimes lines with fur from the original occupant victim. The species evades pursuers by squeezing into tight spaces.

American Mink

Neovison vison L 18–24 in (45–60 cm)

The lustrous, warm coat of the American Mink helps it retain heat, especially when it swims. Today, mink farms mainly supply pelts fashioned into coats and other often controversial apparel.

KEY FACTS

Body is brown, often with white patches on chin, chest, and belly; tail tip is black.

+ habitat: Wooded areas, usually near sources of water

+ range: Much of U.S. and Canada, except Southwest and parts of southeastern U.S.

+ food: Small mammals, birds, eggs, fish, crayfish, frogs, and snakes

The voracious, solitary American Mink spends much of its time—by night and at times by day—filling its demanding stomach. It hunts in the water and on land, swimming well with semi-webbed feet and climbing agilely with its streamlined body; it can dive to depths of 20 feet (6 m). The mink's amphibious lifestyle helps it nab a wide variety of prey, including fish and crayfish. Minks mate in the spring; an average of four young with silvery white coats are born in early summer in burrows often appropriated from other mammals, such as Muskrats. When stressed, the species may squeal, hiss, and empty its anal glands.

American Badger

Taxidea taxus L 24–32 in (60–80 cm)

Like a criminal on the lam, an American Badger in summer often doesn't sleep in the same burrow for two consecutive days. It usually holes up for a longer spell in winter.

KEY FACTS

Body is grayish above; underside is yellowish; darker head has white cheeks and a stripe that extends onto the back.

+ habitat: Grasslands, meadows, woodland edges, and deserts

+ range: Southwestern Canada to western and central U.S.

+ food: Rodents, snakes, birds, invertebrates, and carrion

A low-slung, flat body and long, strong claws make the American Badger a digging machine. It excavates burrows announced at the surface by a hole about a foot wide with a mound of dirt in front. When cornered, a badger may retreat into its burrow or hastily dig a new one, flinging dirt at its pursuer. Badgers are canny hunters, often lying in wait in their prey's empty burrow for a meal delivery. The species is mostly nocturnal and solitary, except during mating season. Females bear one to five cubs in the nest area of a burrow; the cubs stay with her about a month before venturing out and learning to excavate.

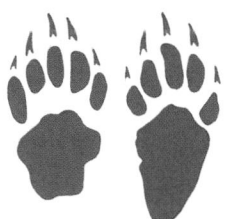

American Hog-nosed Skunk

Conepatus leuconotus L 20–35 in (50–90 cm)

A bare, flexible piglike snout allows the American Hog-nosed Skunk to root for insects, grubs, and worms. Its diggings also resemble the work of hogs.

KEY FACTS

Black body has wide white stripe from head to rump; tail is all white or white on top.

+ **habitat:** Desert valleys, brushy canyons, and agricultural areas

+ **range:** Southern Arizona, New Mexico, and Texas, possibly north to Colorado

+ **food:** Insects, worms, rodents, reptiles, and vegetation

Weighing up to 10 pounds (4.5 kg), this skunk's large size and sturdy claws allow it to make quite a disturbance of soil. In Texas this behavior gives it the popular name "rooter skunk." The Hog-nosed Skunk is seldom distracted by the presence of other species during its nocturnal foragings. It usually keeps its head down and continues to root with its wriggling, naked snout. The Hognose has never been a favorite of trappers because its fur is coarse and somewhat dingy. Even rattlesnakes usually retreat rather than face a pungent barrage. The genus name means "little fox"; it is the only skunk genus known in South America.

Hooded Skunk

Mephitis macroura L 24–30 in (60–75 cm)

A ruff of longer neck hair on this long-tailed skunk gives it its common name. The hood can be black or white; in one pattern, the white of the hood may extend down the skunk's back.

KEY FACTS

Body is black with various white patterns on head and back ranging from white hood to side stripes to wide white central stripe.

+ **habitat:** Streamside brushlands, canyons, grasslands, and deserts

+ **range:** Southern Arizona and New Mexico; western Texas

+ **food:** Insects, rodents, bird eggs, and fruit

Despite varied body markings, the Hooded Skunk often has a characteristic thin white line down the center of its face. It lives a life of nocturnal foraging in the desert Southwest, mainly looking for insects, which form the bulk of its diet. Though mostly solitary, several Hoodeds may come together to eat without incident. The species dens often in rocky crevices, but seldom appropriates abandoned cabins or lives under sheds, unlike the Striped Skunk. A female gives birth to one litter of three to five young each year. Like other skunks, Hoodeds have few predators, although they may fall prey to a Coyote or Bobcat.

Striped Skunk
Mephitis mephitis L 21–34 in (53–86 cm)

An encounter with a Striped Skunk in full defensive posture—spine arched, tail lifted—often leads to a spray of stinky fluid that can travel 15 feet (4.6 m) and be smelled a half mile (0.8 km) away.

KEY FACTS

Black body has forked white stripe extending to rump; face has thin, white central stripe; tail is bushy.

+ **habitat:** Fields, open woodlands, and suburbs

+ **range:** Much of U.S. and Canada, except far northern regions

+ **food:** Vegetation, insects, mice, shrews, bird eggs, and carrion

The Striped Skunk and its close relatives use two scent glands at the base of the tail, with caustic, smelly contents, as an extremely successful defense mechanism. The glands contain about three teaspoonfuls of musk that lasts through five or so discharges. Striped Skunks usually hiss and stamp their feet in warning before a discharge, which they seldom have to deploy; most potential predators learn to give the skunk a wide berth. Basically an easygoing species, the skunk adapts well to human habitation. It is sometimes possible to see a mother skunk out on an evening forage with her four to seven kits following in single file.

Western Spotted Skunk
Spilogale gracilis L 12–17 in (30–43 cm)

North America's smallest skunk is known for the distinctive handstand posture with waving tail that it takes when threatening to spray its oily, stinky musk.

KEY FACTS

Body is black with broken white stripes; a white triangle between eyes and a stripe behind eye; tail tip is white.

+ **habitat:** Deserts, canyons, woodlands, and farmland

+ **range:** Western U.S. into southern British Columbia

+ **food:** Rodents, insects, snakes, lizards, bird eggs, and fruit

This agile skunk sports a silky coat marked with a pattern of broken stripes that resemble spots. The species is an adept tree climber and sometimes feeds—and sleeps—in trees. More often it hunts nocturnally on the ground and dens underground or in rock crevices or hollow trees. It may dig its own burrow or take over one abandoned by another animal. The similar Eastern Spotted Skunk (*S. putorius*) is often a bit bigger, its facial triangle is smaller, and the line behind the eye that runs to the mid-back is narrower. It lives in the midwestern United States, extending slightly into Canada, and in parts of the Southeast.

Ringtail

Bassariscus astutus L 26–31 in (66–79 cm)

Also known as miner's cat, the Ringtail was welcome in western gold mining camps, where it ably kept mice populations in check and was tamed to become a companion to prospectors.

KEY FACTS

Body is brownish gray with very long black-banded tail; eyes are large and ringed in white.

+ **habitat:** Canyons and other rocky areas, mountains

+ **range:** Throughout much of southwestern U.S.; range is expanding

+ **food:** Plants, insects, spiders, and rodents, including ground squirrels

The Ringtail is a late-night creature: It tends to come out to hunt among the rocks and scrub after raccoons and skunks have had their fill. It descends from tree perches and canyon walls with great agility, balanced by a tail that measures as long as its body. This super-agile species also has a hind foot that swivels 180°, aiding climbing and making the contents of high-up bird nests easily accessible to a raiding Ringtail. These animals usually are shy and solitary, although sometimes mated pairs may stay together. A litter of one to four offspring are born in the spring in a rock crevice den, burrow, or tree hollow.

White-nosed Coati

Nasua narica L 3.5–4.5 ft (1–1.4 m)

The White-nosed Coati's long, flexible, and sensitive snout allows it to root successfully for food and gives the southwestern species one of its nicknames: hog-nosed coon.

KEY FACTS

Fur is brownish or reddish; face has dusky mask and white spots around the eyes; tail is long and faintly banded.

+ **habitat:** Woodlands, scrublands, canyons, and mountains

+ **range:** Extreme southwestern United States

+ **food:** Insects and other invertebrates, eggs, small vertebrates, and fruit

Unlike its Raccoon and Ringtail cousins, the White-nosed Coati is most active during the day. It lives in bands of 30 or more females and young, sleeping in trees and coming down at dawn to forage. Adult males are usually solitary (another nickname is *gato solo*—"lone cat"), and are welcomed by females only during mating season. Females give birth to litters of two or more in early summer. A coati walks with its long tail upright; the tail helps it balance when climbing trees, but is not prehensile. Primarily a Central American species, the White-nosed Coati ranges into southern Texas, New Mexico, and Arizona.

Raccoon

Procyon lotor L 22–39 in (56–100 cm)

Humanlike hands with thin, dexterous fingers and a delicate sense of touch give the Raccoon a great advantage when searching for food on land, in the water—or in your latched garbage can.

KEY FACTS

Body is grizzled gray; nose and face mask are black; tail is striped. Color varies regionally.

+ **habitat:** Woods, wetlands, suburban and urban areas

+ **range:** Most of U.S. (except parts of West) and southern Canada

+ **food:** Berries, nuts, seeds, crayfish, crabs, fish, turtles, eggs, small mammals, human food

The Raccoon has a large skill set that allows it to adapt easily to urbanization. The mostly nocturnal species typically spends its days in a tree, coming down at night to pursue an omnivore lifestyle. It often forages for prey in water, using fingers to manipulate, but not wash its catch. Raccoons mate in early spring, and by April or May the two to seven kits are born in a tree hollow den—or perhaps in your attic. Raccoons sometimes den communally and do not hibernate. They can carry rabies and parasites; disoriented or aggressive raccoons should always be avoided. Call local animal control to report one.

Burro/Donkey

Equus asinus H at withers 4 ft (1.2 m)

Descended from strays or castoffs from the 16th century onward, about 6,000 burros roam the southwestern United States. These hardy animals have no natural predators in their range.

KEY FACTS

Shaggy fur color is variable, ranging from red to brown to gray; mane is erect; ears are long; face often has white snout and eye rings; tail is tufted.

+ **habitat:** Canyons, mountains, and deserts

+ **range:** Western U.S., especially in California, Nevada, Arizona, and New Mexico

+ **food:** Grasses, shrubs, and cacti

The Burros served as a sure-footed beast of burden in the settlement of the West. Left to fend for itself in the wild, this close relative of the horse had the necessary skills to survive. In deserts, where browse and water are at a premium, it competes with native species—a fact that led in the past to elimination and more recently to relocation from public lands such as the Grand Canyon. The female, or jenny, bears a foal after a gestation of about 12 months. Jennies and young form stable groups. Males, or jacks, may dominate a territory or wander with other bachelors, braying and battling for mating opportunities.

Horse

Equus caballus H at withers 5 ft (1.5 m)

North America's native horse disappeared at least 8,000 years ago, perhaps a result of Stone Age hunting. Wild populations now consist of feral horses descended from imported breeds.

KEY FACTS

Varying colors of brown, bay, black, white, palomino, or pinto are common; animals tend to be small in height and weight.

+ **habitat:** Range-lands and islands

+ **range:** British Columbia and Nova Scotia; western U.S.; southeastern U.S. barrier islands

+ **food:** Grasses and forbs

Feral Horses derive mainly from Spanish imports, U.S. Cavalry retirees, and workhorses. Most herds occur in western states and British Columbia. Three to 20 animals or more usually are headed by a dominant stallion that controls his harem of females and often an older mare. After an 11-month gestation, mares give birth to one colt, which stays with the herd until age three or four. Assateague Island, a barrier island divided between Maryland and Virginia, is home to two populations of wild "ponies" that have sorted themselves into small bands. How to respond to overpopulation of western herds is a controversial issue.

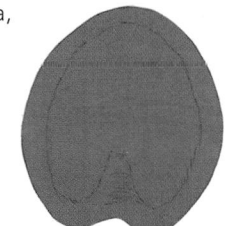

Wild Boar

Sus scrofa H at shoulder 24–42 in (60–107 cm)

This introduced Old World member of the swine family has long interbred with the feral domestic pig, but as a wild species it spends its life mostly unimpeded in much of its range.

KEY FACTS

Body is coarse, brownish or blackish; tusks curve upward; young have stripes and spots.

+ **habitat:** Marshes, forests, brushlands, and mountains

+ **range:** New Hampshire and scattered in U.S. South and West along with feral pigs and hybrids

+ **food:** Roots, tubers, leaves, nuts, fruit, fungi, insects, invertebrates, vertebrates, and eggs

Wild Boars leave many marks on the land, ranging from mud wallows, to extensive swaths of bulldozed earth, to scraped-up trees. They often destroy the habitats of native plants and animals in the process. The boars forage at night, dawn, and dusk, or in the daytime in winter. They are formidable foes, slashing with razor-sharp tusks and sharp hooves. The females and young travel in small bands; males are mostly solitary. Females often bear two litters of 3 to 12 young. Wild Boars rest often, in mud wallows or under a shady canopy they construct by incorporating cut grass with standing vegetation.

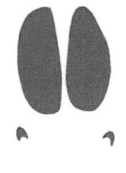 

Collared Peccary/Javelina

Tayassu tajacu or *Pecari tajacu* H at shoulder 15–20 in (38–50 cm)

These piglike hoofed mammals belong to a group that diverged from a common ancestor more than 40 million years ago. Peccaries are native to the Western Hemisphere, pigs to the Eastern.

KEY FACTS

Body is grizzled with pale yellow collar; long snout; downward-pointing tusks.

+ **habitat:** Deserts, canyons, and mixed forest

+ **range:** Southern Arizona, southwestern New Mexico, and southern Texas

+ **food:** Cacti, shrubs, fruit, and nuts

Known as the Javelina for its razor-sharp tusks, the Collared Peccary has a complex stomach well suited to the challenge of one of its favorite foods—Prickly Pear Cactus, which it eats spines and all. Peccaries live in mixed-sex social groups of 5 to 15 individuals. They communicate by means of glands on their rumps that exude an oily, musky fluid. They mark each other with the scent and rub against rocks and trees to delineate a home range. They observe no particular mating season, but males will posture and fight with head-on charges in which their tusked jaws occasionally lock. Twins are usually born four months after mating.

Moose

Alces alces H at shoulder to 7.5 ft (2.3 m)

The largest deer species, the solitary behemoth of the northern woods has a large "Roman" nose and ungainly, high-shouldered appearance. A bull Moose can weigh up to 1,800 pounds (800 kg).

KEY FACTS

Body is blackish brown; legs are usually paler; young are reddish brown.

+ **habitat:** Northern forests, tundra, willow thickets, and swamps

+ **range:** Alaska through Canada and into Rockies, north-central U.S. and New England

+ **food:** Leaves, twigs, bark of trees, aquatic plants

Moose browse on plant parts—the name comes from the Algonquian for "he cuts or trims smooth"—and they also wade out into ponds in summer to feed on pondweed and water lilies. A bull Moose's immense rack of palmate antlers, part of his mating-time appeal, may weigh 70 pounds (32 kg). Rival bulls engage in headfirst shoving matches, but the cows may also lure males with grunting, mooing calls, which hunters often imitate. Cows give birth to one calf, but sometimes twins, after an eight-month gestation. Unlike other young deer, a moose calf lacks spots; its formidable, aggressive mother can keep a wolf at bay, if need be.

Mule Deer

Odocoileus hemiones H at shoulder 3–3.5 ft (0.9–1 m)

Ears that are two-thirds the length of its head give the Mule Deer its name. The large appendages help the western species detect danger at long range.

KEY FACTS

Body is gray to brown to red: rump patches are whitish; slim tail has a black tip.

+ **habitat:** Deserts, brushlands, coniferous and mixed forests, and mountains

+ **range:** Southwest Alaska and southwestern Canada through western U.S.

+ **food:** Twigs, barks, leaves, buds, nuts, and berries

Well adapted to high elevations, broken country, and arid regions, Mule Deer often cover rugged landscapes. They are also known as "jumping deer" because of a method of locomotion called stotting. It consists of bounding stiff-legged with all four feet off the ground at the same time. Adult females and their offspring form groups; bucks usually are solitary. Mating occurs in the fall, and does give birth to one young or twins, depending on their age. Subspecies include the Columbia Black-tailed Deer (*O. h. columbianus*) and the Sitka Deer (*O. h. sitkensis*), which inhabit the Pacific Northwest and British Columbia.

White-tailed Deer

Odocolleus virginianus H at shoulder 2–3.5 ft (0.6–1 m)

In some parts of the continent, White-tailed Deer seem to be becoming as common as Gray Squirrels. The adaptable deer species is pushing ever farther into urban environments.

KEY FACTS

Fur is grayish brown in summer, reddish in winter; underside is white.

+ **habitat:** Deciduous woodlands, brushlands, and scrublands

+ **range:** U.S. and Canada south of Hudson Bay, except parts of southwestern U.S.

+ **food:** Leaves, twigs, nuts, berries, and fungi; crops and ornamental plants

White-tailed bucks grow antlers in the fall that are covered in a blood-rich "velvet" that they scrape off against trees; the antlers fall off after the rut. The species lives in groups of females and young and bachelor males. In winter, larger, mixed groups form in forage areas. Does have a single fawn at first, followed by twins in later years; the spotted young are left in cover while the mother feeds. In many areas, herds are culled regularly to keep populations in check. The tiny Key Deer (*O. virginianus clavium*), an endangered subspecies found only in the Florida Keys, was isolated there at the end of the last ice age.

Caribou

Rangifer tarandus H at shoulder 2.3–4.5 ft (0.7–1.4 m)

Twice a year, tens of thousands of Caribou migrate 900 miles (1,400 km) across the tundra. They head to calving grounds in the spring and return to winter pastures in the fall.

KEY FACTS

Head, body, and legs are brown, neck and rump are pale; male's antlers are large and elaborate, female's are smaller, thinner.

+ **habitat:** Boreal coniferous forest and tundra

+ **range:** Alaska through Canada to extreme northeast Washington and western Idaho

+ **food:** Grasses, sedges, forbs, shrubs, and lichens

When the Caribou isn't migrating, it travels more locally in search of forage in small bands or loose herds of up to several hundred animals. The bulk of its diet, especially in winter, consists of lichens, which it nibbles with small, weak teeth. Adult Caribou eat about 12 pounds (5.5 kg) of the slow-growing lichen each day. Subspecies of Caribou in North America include *R. t. groenlandicus* in Canada and *R. t. granti* in Alaska, known as "barren-ground" Caribou. In spring, Caribou cows in a herd drop their single calves in a synchronized fashion within days of each other, and the precocial young are soon on the move.

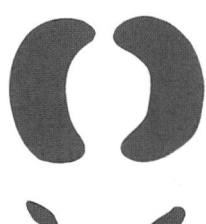

Elk/Wapiti

Cervus elaphus H at shoulder up to 5 ft (1.5 m)

Shawnee Indians called this highly evolved deer "Wapiti." Largely a western species now, the Elk is adapted to grazing and browsing. At least one herd member keeps watch while the others eat.

KEY FACTS

Body is light brown; head, neck, and legs are dark brown; rump patch and tail are pale; male grows antlers.

+ **habitat:** Plains, meadows, woodlands, and mountains

+ **range:** Western U.S. to southwestern Canada; introduced or reintroduced elsewhere

+ **food:** Grasses, forbs, and woody plants

Like a lot of large mammals in North America, the Elk once roamed a larger area; it covered much of the United States and Canada. This stately deer establishes large herds containing hundreds of animals in open areas and smaller numbers in woodlands. Bull Elk join the cows and young during the fall rut when they spar with other males for mating opportunities. They posture and challenge rivals, and when they meet head-on, their antlers—with their single main branches—seldom lock. Females give birth to a single calf in the late spring. The Tule Elk is a small and pale subspecies that lives in east-central California.

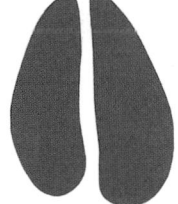

Pronghorn

Antilocapra americana H at shoulder 2.5–3.5 ft (0.7–1 m)

Antilocapra means "antelope goat," but the Pronghorn does not represent either family. It is a remnant of a group of horned mammals that arose in North America some 50 million years ago.

KEY FACTS

Tan with white bands on cheeks, sides, and underside; black markings on male's head and mane.

+ habitat: Grasslands, sagebrush, and desert

+ range: Scattered populations in southwestern Canada and western U.S.; reintroduced elsewhere

+ food: Grasses, forbs, cacti, and shrubby browse

The fastest Northern American mammal, the Pronghorn can reach speeds of more than 50 miles (80 km) an hour for short spurts. It can raise and fan the white hairs of its rump patch as a warning to others. The species has a cruising speed of about 25 miles (40 km) an hour. Adult males sport forked horns, not antlers, and shed only the outer sheath; females have a prongless version shorter than their ears or no horns at all. Dominant males control a territory and the females within it. Usually twins are born to the doe in the spring, and within days they can run faster than a human. Pronghorns once roamed the plains in the millions.

American Bison/Buffalo

Bos bison or *Bison bison* H at shoulder up to 6 ft (1.8 m)

This iconic, cowlike behemoth of the Great Plains was slaughtered almost to extinction by 1900. Today, American Bison live mostly in protected herds.

KEY FACTS

Adult is dark brown, curly on head; calf is reddish; both sexes have short, curved horns.

+ habitat: Plains, prairies, and woodlands

+ range: Small, very scattered populations from Alaska through western Canada and western U.S.

+ food: Grasses and browse

The American Bison, erroneously known as the Buffalo, once roamed the plains in vast herds. Native peoples hunted them and used every part for various needs. The bison is front loaded with enormous shoulders and a massive head with pointed horns. Bulls weighing about a ton (900 kg) join herds of females and subadults during the rut and may lose 300 pounds (130 kg) by its end. Calves are born nine months later and are mobile after a few hours. The Wood Bison (*B. bison athabascae*) of Canada is larger and darker than the plains species, and it lives in protected herds that are kept separate to preserve genetic differences.

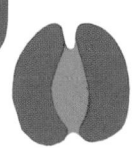

Mountain Goat

Oreamnos americanus H at shoulder 3–4 ft (0.9–1.2 m)

Casually jumping from ledge to ledge on rock faces, steadied by flexible, rubbery, shock-absorbing hooves, the Mountain Goat owns the high crags of northwestern North America.

KEY FACTS

Body is white; eyes, nose, horns, and hooves are black; males, females, and young have stiletto horns.

+ **habitat:** Steep mountain areas

+ **range:** Southern Alaska and western Canada and to Washington, Wyoming, and Idaho

+ **food:** Alpine plants and shrubs

In its rocky realm, the Mountain Goat is mostly safe from predators, apart from an occasional airborne eagle snatching a kid. The landscape is a bigger threat: Avalanches and rockslides take a toll on the species. Forays into mountain meadows also make it vulnerable. The adult female Mountain Goat is a formidable foe, fending off any threat to her kids. A male often takes a juvenile approach to courting, crawling on his belly and squeaking like a baby to win a nanny's favor. After mating, he retreats. One to three kids are born after six months; they're immediately agile and rock worthy, and stay with their mother for about a year.

Muskox

Ovibus moschatus H at shoulder 3–5 ft (0.9–1.5 m)

The Muskox's scientific name means "musky sheep ox," but it's a misnomer. Muskoxen are neither sheep nor oxen, and they lack musk glands, although bull urine is pungent.

KEY FACTS

Body is shaggy dark brown; shoulders and rump are humped; both sexes have large, curved horns.

+ **habitat:** Arctic tundra

+ **range:** Arctic islands and coastal plains of Northwest Territories in Canada; reintroduced elsewhere

+ **food:** Grasses, woody plants, willows, mosses, and lichens

When threatened, Muskox herds form a defensive ring, with their formidable heads facing out and the calves secured in the center behind a rampart of adult bulk. If the threat escalates, an adult will rush to gore an intruder and if successful, the rest may trample it. This tactic works well with wolves, but has made the animals an easy target for human hunters from prehistoric times. Muskoxen graze on the tundra, moving between summer and winter ranges. Their underwool is featherlight and incredibly warm. The Inuit wove mosquito nets from the long guard hairs, which can measure 24 inches (61 cm) on an adult male.

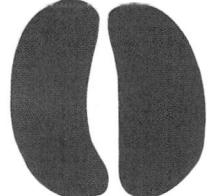

Bighorn Sheep

Ovis canadensis H at shoulder 3.5 ft (1 m)

The Bighorn Sheep lacks the pinpoint agility of the Mountain Goat, but it still takes command of rock ledges in western mountains. A 2-inch (5 cm) path is all it needs for travel.

KEY FACTS

Body is tan to dark brown; male has massive, tightly curled horns; female is smaller with narrower and straighter horns.

+ habitat: Mountain areas

+ range: Rocky Mountains from Canada south to New Mexico and western Texas

+ food: Grasses, sedges, and woody plants

Few animal encounters are as dramatic as the challenge of two Bighorn bucks for dominance. Posturing leads to the head-on crash of two heavy-horned and reinforced skulls, a sound that can be heard from long distances. They continue until one concedes—or dies. With victory comes the opportunity to mate. The males' horns can weigh 30 pounds (14 kg) and show annual growth rings. These sheep travel in herds; females and young stay together as do bachelor males. Desert Bighorns of the southern Rockies and western Texas are smaller and leaner. They obtain moisture from succulent plants and are well adapted to arid lands.

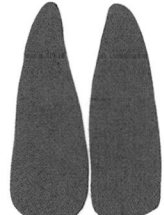

Dall's Sheep

Ovis dalli H at shoulder 3.5 ft (1 m)

Dall's Sheep living in rugged mountain terrain are wary of predators, but in protected areas, they show some of the inquisitiveness that may have led to the early domestication of other sheep.

KEY FACTS

Body is pure white; male has large, darker, spiraled horns; female horns are thinner and straighter.

+ habitat: Mountainous regions

+ range: Alaska, Yukon, western Northwest Territories, and northern British Columbia

+ food: Grasses, sedges, forbs, and woody plants

Dall's Sheep roam in bands separated by sex: Bachelor groups composed of old and young males travel separately from females and juveniles. Young males appease the established adults by behaving like females. The sexes come together only during the mating season. Dall's Sheep move between grazing patches that may be 40 miles (64 km) apart. Young animals learn the winding routes from their elders. The species belongs to the "thinhorn" branch of the genus *Ovis*. Their horns spiral widely away from the head. Stone Sheep (*O. d. stonei*) is a darker subspecies that overlaps with Dall's in the southern part of the range.

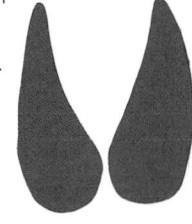

Red-tailed Hawks are North America's most widespread raptor, found from the desert Southwest to the Arctic tundra.

Birds

An Invitation to Birding

"Earth's the right place for love," wrote Robert Frost. Earth is also the right place for birds—masters of land, sea, and air, everywhere from Antarctica to beyond the Arctic Circle. Of warm-blooded creatures on the planet, there exist roughly twice as many bird species as mammal; except for whales in the ocean depths, birds can be found everywhere mammals live and beyond.

People who take up bird-watching as adults are often amazed to discover just how many birds are out there. Most of us will spot no more than a fraction of the almost 10,000 bird species, but the species visiting your own backyard represent a good beginning. Indeed, the easiest place to watch birds is right at home, especially if you have a good assortment of native trees and shrubs, a bird feeder, and a birdbath. A surprising number of birds can even be attracted to an apartment balcony in the city. Start looking, and you will see: Birds are everywhere, a delight to behold, a pleasure to learn to identify.

■ Why Learn Bird Identification?

Recognizing birds may not be necessary to individual survival, but learning about the creatures with which we share this planet, especially our own little corner of it, enhances our daily lives, giving us grace notes of joyful recognition as we travel to new places or look out our own windows, whether the birds we see are comfortably familiar or exciting new discoveries.

Using your senses in new ways enhances your ability to detect birds, and as your awareness grows, your progress accelerates. People with vision or hearing impairments can be excellent birders, locating and identifying birds entirely by either sound or visual cues. Those of us who can use both senses can train our eyes and ears to more easily detect birds with practice. The more birds we see and identify, the simpler it is to find and identify more.

■ Where to Find Birds

+ Feeding stations and birdbaths can offer long, leisurely looks at songbirds.
+ Ponds and other park waterways attract waterfowl and gulls.
+ Many states publish a birding guide or birding trail map with directions to the best birding locales.
+ State and local bird clubs and ornithological societies offer free or inexpensive local trips guided by experienced birders to birding spots they know best.
+ Birding festivals and bird club conventions can take you farther afield, to find more exotic

species with the help of expert guides.

+ Books such as National Geographic's *Guide to Birding Hot Spots of the United States* can give you travel ideas to broaden your birding knowledge.

■ Optics

Binoculars are an essential tool for bird-watchers, especially when you want to see details on birds beyond the window feeder. Optical quality improves with cost, so buy the best pair that you can comfortably afford. Ten-power binoculars have the best magnification but are the hardest to hold steady, provide less light than the same size model in a lower power, and have a smaller field of view, making it trickier to locate birds in the first place.

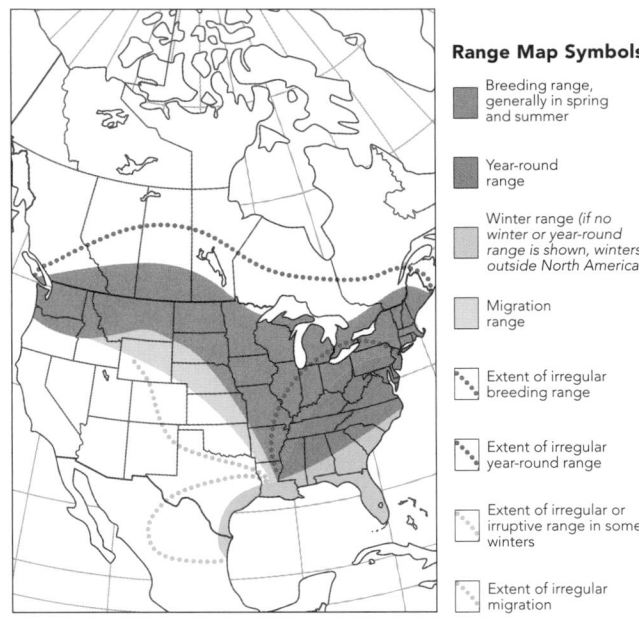

Range Map Symbols

Breeding range, generally in spring and summer

Year-round range

Winter range (*if no winter or year-round range is shown, winters outside North America*)

Migration range

Extent of irregular breeding range

Extent of irregular year-round range

Extent of irregular or irruptive range in some winters

Extent of irregular migration

KEY IDENTIFYING FEATURES

Identifying birds is the first step. Here are the key features that distinguish the many bird species from one another. Once you've identified a bird, keep a record of the species, as well as when and where you spotted it. This can be the beginning of your life list.

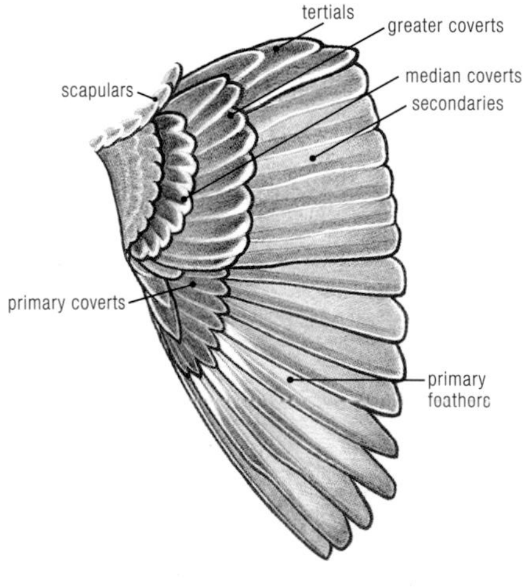

tertials
greater coverts
scapulars
median coverts
secondaries
primary coverts
primary feathers

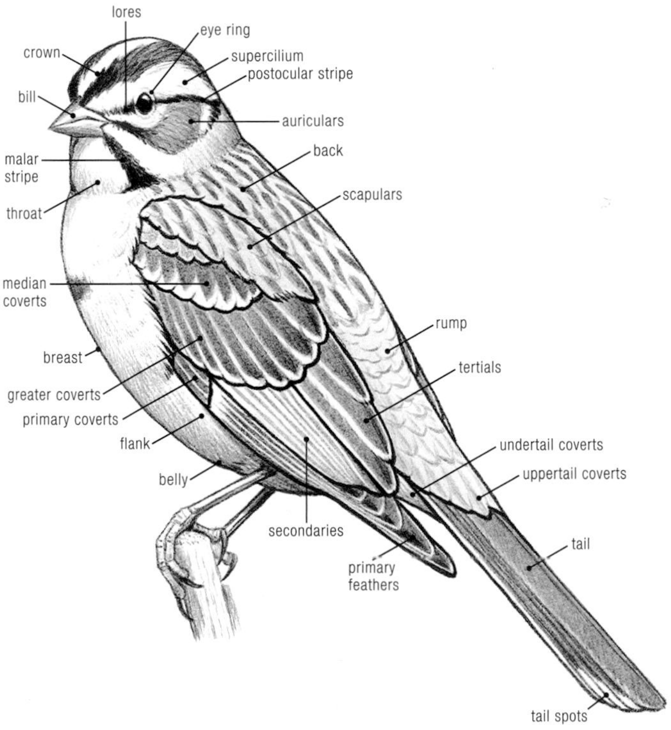

lores
crown
eye ring
supercilium
postocular stripe
bill
auriculars
malar stripe
back
throat
scapulars
median coverts
breast
rump
greater coverts
tertials
primary coverts
flank
undertail coverts
belly
uppertail coverts
secondaries
tail
primary feathers
tail spots

Most birders find that seven- or eight-power binoculars offer more than enough magnification. Because compact models are lighter and often less expensive, they are a popular choice.

Binoculars can take some getting used to. If you've never used binoculars before, practice on nonmoving objects first. Sight an object with your eyes, then train yourself to lift the binoculars and point them in the same direction. When you spot a bird, keep your eyes on it as you lift the binoculars up to your eyes.

Birding festivals and other wildlife events offer opportunities to compare binocular models out-doors, so you can see which work best for your eyes, hand size, and other needs. Many orga-nized field trips will also give you chances to see birds through a spotting scope before you buy your own.

Note: If you wear glasses to see distances, keep them on as you use your binoculars, but make sure that the eyecups are folded or pushed down; otherwise, keep them extended so they frame your eyes. The extra time you gain by not lifting your eyeglasses can be critical to getting a good look before a bird flies off.

■ Using This Guide

This chapter is a sampler of the rich array of North American birds. We have selected 160 species in total, including both the birds most commonly observed throughout North America and a few birds less easily seen but considered iconic, such as the Bald Eagle.

Here are a few pointers that will help you get the most out of this chapter.

+ Look up every bird you see. Many will fly away before you can identify them, but with practice, you'll become a quicker observer.
+ Notice the length and shape of the bill, relative length of the tail, body posture, and markings

Identifying this Northern Mockingbird is just the first step. It's great fun to observe feeding behaviors and how they balance in tricky positions.

‖‖‖

A light broke in upon my brain,—
It was the carol of a bird;
It ceased, and then it came again,
The sweetest song ear ever heard
—LORD BYRON, "THE PRISONER OF CHILLON," 1816

‖‖‖

on the wing and head. These features are among the best clues for identifying a species.
+ Use the Key Facts to focus on each bird's most important field marks—the physical characteris-tics distinctive to a particular species.
+ Use the range maps to check on whether a spe-cies is likely to be found in your area. The key is on page 96.
+ Keep the book handy. Pick it up now and then to scan a few pages, even if you're not looking up a specific bird. That way, you can become familiar with the species you are likely to see.

+ Whenever you identify a new bird—and even when you encounter a familiar one—look it up in the chapter. Look at the illustrations and read the text for pointers on all the important features that distinguish that bird species from others.

The more you use this chapter, the more proficient you'll be at finding new birds in it. By becoming familiar with it, you are learning the language of birders, and soon, even without this chapter in hand, you will have enough knowledge to place an unfamiliar bird in the right family and to notice details that can help you identify it later.

■ Taking It to the Next Level

Identifying birds is the first step. More enjoyment is in store if you watch behaviors, provide birdhouses and feeders, plant bird-attracting flowers and shrubs, monitor nest boxes, track which species come and go in different seasons, and look for personality quirks in individuals of a single species. Such challenges and pleasures will enhance your bird-watching in the backyard, your neighborhood, your local park, and on vacation in the mountains or at the shore. You'll be connected with nature in powerful, soul-satisfying ways. Come join us!

OFFICIAL BIRDS OF THE U.S. STATES AND CANADIAN PROVINCES

+ Alabama: Northern Flicker (Colaptes auratus)
+ Alaska: Willow Ptarmigan (Lagopus lagopus)
+ Arizona: Cactus Wren (Campylorhynchus brunneicapillus)
+ Arkansas: Northern Mockingbird (Mimus polyglottos)
+ California: California Quail (Callipepla californica)
+ Colorado: Lark Bunting (Calamospiza melanocorys)
+ Connecticut: American Robin (Turdus migratorius)
+ Delaware: Blue Hen Chicken (Gallus gallus)
+ Florida: Northern Mockingbird (Mimus polyglottos)
+ Georgia: Brown Thrasher (Toxostoma rufum)
+ Hawaii: Nēnē or Hawaiian Goose (Branta sandvicensis)
+ Idaho: Mountain Bluebird (Sialia currucoides)
+ Illinois: Northern Cardinal (Cardinalis cardinalis)
+ Indiana: Northern Cardinal (Cardinalis cardinalis)
+ Iowa: American Goldfinch (Spinus tristis)
+ Kansas: Western Meadowlark (Sturnella neglecta)
+ Kentucky: Northern Cardinal (Cardinalis cardinalis)
+ Louisiana: Brown Pelican (Pelecanus occidentalis)
+ Maine: Black-capped Chickadee (Poecile atricapillus)
+ Maryland: Baltimore Oriole (Icterus galbula)
+ Massachusetts: Black-capped Chickadee (Poecile atricapillus)
+ Michigan: American Robin (Turdus migratorius)
+ Minnesota: Common Loon (Gavia immer)
+ Mississippi: Northern Mockingbird (Mimus polyglottos)
+ Missouri: Eastern Bluebird (Sialia sialis)
+ Montana: Western Meadowlark (Sturnella neglecta)
+ Nebraska: Western Meadowlark (Sturnella neglecta)
+ Nevada: Mountain Bluebird (Sialia currucoides)
+ New Hampshire: Purple Finch (Carpodacus purpureus)
+ New Jersey: Eastern Goldfinch (Carduelis tristis)
+ New Mexico: Chaparral Bird or Greater Roadrunner (Geococcyx californianus)
+ New York: Eastern Bluebird (Sialia sialis)

+ North Carolina: Northern Cardinal (Cardinalis cardinalis)
+ North Dakota: Western Meadowlark (Sturnella neglecta)
+ Ohio: Northern Cardinal (Cardinalis cardinalis)
+ Oklahoma: Scissor-tailed Flycatcher (Tyrannus forficatus)
+ Oregon: Western Meadowlark (Sturnella neglecta)
+ Pennsylvania: Ruffed Grouse (Bonasa umbellus)
+ Rhode Island: Rhode Island Red (Gallus gallus)
+ South Carolina: Carolina Wren (Thryothorus ludovicianus)
+ South Dakota: Ring-necked Pheasant (Phasianus colchicus)
+ Tennessee: Northern Mockingbird (Mimus polyglottos)
+ Texas: Northern Mockingbird (Mimus polyglottos)
+ Utah: California Gull (Larus californicus)
+ Vermont: Hermit Thrush (Catharus guttatus)
+ Virginia: Northern Cardinal (Cardinalis cardinalis)
+ Washington: American Goldfinch (Spinus tristis)
+ West Virginia: Northern Cardinal (Cardinalis cardinalis)
+ Wisconsin: American Robin (Turdus migratorius)
+ Wyoming: Western Meadowlark (Sturnella neglecta)

Canadian Provinces and Territories

+ Alberta: Great Horned Owl (Bubo virginianus)
+ British Columbia: Steller's Jay (Cyanocitta stelleri)
+ Manitoba: Great Grey Owl (Strix nebulosa)
+ New Brunswick: Black-capped Chickadee (Poecile atricapillus)
+ Newfoundland and Labrador: Atlantic Puffin (Fratercula arctica)
+ Northwest Territories: Gyrfalcon (Falco rusticolus)
+ Nova Scotia: Osprey (Pandion haliaetus)
+ Nunavut Territory: Rock Ptarmigan (Lagopus mutus)
+ Ontario: Common Loon (Gavia immer)
+ Prince Edward Island: Blue Jay (Cyanocitta cristata)
+ Quebec: Snowy Owl (Nyctea scandiaca)
+ Saskatchewan: Shark-tailed Grouse (Tympanuchus phasianellus)
+ Yukon Territory: Common Raven (Corvus corax)

BIRDS

Canada Goose

Branta canadensis L 30–43 in (76–109 cm)

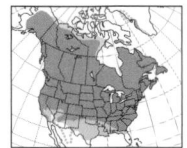

Canada Geese were, until the 1970s, seen in many places only on migration and in winter.

KEY FACTS

Large dark goose; black neck; white chin strap; paler below.

+ **voice:** Call is deep but nasal *honk-a-lonk.*

+ **habitat:** Common and familiar. Feral birds frequent suburban areas, such as golf courses, parks, reservoirs; wild flocks migrate in V-formation.

+ **food:** Grazes on grass; eats a variety of pond life and waste grain in rural areas.

Geese are among a handful of birds capable of digesting grass, and so are drawn to expansive lawns. "Honkers" are extremely sociable. Once paired, they usually remain with their mate year-round for life. In winter, they may also maintain social bonds with young from previous seasons. They learn their migratory routes from their parents. When a pair of geese becomes urbanized and remains in a city for the winter, their young often follow suit. Adults undergo a flightless period while molting flight feathers in summer, before their goslings can fly. Goslings can swim and even dive to evade predators within hours of hatching.

Adult

Snow Goose

Chen caerulescens L 26–33 in (66–84 cm)

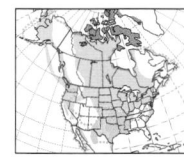

White farm geese, domesticated in Eurasia, are not the same species as wild American geese.

KEY FACTS

Medium-size goose with two color morphs: White morph entirely white with black wing tips; blue morph has dark gray body, white head. Juveniles are dingy gray.

+ **voice:** Gives single, high-pitched calls.

+ **habitat:** Nests in Arctic. Winters on marshes and open fields, often in large flocks.

+ **food:** Vegetarian; eats various grasses, weeds, and tubers.

"Waveys" got their nickname from their long, undulating lines on migration. The two color morphs (white and blue) were considered separate species until 1983. Birds of both types usually select mates with the same plumage; the blue morph is dominant genetically, like brown eyes in humans. These vegetarians are gregarious except during the breeding season. Goslings remain with their parents until their second or third year. Populations are at historical highs, and many biologists are concerned about habitat destruction on the tundra because Snow Geese grub for underground tubers and roots, destroying large swaths of delicate habitat.

Blue-morph adult

White-morph adult

Tundra Swan

Cygnus columbianus L 52 in (132 cm)

Feral Mute Swans, brought here from Europe, displace native Tundra and Trumpeter Swans.

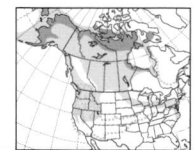

KEY FACTS

Larger and longer necked than geese. Entirely white; black bill with yellow spot near eye. Similar Mute Swan (not shown) has orange bill with black knob at base.

+ voice: Call is a single or double *honk* like an old car horn.

+ habitat: Nests in Arctic. Winters in large flocks in wetlands.

+ food: Aquatic plants and tubers, some mollusks, arthropods

Usually seen in flocks during migration and winter, America's "whistling swan" breeds in the Arctic and winters mostly near the Pacific and mid-Atlantic coasts of the United States. They once fed mostly on submerged aquatic vegetation, but habitat losses on their wintering grounds led them to depend increasingly on grain fields. Some wild Tundra Swans have lived more than 23 years. They remain with their mate for life. Their chicks, called cygnets, can follow their parents within hours of hatching. Recently hatched cygnets are too buoyant to feed easily while swimming; they feed more often on land than the young of other swans.

Juvenile

Adult

Wood Duck

Aix sponsa L 18.5 in (47 cm)

The Wood Duck is one of the very few ducks that perches in trees and nests in tree cavities or nest boxes.

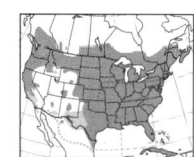

KEY FACTS

Glossy, colorful male is unmistakable; female is gray-brown with speckled breast, white belly. Juvenile similar to female.

+ voice: Female's flight call is loud squealing *oo-eek;* male gives a soft whistle.

+ habitat: Quiet rivers, wooded swamps, bottomlands; also along streams and rivers

+ food: Aquatic plants and pond life; some acorns and grain

This spectacular species declined dangerously in the late 1800s from overhunting for food and its beautiful feathers. Protection and nest box programs helped it recover. It takes two weeks or more for a female to lay about 13 eggs, but the entire clutch hatches out together within about 6 to 18 hours. The mother and chicks synchronize hatching by calling to one another while the ducklings are still inside the eggs. When it seems safe the morning after they hatch, the female calls to her brood from the ground, and one by one they jump safely down to her, as far as 50 feet (15 m) below.

Mallard

Anas platyrhynchos L 23 in (58 cm)

Dabbling ducks such as the Mallard are buoyant, with relatively large wings. Most take off from water in a leap.

KEY FACTS

Male has green head and yellow bill; female is mottled brown with darker bill.

+ **voice:** Female goes *quack-quack-quack;* male makes a raspy *kreep.*

+ **habitat:** Abundant denizens of park ponds; wilder birds live in a variety of shallow-water wetlands.

+ **food:** Tips forward to reach underwater plant and animal life; also forages on land.

The handsome Mallard, America's most abundant duck, is the parent species of most domesticated ducks. Farm ducks often mate with wild Mallards, producing a variety of hybrids. If hybrids continue to breed with wild Mallards, over generations, their offspring become more Mallard-like, though often larger and with a wider white neck band. Each female chooses her mate on her winter range; he follows her back to where she originally hatched. Even devoted males often mate with other females, so a single brood of ducklings may have more than one father. Mallards have few taste buds on their tongue, but many under the tip of their bill.

Northern Pintail

Anas acuta L 21 in (53 cm) + 4 in (10 cm) tail of male

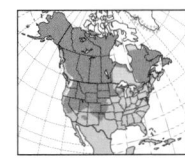

Shallow wetlands from the prairies to the tundra are ideal breeding grounds for pintails.

KEY FACTS

Elegant and long necked. Male has brown head with white stripe, long tail; female is mottled brown with paler head (compare to female Mallard).

+ **voice:** Female quacks; male whistles.

+ **habitat:** Found in marshes and open areas with ponds; more common in West than East.

+ **food:** Tips forward to reach underwater plant and animal life; also forages on land.

The Northern Pintail is the greyhound of ducks—elegant, slender, and fast. From a distance, its delicate silhouette makes it seem smaller than it really is. Pintails nest almost as soon as ice goes out in northern areas. Each male remains with his mate until she has laid a full clutch and begins incubating. Nests may be a mile from the nearest pond or stream; as soon as ducklings hatch, they follow their mother to water. After recovering from botulism, one male banded in Utah in 1942 was found 82 days later in a flock of other exhausted pintails on Palmyra Island, an amazing 3,600 miles (5,800 km) from where he was banded.

Green-winged Teal

Anas crecca L 14.5 in (37 cm)

Teal are highly sociable, short-necked dabbling ducks that fly very fast in large groups.

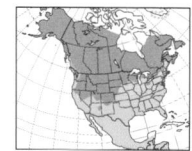

KEY FACTS

Small, compact, and fast flying; green wing patch; yellowish under tail. Male has chestnut head, green patch through eye, white shoulder bar; female mottled brown

+ **voice:** Male call, a liquid *krick;* female, a shrill *quack*

+ **habitat:** Aquatic habitats, such as coastal marshes in winter

+ **food:** Dabbles in shallows for seeds and small aquatic animals.

Tiniest of our puddle ducks, Green-winged Teal have been called feathered minnows for their ability to twist, turn, and bank in high-speed flight and even dive into the water in tight formations in perfect unison. This is the only species of duck known to scratch itself while flying. The young can swim, dive, walk, and feed themselves within hours of hatching, but cannot regulate their body temperatures well when air temperature is less than about 50°F (10°C), so the mother may brood them both at night and during cool days for a week or more. The color teal was named for the shade of this bird's facial patch in some light.

Ring-necked Duck

Aythya collaris L 17 in (43 cm)

Diving ducks such as the Ring-necked Duck dive to feed and elude predators. They run on the water's surface to take off.

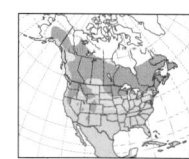

KEY FACTS

Small diving duck with peaked head, bold white ring around bill. Male purple-black with pale gray flanks; female brown with gray face, white eye ring

+ **voice:** Usually silent; soft growly calls

+ **habitat:** Lakes and ponds; also coastal marshes, rivers in winter

+ **food:** Dives for aquatic plants and insects; sometimes tips up like a dabbling duck.

The inconspicuous chestnut ring around the male's neck is visible only in certain light or at very close range, so hunters call this species the "ring bill." Like other diving ducks, ring-necks can be hard to track while they're actively feeding because they disappear from sight when they dive, often popping up some distance away. During migration, they often associate with other diving ducks, especially scaup, but the male's angular head, black back, and white vertical patch behind the black breast make this species easy to pick out. During the 1980s and early 1990s, when many other duck populations were declining, Ring-necked Ducks expanded their range.

Bufflehead

Bucephala albeola L 13.5 in (34 cm)

The Bufflehead, a "sea duck," has dense plumage, extra fat reserves, and special glands to excrete salt.

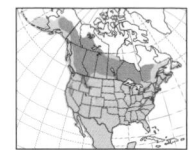

KEY FACTS

Tiny diving duck with a large, puffy head. Very active when feeding. Male has large white patch on glossy black head; female has brown head with white cheek.

+ voice: Usually silent

+ habitat: Fairly common; nests in woodlands near small ponds; in winter, found on sheltered bays, rivers.

+ food: Dives for insects, crustaceans, mollusks; some plants.

The Bufflehead takes its name, which means "buffalo head," from its large head and tiny body. It nests in abandoned flicker holes near freshwater lakes and rivers. Courting males make short flights over a female and land feet first, "skiing" on the water with white head feathers erect and bright pink feet fully exposed. Males frequently bob their head up and jerk it back in a dramatic courtship display. Unlike most ducks, pairs of Bufflehead sometimes return to mate year after year. These cavity nesters seldom walk on land except when females lead newly hatched ducklings to water. Hunters call them "butterballs" for their plumpness.

Common Merganser

Mergus merganser L 25 in (64 cm)

Mergansers have narrow, serrated bills to grasp prey. Many nest in cavities and nest boxes.

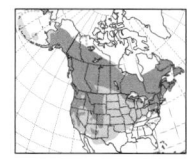

KEY FACTS

Large diving duck with thick-based, red bill. Male has dark green head, bright white sides. Female has chestnut head, white chin, gray sides.

+ voice: Usually silent

+ habitat: Prefers fresh water at all seasons; clear rivers and wooded lakes for breeding; winters on large rivers and lakes.

+ food: Pursues fish underwater; also mollusks, crustaceans.

These large, striking ducks often swim with their head under water up to the eyes as they watch for fish. They can slowly submerge without a ripple, or leap into a graceful dive. They prefer fishing in slow-moving, clear rivers and streams. Smaller, but shaped something like loons, males are easy to distinguish from loons by their color pattern, females by their "bad hair day" appearance. Males also have long head feathers but don't erect them. Chicks feed on aquatic insects for the first week or two, adding fish to their diet as they get bigger and more skilled. Few hunters target mergansers because of their strong fishy taste.

California Quail

Callipepla californica L 10 in (25 cm)

Quail aren't closely related to grouse and pheasants. They live in groups called coveys.

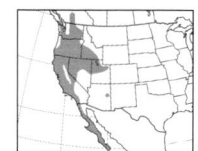

KEY FACTS

Ground-dwelling, chunky bird seen in groups; can burst into flight but usually runs for cover. Spiffy head pattern and topknot in male; female is plainer.

+ voice: Main call is loud *chi-CA-go.*

+ habitat: Brushy chaparral and scrubby lowlands in the West

+ food: Vegetable matter and insects; mainly seeds and small fruits in winter

California's state bird is shy and retiring, more often seen in cartoons than in its natural range, but its loud call is often heard. A paired male and female alternate calls in a neatly arranged duet. Families usually associate in coveys—all the adults care for the young. This bird is well adapted to arid environments, and except during extreme heat can survive without drinking water, extracting fluids from food. It feeds mostly in the morning and evening, roosting in shade during daytime. When approached by a potential predator, it bursts into noisy flight.

Juvenile

Northern Bobwhite

Colinus virginianus L 9.8 in (25 cm)

The bobwhite, like the killdeer, phoebe, and chickadee, is named for the sound of its call.

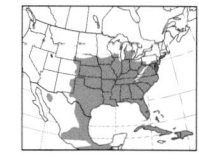

KEY FACTS

Small chicken-like bird; only quail in East. Overall reddish brown with streaks and spots; male has boldly patterned head. Popular game bird

+ voice: Loud whistled call *bob-WHITE!* is heard mostly in spring.

+ habitat: Farm fields and brushlands with good cover. Declining in most states

+ food: Forages on ground for seeds and other plant material.

Bobwhites are sociable quail the size of soccer balls. Males are aggressive toward other males during pair formation, but when both parents start incubating, males become peaceable again. Families often associate with other families soon after the fuzzy chicks have hatched. The species was once considered monogamous, but telemetry studies have revealed that both males and females may incubate and raise broods with more than one mate. Bobwhites were once very common, nesting in fencerows and other small patches, but are now declining in many areas. Captive-bred bobwhites are used to stock private hunting areas and to train retrievers. In many areas, most sightings are of escaped birds.

Ring-necked Pheasant

Phasianus colchicus ♂L 33 in (84 cm) ♀L 21 in (53 cm)

The pheasant, one of the world's most popular game birds, is the state bird of South Dakota.

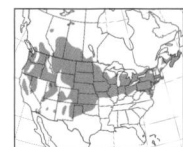

KEY FACTS

Long-tailed game bird introduced from Asia. Large, flashy male is iridescent bronze with fleshy red eye patches and white neck ring. Smaller female is buffy.

+ **voice:** Male's call a loud, harsh *kak-CACK!*

+ **habitat:** Resident in open country, farmland, brushy fields, woodland edges

+ **food:** Forages on ground for seeds, grain, insects.

Native to western Asia and eastern Europe, pheasants were traded and raised for food by ancient Greeks and Romans and introduced to many places over the world. Today, they sometimes escape from breeders and game farms even outside their range. As with other birds that mostly move about by walking or swimming, pheasant legs are powered by red muscle fibers—the dark meat—that are virtually tireless. The white muscle fibers powering their wings—the breast meat—produce rapid, strong bursts of flight, but tire quickly. Males may acquire a "harem" of several females.

Wild Turkey

Meleagris gallopavo ♂L 46 in (117 cm) ♀L 37 in (94 cm)

Wild Turkeys are strong short-distance fliers, and usually roost in trees at night.

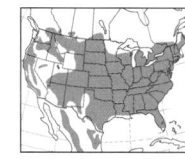

KEY FACTS

Very large game bird with iridescent, bronze plumage and bare-skinned head. Male larger than female with glossier plumage and breast tuft. Male displays in spring.

+ **voice:** Male's gobbling call may be heard a mile away.

+ **habitat:** Resident in open hardwood forest, woodland edges

+ **food:** Forages on the ground for seeds, nuts, acorns, insects.

Audubon placed the turkey first in *Birds of America*, and Benjamin Franklin wanted it on our national emblem. Turkeys had been domesticated by the Aztec, brought to Spain by conquistadors, and appeared on English tables by the 1540s. Back then, the wild birds fed on American chestnuts, walnuts, and other riches of the eastern forest. By the mid-1800s, turkeys had vanished from much of their range due to chestnut blight, deforestation, and overhunting. Now they've been successfully reintroduced over much of their former range and even beyond.

Ruffed Grouse

Bonasa umbellus L 17 in (43 cm)

The Ruffed Grouse may be red or gray, the color determined genetically like humans' hair color.

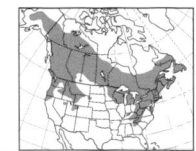

KEY FACTS

Chicken-like bird with mottled gray or reddish-brown plumage that matches the forest floor. Black ruff feathers and banded tail conspicuous only during display

+ **voice:** Male's drumming sounds produced by rapidly beating wings

+ **habitat:** Fairly common resident of deciduous and mixed woods

+ **food:** Buds, twigs, leaves, nuts, berries

A deep, resonant *thump, thump, thump* starts up in an aspen forest. Like a bouncing ball, it speeds up until ending in a muffled roar. To make this drumming sound, the male Ruffed Grouse cups air in his "wing pits" and flaps hard. Females, unassisted by males after mating, produce up to 16 young in a single brood. The chicks first fly when less than a week old. In winter, grouse feet grow comblike "snowshoes" that help them grip icy branches as they feed on aspen buds. Their intestines also change dramatically in winter to digest this woody tissue. The state bird of Pennsylvania is a popular game bird.

Displaying ♂

♀

Common Loon

Gavia immer L 32 in (81 cm)

Sharp backward projections on upper palate and tongue help the Common Loon hold slippery prey.

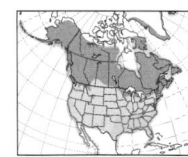

KEY FACTS

Large, low-riding waterbird. Breeding bird has checkerboard back, dark green head, striped collar; winter bird is dark gray with white throat and half collar.

+ **voice:** Iconic yodeling call heard mostly in breeding season

+ **habitat:** Nests on northern lakes; winters in coastal waters and large ice-free lakes.

+ **food:** Mostly fish

The state bird of Minnesota has oversize webbed feet that flatten on the forward stroke. These visual hunters choose lakes based on water clarity, and swallow prey headfirst. Their eyes are brightest red during breeding season. Loons nest at the shoreline or on floating islands and platforms. Their long, haunting "wail" is a contact call between family members. They make the laugh-like "tremolo" when stressed or in flight. Only territorial males make the nighttime "yodel" calls. They winter in salt water. A gland between their eyes excretes excess salt through their nostrils.

Breeding adult

Winter adult

Pied-billed Grebe

Podilymbus podiceps L 13.5 in (34 cm)

Grebes, nicknamed "hell divers" for their long dives, have lobed toes, not webbed feet.

KEY FACTS

Brown, duck-like, diving bird with lobed toes and thick, white bill. Black ring on bill and black throat; both lost in winter. Can hide by slowly sinking until only its head shows.

+ **voice:** Loud series of gulping notes heard on breeding grounds

+ **habitat:** Nests around marshy ponds and sloughs; winters on fresh or salt water.

+ **food:** Dives for fish, crustaceans, insects.

The Pied-billed Grebe is a tiny, somewhat secretive but widespread diving bird. It takes its name from the black-and-white, or "pied," bill of breeding adults. It may sink slowly into the water by expelling air from internal air sacs and from between its feathers, or may dive quickly. A nesting pair produces about four to six eggs. Zebra-striped chicks have red faces; they can swim soon after hatching but cannot maintain body temperature at first so often ride on their parents' backs. These nocturnal migrants are seldom seen in flight but have wandered as far as Hawaii and Europe.

Breeding adult

Downy young

Winter

Western Grebe

Aechmophorus occidentalis L 25 in (64 cm)

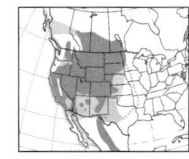

Two similar grebes, Clark's and Western Grebes, were long believed to be the same species.

KEY FACTS

Large black-and-white grebe with long, thin neck and yellow-green bill. Very similar Clark's Grebe has more white around eye and orange bill.

+ **voice:** Call is a loud, two-note *crick-kreek.*

+ **habitat:** Nests in reed beds of fresh-water lakes; winters along Pacific coast and on large inland bodies of water.

+ **food:** Dives for fish and some crustaceans.

This beautiful, slender-necked grebe is most famous for its "rushing" dance, when pairs or competing males rush toward each other and rise out of the water to run on the surface in synchrony. Mated pairs also engage in a "weed dance" during which each bird dives to grab vegetation, swims toward the other, and rises vertically out of the water in erect posture, bill lifted. Western Grebes usually breed in colonies. They lay three to four eggs in a nest set in floating vegetation hidden among plants. They fly only during migration. After arriving on the breeding grounds, their flight muscles atrophy for the season, growing again in time for fall migration.

Adult

Downy young

Double-crested Cormorant
Phalacrocorax auritus L 32 in (81 cm)

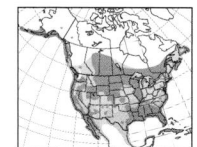

The double crest of this cormorant is held from spring to early summer, and varies from white to black.

KEY FACTS

Large, all-black water bird with yellow-orange facial skin; breeding adult has wispy crest. Young bird has pale breast. Often seen holding wings spread out to dry.

+ **voice:** Some grunting noises; usually silent

+ **habitat:** Common and widespread along coasts, lakes, and rivers

+ **food:** Pursues fish under water, propelled by large webbed feet.

Cormorants look like dark loons when swimming and like dark geese in flight. Wettable flight feathers decrease their buoyancy while diving. They are very sociable and nest in colonies with other cormorants and sometimes herons, on rocks and cliffs or in trees. All four toes are webbed. To incubate, they roll eggs atop their feet and rest the abdomen and breast on top of them. Hatchlings look like rubber toys, with naked dark brown, translucent skin. Within two weeks, they're covered with woolly black down. Clumsy on land, chicks and adults use their hooked bill tip to balance and walk.

Juvenile

Winter adult

Anhinga
Anhinga anhinga L 35 in (89 cm)

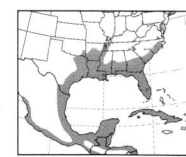

Nicknamed the "snake bird," the Anhinga often swims with only its neck and bill above water.

KEY FACTS

Long, snakelike neck and slender bill; long, broad tail. Male is black with silvery white patches on wings and upper back; female has buffy neck and breast.

+ **voice:** Mostly silent

+ **habitat:** Prefers shallow, slow-moving waters and swamps. Often seen soaring high overhead.

+ **food:** Stalks fish under water, spearing them with its long, pointed bill.

The only species in its family, the Anhinga's long tail counterbalances its long narrow neck and gives it a distinctive cross-like shape when soaring. Anhingas have dense bones and wettable feathers, allowing them to stay submerged while chasing and spearing fish. They lose body heat while under water, and must perch and spread their wings to dry their feathers and to warm their bodies in the sun. Heat loss during swimming restricts Anhingas to warmer climates than cormorants. They nest in loose colonies.

Breeding adult

American White Pelican

Pelecanus erythrorhynchos L 62 in (158 cm)

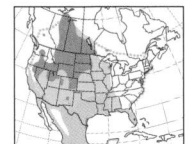

The huge American White Pelican doesn't dive, but rather scoops up fish as it swims on the surface.

KEY FACTS

Huge waterbird with massive orange bill and throat pouch. White with black outer wing; juvenile with dusky marks on wings, neck, and head. Often soars.

+ **voice:** Mostly silent

+ **habitat:** Breeds on inland lakes throughout the mountain West and northern Great Plains; winters along the coast.

+ **food:** Birds corral fish, then dip their open bills to catch them.

Pelicans are gangly on land, graceful in water and air. These gregarious birds hunt in groups, chasing fish to one another, herding schools of fish into shallow water, or encircling a large school of fish and closing in to concentrate them. Their lower mandible is flexible, widening enormously to serve as a scoop. This is the only pelican that grows a plate on the bright yellow upper bill during the breeding season. White Pelicans nest in large inland colonies. Both sexes build the nest on the ground and incubate the eggs under their huge webbed feet. Populations are growing, possibly due in part to fish farms.

Breeding adult

Brown Pelican

Pelecanus occidentalis L 48 in (122 cm)

The Brown Pelican is the only pelican species in the world that plunge-dives for food.

KEY FACTS

Very large, dark waterbird. Adult is gray-brown with blackish belly; top of head white to yellowish. Neck brown in summer, white in winter. Juvenile is dirty brown with pale belly.

+ **voice:** Mostly silent

+ **habitat:** Saltwater. Flocks often seen gliding low over the water.

+ **food:** Plunge-dives for schooling fish and traps them in its ballooning pouch.

This sociable saltwater species often approaches people for handouts. In turn, gulls often grab fish from Brown Pelicans. Brown Pelicans fly in long lines barely above the water's surface, seeming to disappear behind large waves. They alight on water while extending their huge webbed feet as skids. A pelican's bill holds more than its belly can—after filling its pouch, it drains as much as 3 gallons of water before swallowing the fish. Brown Pelicans nest on the ground or in trees. The male selects the site and brings nesting materials; the female builds the nest.

Immature

Non-breeding adult

Breeding adult

Great Blue Heron

Ardea herodias L 46 in (117 cm)

Herons, unrelated to cranes, roost and nest in trees. Nest building can take up to two weeks.

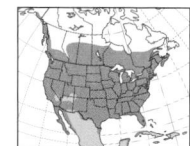

KEY FACTS

Large, gray-blue heron; white crown with black stripe over eye; ornate head and neck plumes when breeding. Juvenile has black crown.

+ voice: Occasional deep croaks

+ habitat: Common and widespread in fresh- and saltwater locations; also hunts in fields.

+ food: Includes fish, frogs, snakes, and small mammals.

This huge bird has powerful neck muscles that drive its spear-like bill with amazing speed and precision. It swallows fish headfirst, usually waiting for larger fish and those with spines to die before swallowing. The bill and neck are relatively heavy, so as it takes off in flight, it kinks its neck to draw the weight nearer its wings. Like other herons, it does not carry prey, but feeds the young by regurgitating fish into the nest. Most herons nest in colonies (rookeries), but some pairs are territorial. Many select new mates each year, but some pairs reunite. In southern Florida, some are pure white or are pale with a white head.

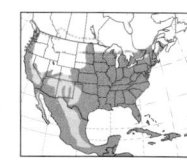

Green Heron

Butorides virescens L 18 in (46 cm)

This handsome, not-very-green heron's yellow legs turn orange during the breeding season.

KEY FACTS

Small, dark heron with short legs and dark bill. Adult has dark cap, chestnut neck, and greenish-blue back; juvenile is brownish above and streaked below. Usually solitary.

+ voice: Loud, sharp *kyowk*, often given when taking flight

+ habitat: Wooded streams, ponds, and marshes

+ food: Patiently stalks small fish and other aquatic life.

The Green Heron can stand absolutely still for several minutes along the water's edge until, with an explosive strike, it grabs or stabs prey. It sometimes baits fish, dropping bits of bread, insects, earthworms, twigs, or feathers onto the water's surface to lure them in. Like other herons, it sometimes fishes by night under moonlight. It usually nests in isolation but sometimes in loose colonies in swampy freshwater thickets, and often winters in coastal areas. Two nicknames, "shitepoke" and "chalkline," were given for its habit of conspicuously excreting a long stream as it takes off in flight. It's noisier than most herons, often squawking in flight.

Great Egret

Ardea alba L 39 in (99 cm)

Egrets are herons with specialized nuptial feathers called aigrettes.
Most are white.

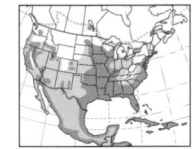

KEY FACTS

Large white heron with heavy yellow bill, blackish legs and feet. In breeding plumage, long, wispy plumes trail from the back.

+ **voice:** Occasional deep croaking *kaaark*

+ **habitat:** All types of wetland habitats, including damp fields

+ **food:** Fish, invertebrates, reptiles, and small mammals. Walks slowly or waits for prey, then lunges with bill.

Great Egrets were hunted excessively for their beautiful breeding plumage, and became a symbol of the Audubon Society and conservation in general. They are widespread, found on every continent except Antarctica. They look huge but weigh barely 2.5 pounds (1.1 kg), less than half a Great Blue Heron's weight. They're aggressive toward other herons but surprisingly unaggressive toward potential nest predators— they've even been known to stand next to their nests without reacting as Fish Crows fly in and remove eggs. Some mooch for handouts at Florida theme parks. One lived more than 22 years.

High breeding adult

Non-breeding

Snowy Egret

Egretta thula L 24 in (61 cm)

The Snowy Egret's exquisite nuptial plumes are flaunted in spectacular
breeding displays.

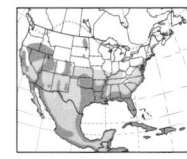

KEY FACTS

Small white heron with black bill and yellow feet ("golden slippers"). Breeding plumes on neck, head, and lower back

+ **voice:** Low, raspy note, mostly near nest

+ **habitat:** Found in salt- and freshwater habitats.

+ **food:** Small fish and crustaceans. More active than most herons; also uses feet to stir up prey in shallow water.

Snowy Egrets often run around in comically frantic pursuit of prey, wings held aloft. They are very sociable, often hunting in aggregations that include other wading birds, gulls, and terns, and nest in large colonies that usually include several species of waders. Both parents incubate the three to five eggs and feed and brood the young. After the breeding season, many Snowy Egrets wander well north of their breeding range. In 1886, this bird's plumes sold for $32 an ounce, double the price of gold at the time. The species declined until enactment of the Migratory Bird Act, which ended the feather trade.

High breeding adult

Breeding adult

Black-crowned Night-Heron

Nycticorax nycticorax L 25 in (64 cm)

Night-herons hunt mostly at twilight and night. By day, they are often seen roosting.

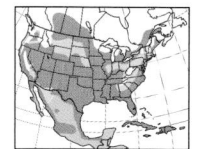

KEY FACTS

Stocky heron with short neck, big head, and yellow legs. Adult has black crown and back. Juvenile is brown with white spots on upperparts, streaks below.

+ voice: Loud, harsh *wok*

+ habitat: Fairly common in salt- and freshwater wetlands, around ponds and nearby fields

+ food: Fish, aquatic invertebrates, rodents, lizards, snakes

These herons roost in trees or dense vegetation near water, even in large cities, and are one of the most widespread of all herons worldwide. Some winter in the north, feeding on mice when ice covers fishing areas. Like other herons, they nest in colonies, often sharing the same trees with other waders. Chicks leave the nest well before they can fly, climbing through branches and into other nests. They sometimes beg for and receive food from adults other than their parents. Black-crowns are often seen flying between feeding areas and roosts.

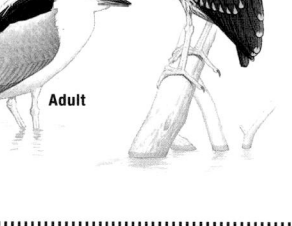

Juvenile

Adult

White Ibis

Eudocimus albus L 25 in (64 cm)

Ibises and spoonbills are large wading birds with specialized bills. They are strong fliers.

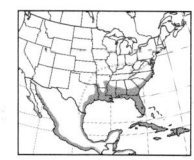

KEY FACTS

White wading bird (not a heron) with decurved red bill, red face and legs. Juvenile is dark brown above; older birds are mottled brown and white.

+ voice: Occasional low grunts

+ habitat: Southern swamps, coastal wetlands, beaches, lawns. Often seen in flocks.

+ food: Probes underwater for crayfish, crabs, and other aquatic life.

White Ibises are abundant and conspicuous in the Deep South, where they are mostly associated with marshes, mangrove swamps, and other estuary and inland wetlands. Foraging in shallow water, they use their sensitive bill tip to locate prey by touch, but when an injured ibis's bill tip is replaced with a prosthetic, it can still forage successfully. They also feed on lawns and mooch for food at Florida theme parks. They nest in large colonies, which may lose up to 44 percent of all eggs to predation. Independent young may join other juveniles away from adults.

Adult

Juvenile

Roseate Spoonbill
Platalea ajaja L 32 in (81 cm)

The Roseate Spoonbill, unrelated to the flamingo, is a conspicuous wader that feeds in tight flocks.

KEY FACTS

Flamboyant, pink wading bird with long, flat, spoon-shaped bill, unfeathered greenish head, white neck, and reddish legs. Juvenile much paler; takes 3 years to reach adult color.

+ **voice:** Mostly silent

+ **habitat:** Fairly common in shallow water of swamps and marshes along Gulf Coast

+ **food:** Mostly small fish, shrimp, and aquatic insects

Roseate Spoonbills feed by walking forward, sweeping their head side to side in wide semicircles, the sensitive bill slightly open until it detects prey items and snaps shut. Flocks may serve as "beaters," stirring up prey items for other waders that gravitate to them while feeding. They often roost in trees. When one flies over, a perched flock sometimes "sky-gazes," all extending their necks to point their bills skyward for a few seconds. They nest in colonies with other waders. Chicks insert their bill down a parent's throat to be fed. As they mature, they lose their head feathers and the bald skin gets more colorful.

Juvenile

Breeding adult

Wood Stork
Mycteria americana L 40 in (102 cm)

The Wood Stork, America's only stork, feeds by touch as well as sight, finding food easily in murky water.

KEY FACTS

Very large, white wading bird with heavy, downcurved bill and unfeathered gray head. Black flight feathers and tail; flies with neck extended, unlike herons. Juvenile has brown neck.

+ **voice:** Mostly silent

+ **habitat:** Southern swamps, shallow pools, and coastal marshes

+ **food:** Mostly fish. Walks slowly through shallow water with open bill submerged.

The "ironhead" or "flinthead" has been declining in South Florida since the 1960s, but small populations are expanding in the northern part of their range. Wood Storks nest in colonies with Anhingas and herons. To cool their chicks, parents regurgitate water over them. European storks nest on roofs near chimneys, perhaps why folklore credits them with delivering babies. Wood Storks weigh 4.5 to 6 pounds (2–2.7 kg), far too light to lug a newborn. Storks often spend hours on sunny days soaring on thermals.

Juvenile

Adult

Turkey Vulture
Cathartes aura L 27 in (69 cm)

The Turkey Vulture has a distinctive bald head, easy to clean after being plunged into carrion.

KEY FACTS

Blackish-brown with red head and pinkish legs; juvenile has grayish head, legs. In flight, note silver and black underwings and side-to-side rocking motion.

+ **voice:** Usually silent

+ **habitat:** Common and widespread; favors wooded areas for breeding; open areas for foraging

+ **food:** Animal carcasses, including roadkill

Turkey Vultures locate carrion by scent. Power companies add ethyl mercaptan to natural gas; the smell attracts Turkey Vultures, which circle above ruptured pipes. Their gastric juices destroy disease organisms, but, given a choice, they prefer fresh food. They can devour skunks without piercing the scent glands. Their wing surface area is huge relative to their weight, so they can stay aloft on rising air currents with very little flapping. When trapped, they vomit toward potential predators, and urinate on their legs to cool themselves.

Adult

Juvenile

Osprey
Pandion haliaetus L 24 in (61 cm)

Ospreys build stick nests at the very top of a tree or power pole or on a nest platform.

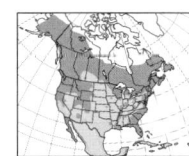

KEY FACTS

Large, eagle-like raptor. Brown above, white below with prominent dark eye stripe. Juvenile marked with pale fringes above. Flies with marked kink in wings.

+ **voice:** Series of loud, whistled *kyew* notes

+ **habitat:** Fresh- or saltwater habitats with clear water so that fish are visible

+ **food:** Large fish caught by spectacular, feet-first dives

Osprey talons are designed for hanging on to slippery fish. One front toe rotates backward to balance the two normal front toes. Spiny scales under the toes make the hold even more secure. When a fish is heavy, an Osprey may paddle to shore with its wings. Bald Eagles sometimes steal fish from them, one reason Ben Franklin disapproved of eagles. A courting male performs a "fish flight" sky dance. Screaming and holding a fish in his dangling legs, he alternates hovering and steep ascents.

Adult

Juvenile

Mississippi Kite

Ictinia mississippiensis L 14.5 in (37 cm)

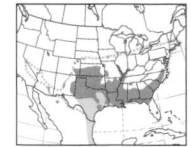

Kites are graceful, acrobatic hawks adapted for capturing small, erratically flying prey.

KEY FACTS

Graceful, mostly gray raptor with long, pointed wings and dark tail. Juvenile is heavily streaked below with barred tail and mottled underwings.

+ **voice:** Thin, high-pitched whistle

+ **habitat:** Summer breeder in southern woodlands and shelterbelts; sometimes wanders north.

+ **food:** Mainly large insects caught in flight

This crow-size hawk with long, pointed wings can be mistaken for a falcon in fast flight. It's easier to recognize while soaring with wing and tail feathers spread or engaged in aerial acrobatics. Some nest in urban parks and golf courses where they defend nests aggressively, often joined by mockingbirds. Small songbirds sometimes mob them. Some build nests on wasp nests, which may protect them from climbing predators. The two chicks seldom squabble. Kites are most active at midday and late afternoon.

Northern Harrier

Circus cyaneus L 18 in (46 cm)

Harriers, named for harrying prey, have owl-like facial discs to detect prey by hearing.

KEY FACTS

Long-winged, long-tailed raptor. Male is gray above, mostly white below; female is brown above, streaked below, with barred wings; juvenile resembles female but with rusty underparts.

+ **voice:** Various high-pitched calls

+ **habitat:** Open wetlands, grasslands, pastures, and fields

+ **food:** Flies low searching for rodents, reptiles, frogs.

Northern Harriers are buoyant, low-flying hawks that often hover while hunting. They sometimes follow large predators, seizing prey scared up by them, or along the edge of advancing wildfires, grabbing animals trying to escape. Adult males and females have different plumages. Females are mottled brown; males are nicknamed "gray ghosts" for their beautiful white and gray plumage accentuated by black wing tips. Males provide food and protection for their mate and young. In years of abundant prey, males may have a "harem" of 2 to 5 mates.

Bald Eagle

Haliaeetus leucocephalus L 34 in (86 cm)

The Bald Eagle isn't really bald; the reference is to the adult's shining white head feathers.

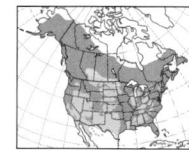

KEY FACTS

Huge eagle. Adult has white head and tail, huge yellow bill. Juvenile is dark brown with mottled tail and underwings; birds attain adult plumage when about 5 years old.

+ **voice:** Series of high-pitched twitters

+ **habitat:** Prefers seacoasts, large rivers, and lakes. Builds large stick nest in tall tree.

+ **food:** Mostly fish and waterfowl, also carrion

The Bald Eagle's imposing appearance led John Adams and Thomas Jefferson to use it on the national emblem, perhaps to symbolize America's military will and might. Eagle feet have three front toes and one rear toe. The imbalance of six toes on one side of prey and only two on the other may allow fish to thrash out of their grasp. In 1987, a Bald Eagle dropping its prey caused a midair collision between an Alaska Airlines jet and a fish; luckily, the only fatality was the fish. The oldest wild eagles known lived to their early 30s.

Juvenile

Adult

Adult

Cooper's Hawk

Accipiter cooperii ♂L 15 in (38 cm) ♀L 18 in (46 cm)

Accipiters have short, rounded wings and a long rudder-like tail to maneuver in woodlands.

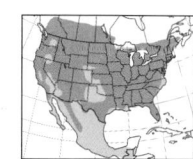

KEY FACTS

Long-tailed, woodland hawk. Adult has dark gray upperparts, rufous bars below, and banded tail. Juvenile is brownish above, streaked below. Very similar Sharp-shinned Hawk is smaller.

+ **voice:** Nasal series of *kek* notes

+ **habitat:** Woodlands, even in suburbs and towns

+ **food:** Preys on small birds; often attacks birds at feeders.

Cooper's Hawks have become urbanized, nesting in large shade trees, but are secretive and seldom noticed outside migration except when hunting in backyards or defending a nest. Males build the nest and provide food for their mate from the time they begin courting through the time the chicks are 12 to 14 days old. Only females incubate eggs and brood and feed chicks. Females are significantly larger than males. They may chase prey on foot through dense undergrowth.

Juvenile ♀

Adult ♂

Red-shouldered Hawk

Buteo lineatus L 17 in (43 cm)

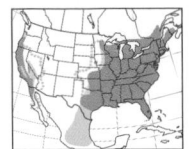

Buteos have long, broad wings and a short, broad tail and can float on rising air currents.

KEY FACTS

Medium-size hawk with distinctive black-and-white pattern on wings and tail, reddish bars on underparts, shoulders, and wing linings. Juvenile is browner with streaked breast.

+ **voice:** Loud, plaintive *kee-yeer*, often given in flight; imitated by Blue Jays.

+ **habitat:** Mixed woodland, particularly near water, swamps

+ **food:** Small mammals, frogs, snakes

This hawk has widely separated ranges. In the East, it is found in swamps, bottomland hardwood forests, and wet woodlands. In the West, it's found in riparian and oak woodlands, eucalyptus groves, and suburbs. It drops down from a perch to seize prey. Red-shouldered Hawks often mate for life. Many chicks are taken by Great Horned Owls. Turning the tables, one Red-shouldered Hawk chased a Great Horned Owl while the hawk's mate seized and ate the owl's chick. Migrating Red-shouldered Hawks join other hawks in large groups, called kettles, circling on the same thermals.

Juvenile

Adult

Red-tailed Hawk

Buteo jamaicensis L 22 in (56 cm)

In *Oklahoma*, the "hawk making lazy circles in the sky" is a buteo, most likely a Red-tail.

KEY FACTS

The "default" hawk in North America, often seen soaring. Most adults have rufous tail and dark bar on front edge of underwings. Juvenile has banded (not rufous) tail and is heavily streaked below.

+ **voice:** Harsh, descending *kee-eerrrr*

+ **habitat:** Varied, from mountains and forests to prairies and deserts

+ **food:** Small mammals, reptiles, birds

Conspicuous Red-tailed Hawks live near humans even in large cities, and are also found in wilderness. They usually hunt from a perch until they spy prey, but can drop down to seize animals spotted while circling high above. Adults usually return to the same mate and nest year after year; nonmigratory pairs may remain on or near their territory year-round. Both adults incubate eggs and brood and feed the young. One famous Red-tail, "Pale Male" (see photo), constructed a nest on a building overlooking Central Park in New York City in 1992, and has raised young there for over two decades. Some wild Red-tails have lived more than 30 years.

Adult

American Kestrel

Falco sparverius L 10.5 in (27 cm)

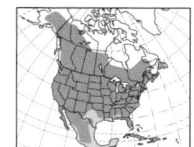

Despite similarities to hawks, falcons are more closely related to parrots and songbirds.

KEY FACTS

Petite, colorful falcon. Male has bluish wings, barred chestnut back and tail, apricot underparts with black spots; female is reddish brown above with streaked underparts.

+ **voice:** Shrill, rapid *killy-killy-killy*

+ **habitat:** Open country, farmland, roadsides. Nests in old woodpecker hole, birdhouse.

+ **food:** Mice, insects, and other small prey

American Kestrels perch on power lines along highways; long wings make them seem to swell in size when taking off in flight. Kestrels often hover in place, studying the ground to find prey. They can see ultraviolet light, allowing them to detect urine-marked rodent trails. Unlike most raptors, males and females have different plumage, and females are larger and more aggressive. They nest in cavities—setting out nest boxes has been shown to help them spread into new areas. Nestlings shoot streams of droppings onto the walls of the nest cavity, giving the nest a strong odor of ammonia and probably killing bacteria.

Adult ♂ **Adult** ♀

Peregrine Falcon

Falco peregrinus L 18 in (46 cm)

The majestic Peregrine Falcon can be found in almost all habitats on every continent except Antarctica.

KEY FACTS

Powerful falcon. Head blackish, appearing helmeted; blue-gray above; pale underparts with bars and spots. Juvenile brown above with heavily streaked underparts

+ **voice:** Harsh *cack* notes around nest site

+ **habitat:** Nests on tundra, coastal cliffs, also in cities; frequents open wetlands. Population is increasing.

+ **food:** Birds, hunted in flight at high speed

The fastest known bird has been recorded flying over 200 mph (322 kph) in vertical dives, or stoops; up to 70 mph (113 kph) while directly chasing prey; and 30 mph (48 kph) while traveling along. Peregrines hunt nocturnal migrants at lighted skyscrapers and oil rigs. They were extirpated from the United States east of the Rockies by the 1960s, but reintroduction programs have been widely successful. Most Peregrines historically nested on cliffs, but a pair nested on the Sun Life Building in Montreal from 1936 until 1952, perhaps providing the original inspiration to build nest platforms on tall buildings.

Juvenile **Adult**

Sora

Porzana carolina L 8.75 in (22 cm)

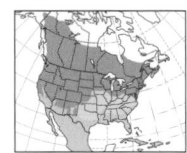

Rails such as the Sora are secretive marsh birds that seem weak in flight but may migrate thousands of miles.

KEY FACTS

Most common rail in North America, but small and secretive. Thick yellow bill, black face and throat, streaked above. Juvenile lacks black feathering.

+ **voice:** Descending whinny and high-pitched *keek*

+ **habitat:** Fresh and brackish marshes; saltwater marshes in winter

+ **food:** Seeds and aquatic invertebrates

This common rail makes a distinctive whinny from within marsh vegetation. "Meadow chickens" provide food for harriers, Coyotes, and other predators, and in fall, fattened on wild rice, they're hunted in 31 states and 2 provinces. To elude detection, these tasty, defenseless birds flatten their bodies laterally to slip between cattails without rustling them. A pair may produce up to 18 eggs in a single clutch. They build "dummy" nests to use as resting platforms or for brooding first-hatched chicks away from remaining eggs. Soras have unwebbed feet but are fine swimmers, even right after hatching. The word *Sora* was derived from a Native American name for this bird.

Juvenile

Breeding adult

American Coot

Fulica americana L 15.5 in (39 cm)

Coots and their relatives the gallinules are often mistaken for ducks, but they are not related.

KEY FACTS

Duck-like bird with blackish plumage and stubby white bill. Juvenile is paler below with darker bill. Lobed toes for swimming; requires long, running takeoff.

+ **voice:** Grunting and clucking calls

+ **habitat:** Common. Nests in freshwater habitats; winters in both fresh and salt water.

+ **food:** Aquatic plants; often seen in flocks, grazing (or loafing) on lawns.

"Cute coots with their white snoots" swim in flocks sometimes numbering in the thousands. They also feed on land, usually in smaller groups. "Mud hens" are awkward in flight, especially on takeoff—a flock may noisily patter on water, beating their wings, for a long distance before getting airborne. Their long toes are lobed for swimming and also for supporting their weight when standing in muck. The lobes fold as the foot is lifted while walking and in the forward stroke as they swim. Most birds in this family have a tiny, triangular tail with fluffy white feathers beneath. Some Bald Eagles specialize in hunting coots.

Adult

Sandhill Crane

Grus canadensis L 41–46 in (104–117 cm)

Cranes, a small family of large birds—15 species worldwide—are unrelated to herons.

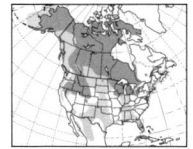

KEY FACTS

Very tall, gray bird with long black legs and red crown. Flies with neck extended and has feather "bustle," unlike any heron.

+ **voice:** Very loud, bugling call *gar-oo-oo*

+ **habitat:** Locally common. Breeds in marshes and on tundra; winter flocks forage in grain fields and wetlands.

+ **food:** Mostly grain and seeds, some insects

Huge numbers of Sandhill Cranes gather along Nebraska's Platte River in early spring to fatten up before completing spring migration. Cranes pair off when 2 to 7 years old, and remain together for life. Courtship dancing synchronizes their breeding readiness. Bare, bumpy skin on the crown gets more intensely red during territorial and courting activities. Chicks leave the nest soon after hatching. They remain with their parents for almost a year. Sandhill Cranes rub mud into their feathers while preening to stain them the color of local soil.

Adult

Stained adult

Black-bellied Plover

Pluvialis squatarola L 11.5 in (29 cm)

Plovers—short-billed shorebirds—run and pause while feeding. Some live far from water.

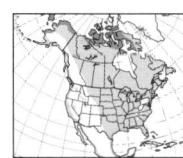

KEY FACTS

Large, chunky shorebird with short bill. Breeding adult has striking black face and belly; winter adult and juvenile are grayish above and streaked below.

+ **voice:** Rich, three-note whistle *pee-ooo-whee*

+ **habitat:** Nests on Arctic tundra; winters on mudflats and beaches.

+ **food:** Marine invertebrates; mostly insects on tundra

Inconspicuous when not in breeding plumage, these handsome birds are among the most wide-ranging of all shorebirds. They winter along coasts of every continent except Antarctica and may appear just about anywhere during migration. Their large size and ability to change feeding methods help them adapt to extreme conditions; large eyes help them feed at night. Both parents incubate their four eggs in a lichen-lined ground nest in the high Arctic, and both raise the young. They call loudly at the first sign of danger, providing sentinel services for other shorebirds. Their rapid flight and wariness kept them common when market hunting decimated other shorebirds.

Winter

Breeding ♂

Semipalmated Plover

Charadrius semipalmatus L 7.25 in (18 cm)

This small wader can swim short distances, aided by partially webbed (semipalmated) feet.

KEY FACTS

Small, brown-backed shorebird with single black breast band and white underparts. Short, black-tipped, orange bill and orange legs

+ **voice:** Unique, up-slurred whistle *chu-weet*

+ **habitat:** Beaches, lakeshores, and tidal flats; seen across the continent on migration.

+ **food:** Small invertebrates found in wet areas

This shorebird, often described as "adorable," is usually seen in flocks with other waders during migration. When not feeding near the water's edge on mudflats or sandy beaches, these plovers rest in small flocks safely above the high watermark. They have few defenses except subterfuge to protect their exposed ground nest, set in a shallow depression lined with debris, leaf litter, and other camouflaging items. Adults frequently sit low as if incubating eggs away from the nest, probably to confuse predators. They also lure away potential nest predators with a broken-wing display. They've recently expanded their Arctic breeding range southward.

Juvenile

Breeding ♂

Killdeer

Charadrius vociferus L 10.5 in (27 cm)

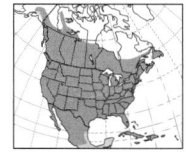

The Killdeer was given both its name and its scientific name, *vociferus*, for its loud, piercing calls.

KEY FACTS

Large, dark plover with two black breast bands. Horizontal stance with long tail; orange rump (best seen in flight) and white eyebrow.

+ **voice:** Loud, piercing *kill-dee*

+ **habitat:** Common in open fields (also lawns and parking lots) and on shores, but not tied to water

+ **food:** Invertebrates, especially earthworms, grasshoppers, beetles

Our most widespread plover nests on the ground in short-grass meadows and burned-over tracts, and also on construction sites, gravel roads, and driveways, often far from water, sometimes even on flat rooftops. When a nesting Killdeer spots a potential predator, it performs an injury-feigning display. But when a grazing, nonpredatory animal such as a deer or cow approaches a nest, the Killdeer rushes toward it screaming loudly, which may turn the animal away. Chicks can see and follow their parents to food within hours of hatching. Killdeer follow plows to take churned-up worms and insects. They are fine swimmers.

Adults

American Avocet

Recurvirostra americana L 18 in (46 cm)

Avocets belong to the family Recurvirostridae, the name referring to the up-tilted bill.

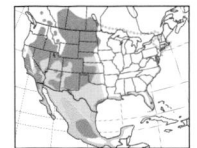

Black-and-white shorebird with upturned bill (straighter in male) and long legs. Head and neck are rusty in breeding plumage, gray in winter.

+ voice: Loud *kleek, kleek, kleek*

+ habitat: Fairly common in shallow alkaline and saltwater wetlands

+ food: Sweeps its long bill across the water (like a scythe) to catch small invertebrates.

Avocets sweep their bill through shallow water, detecting food by touch as well as sight. The bill is so sensitive that a bird in the hand recoils at the gentlest touch. Colonies breed in temporary wetlands of the arid West, where selenium contamination in irrigation water may cause embryo deformities. Incubating birds sit tight when aerial predators approach, but when off the nest aggressively strike at them. When a ground predator approaches, an avocet may fly at it while making changes to the pitch of its call. The Doppler effect makes the avocet seem to be approaching faster.

Winter ♂

Breeding ♀

Spotted Sandpiper

Actitis macularius L 7.5 in (19 cm)

The "teeter-peep" has the most widespread breeding range of any American sandpiper.

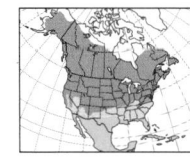

Small sandpiper with a teeter-tottering walk; flies with stiff, stuttering wing beats. Breeding bird has spotted underparts and pink-orange bill. Juveniles and winter birds lack spots.

+ voice: Shrill *peet-weet*; in flight, series of *weet* notes

+ habitat: Widespread in freshwater habitats, also seacoasts

+ food: Probes and picks for invertebrates.

This common sandpiper is found along lakes and rivers, urban waterfronts, and agricultural ponds, bobbing its tail wherever it goes. It can dive underwater and swim or walk on the bottom to elude predators. When surprised by ground predators, it makes a display called a "rodent run," squealing as it crawls low to the ground, wings flapping and tail spread. Males usually provide more care for young than females do. The chicks stay with their father for at least four weeks, sometimes joining flocks with other families. Females often leave their chicks with their mate to start a new nest with another male.

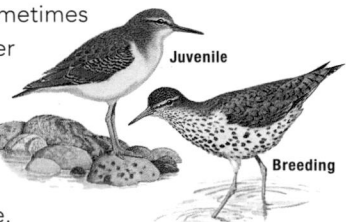

Juvenile

Breeding

Greater Yellowleg

Tringa melanoleuca L 14 in (36 cm)

The "Tattletale" is a tall shorebird that is easy to find, making piercing alarm calls at the slightest disturbance.

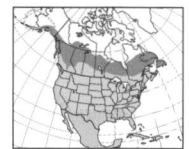

KEY FACTS

Long, bright yellow to orange legs. Breeding bird has streaked breast and barred flanks; juveniles and winter birds less heavily marked. Very similar to Lesser Yellowlegs but obviously larger

+ **voice:** Strident series of *tew* notes

+ **habitat:** Freshwater ponds and tidal marshes

+ **food:** Small invertebrates, minnows

Greater Yellowlegs, the size of small ducks, are widespread and easy to observe in wetlands during migration, and near coasts in winter. Small groups may work together to maneuver fish to where they can catch them easily. They spend as short a time as possible in the swampy muskeg habitats of Canada and Alaska where they breed. Little is known about their nesting habits because their breeding areas are so inaccessible and unpleasant, ridden with mosquitoes and black flies. Flukes (trematodes) infest Greater Yellowlegs so frequently and severely that parasites are a primary reason why they start migrating south in June.

Willet

Tringa semipalmata L 15 in (38 cm)

The Willet breeds in two very different locations: prairie wetlands and salt marshes.

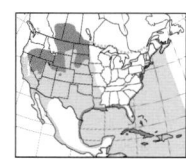

KEY FACTS

Large, plump shorebird. Plain and grayish overall, until it takes flight, revealing striking black-and-white wing pattern. Breeding bird is barred below; bars absent in winter.

+ **voice:** Loud, territorial call *pill-will-willet*

+ **habitat:** Common along coasts in winter; breeds in Atlantic and Gulf salt marshes and western prairies.

+ **food:** Probes for invertebrates.

Willets were named for their ringing *pill-will-willet* calls. Eastern and western populations vary in important ways. Eastern birds usually return to the same mate as long as both birds survive; western birds don't maintain pair bonds consistently. Eastern birds call more frequently and at a higher pitch, probably due to different ambient sounds. On their breeding grounds, Willets aggressively chase predators, but in winter they rise up in a flock to evade them. Scientists once placed them in their own genus, but DNA studies show they are fairly closely related to yellowlegs and some other sandpipers.

Whimbrel
Numenius phaeopus L 17.5 in (45 cm)

Curlews, including the Whimbrel, are large shorebirds with long, slender bills.

KEY FACTS

Large, streaky brown and buff shorebird with long downcurved bill. Bold dark stripes on crown and through eye; long grayish legs

+ **voice:** Fast series of hollow whistles on one pitch

+ **habitat:** Nests on Arctic tundra; along coasts in winter and during migration, when found almost worldwide.

+ **food:** Probes for invertebrates.

This large wader's scientific name, *Numenius*, comes from Greek for "new moon," referring to the crescent shape of the bill, which matches the curve of fiddler crab burrows. On their wintering grounds, Whimbrels reach deep into these burrows for crabs. During breeding season and on migration they usually pick at berries, insects, and other food items with the bill tip. To swallow, they simply jerk the head back and catch the item in the back of the throat. Some Whimbrels make a nonstop flight from southern Canada or New England all the way to South America, a treacherous 2,500-mile (4,000 km) journey. Pairs reunite on their breeding territory.

Adult

Ruddy Turnstone
Arenaria interpres L 9.5 in (24 cm)

Turnstones have a bill specially adapted for flipping stones to reveal food beneath.

KEY FACTS

Breeding male is striking, with black-and-white head, black-and-chestnut back, and short orange legs. Female is duller, and both sexes are less colorful in winter. Bold flight pattern

+ **voice:** Low-pitched, guttural rattle

+ **habitat:** Nests on Arctic tundra; along coasts in winter.

+ **food:** Aquatic invertebrates and insects

Ruddy turnstones are exceptionally adaptable. Their short, sturdy legs and flattened toes, spiny beneath and armed with sharp claws, are adaptations for running on slippery wet rocks and holding their stance securely when prying into crevices and turning over stones and debris. During breeding season on the Arctic tundra, they feed almost entirely on flies. The rest of the year, they have an extremely varied diet that includes handouts from beachgoers. Breeding pairs are monogamous and territorial, but associate with other shorebirds elsewhere. They jab at birds approaching too close.

Breeding ♂

Winter

Sanderling
Calidris alba L 8 in (20 cm)

These iconic sandpipers are found worldwide on temperate and tropical sandy beaches.

KEY FACTS

Small sandpiper. Winter plumage is very pale gray. Breeding bird is variably rufous. Bold wing stripe in flight

+ **voice:** Series of *kip* notes. Flocks twitter.

+ **habitat:** Nests on Arctic tundra; along coasts in winter and during migration, when found almost worldwide.

+ **food:** Aquatic invertebrates and insects

Sanderlings run ahead of advancing waves and chase retreating ones, giving them a frenzied, manic aspect. During the breeding season on the high Arctic, they are aggressively territorial. Either parent may incubate eggs and raise the young, but often only one does, the other finding a new mate and starting a new brood. Away from the breeding grounds, they associate in flocks that average 5 to 30 birds but may reach 2,500, especially along the Pacific coast. When chased by falcons, they fly in dense flocks moving erratically. When an individual is cut off from the flock, it may dive into the ocean.

Winter   Breeding

Least Sandpiper
Calidris minutilla L 6 in (15 cm)

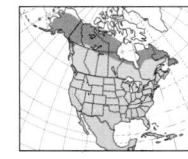

"Peeps" are small, similar-looking sandpipers that test birding skills. The Least Sandpiper is the smallest of the sandpipers.

KEY FACTS

The Least Sandpiper has yellow legs—unlike other sandpipers—and prefers to forage back from the water's edge rather than in the water.

+ **voice:** High *kreeep*

+ **habitat:** Northern nester; freshwater and saltwater locations in winter and during migration

+ **food:** Tiny invertebrates

The tiny Least Sandpiper is often found with other peeps on mudflats, where muck can coat the diagnostic yellow legs. Quiet, slow-moving birders can make careful comparisons because they often allow close approach. Least Sandpipers return annually to the same territory on the subarctic tundra and northern boreal forest, and often reunite with the same mate as long as they both survive. The eastern population probably flies nonstop from the Northeast to South America, yet is surprisingly long-lived. One bird banded as a juvenile in Nova Scotia was recaptured there, alive, when 15 years old.

Breeding   Juvenile

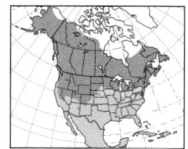

Wilson's Snipe
Gallinago delicata L 10.25 in (26 cm)

Despite mythical "snipe hunts" with a paper bag and two sticks,
the snipe is quite real.

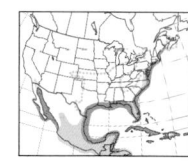

KEY FACTS

Stock shorebird with very long bill, boldly striped head, barred flanks. Very well camouflaged; sits tight when approached, then bursts into flight, and flies off in a zigzag path.

+ **voice:** Harsh *ski-ape* call when flushed

+ **habitat:** Wet meadows, bogs, and swamps for nesting; muddy fields during migration and winter

+ **food:** Insects, worms, some seeds

The snipe is common and widespread but elusive and shy. In England, its name has been in use since at least the 1500s, probably a variant of snout in reference to the bird's long bill. Snipe probe deep into wet, mucky mud for worms and other prey that they grasp with the flexible tip while keeping the rest of the bill shut. Sharpshooting skills are essential to successfully hunt these erratic fliers, leading to the word *sniper*. Snipe make a variety of calls. The haunting winnowing sound, which serves as a territorial "song," is produced not by voice but by airflow over the outer feathers of the outstretched tail.

Laughing Gull
Leucophaeus atricilla L 16.5 in (42 cm)

Some "seagulls" live far from the sea, but the Laughing Gull is seldom
far from salt water.

1st winter

Breeding adult

KEY FACTS

Midsize, dark gray gull, takes 3 years to become adult. Breeding adult has black hood and red bill; lacks hood in winter. First winter is browner and more subtly patterned, has black tail band.

+ **voice:** Crowing series of *hah* notes

+ **habitat:** Common along Gulf and Atlantic coasts; rare inland

+ **food:** Almost anything, including fish and beach handouts

Laughing Gulls are a familiar sight to anyone spending time in their range. They follow boats, mooch from picnickers, visit landfills, and grab fish from pelicans. Named for their flight calls, they are adapted for mobility on land, water, and air, often hovering to feed on flying insect swarms. Less aggressive than other gulls, they seldom take eggs or chicks from terns and shorebirds. Early in the nesting season, large bathing groups may gather for 15 to 20 minutes of dunking and dipping. Chicks aren't buoyant and drown if high tides inundate the nest.

Ring-billed Gull

Larus delawarensis L 17.5 in (45 cm)

Many of these abundant, opportunistic gulls never spend any part of their lives on the sea.

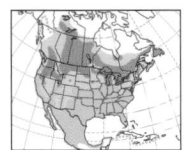

KEY FACTS

Midsize, pale gray gull, takes 3 years to become adult. Adult has yellow bill with black ring, pale eyes, and yellow legs. First-winter bird has pink bill with dark tip, dark eyes, and brownish wings.

+ **voice:** Mewing *kee-ew;* sharp *kyow*

+ **habitat:** Abundant and widespread along coasts and inland

+ **food:** Almost anything, including fish and garbage

This gull has adapted to human land-use patterns. Large flocks gather at dumps, freshly plowed or mowed fields, and fast-food restaurants. They nest on the ground near water in busy harbor and recreation areas as well as wilder spots. Their population expanded greatly between 1976 and 1990, especially in the Great Lakes region, and increased as much as 250 percent during the following two decades. In many areas, they're considered serious pests, but control measures have not curbed the population expansion. It's hard to imagine today, but this gull was nearly wiped out by persecution and habitat destruction between about 1860 and 1920.

1st winter

Breeding adult

Herring Gull

Larus argentatus L 25 in (64 cm)

Several large, white-headed gulls hybridize with and can be confused with this common gull.

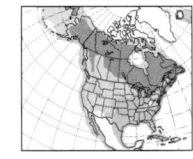

KEY FACTS

Large, pale gray gull, takes 4 years to become adult. Adult has white head (streaked in winter), yellow bill with red spot, pale eyes, pink legs. First-winter bird is mottled brown with dark bill and pink legs.

+ **voice:** Loud *kyow*

+ **habitat:** Widespread along coasts and inland

+ **food:** Marine invertebrates, fish, other birds, carrion, garbage

Herring Gulls drink ocean water when no fresh water is available, their nasal glands excreting excess salt. Away from the coast, they are most common around large lakes and rivers. They catch wild natural foods, scavenge on dead fish, and eat garbage at landfills. To break open hard-shelled prey, they often drop it on a rock or other hard surface from the air. A Herring Gull in Europe was observed using bread to bait goldfish. Males regurgitate food to their mate as part of courtship. Pairs usually remain mated for life, remaining with chicks for up to six months.

1st winter

Winter adult

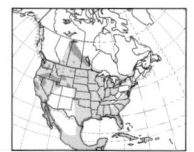

Caspian Tern
Hydroprogne caspia L 21 in (53 cm)

The Caspian Tern is the largest and strongest of all terns, almost gull-like in flight.

KEY FACTS

Very large tern with a coral red bill and black undersides to wing tips. Breeding adult has a black cap, speckled in winter. Juvenile has dark edges to back and wing feathers.

+ voice: Loud, raspy *kowk* and *ca-arr*, often given in flight

+ habitat: Breeds on large lakes (on islands) and in marshes; winters on southern coasts.

+ food: Mostly fish

Caspian Terns fly low over the water's surface searching for food, hovering and diving to take prey. They also snatch fish from smaller terns and gulls. Nesting Caspian Terns are fiercely protective, attacking hawks and inflicting bloody wounds on researchers' heads. Sometimes Herring Gulls and other opportunistic predators dart in and take chicks while the adults are chasing an eagle. Chicks need experience to get good at fishing. Meanwhile, their parents continue to feed them even on their wintering grounds. Some Caspian Terns "play," repeatedly dropping sticks or other objects and catching them before they hit the water.

Winter adult

Forster's Tern
Sterna forsteri L 14.5 in (37 cm)

Many terns are nicknamed "sea swallows" for their graceful flight and long, forked tail.

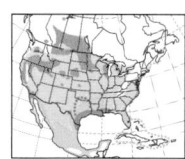

KEY FACTS

Midsize tern with a long tail; pale gray above, pure white below. Breeding adult has black cap and orange bill with black tip; winter adult has black face mask and dark bill.

+ voice: Hoarse *kyarr* notes, often in a series

+ habitat: Fresh- and saltwater marshes, beaches, lakes, and rivers—the most likely tern in many places

+ food: Mostly fish; some insects

This delicate-looking tern fishes in the deeper areas of fresh- and saltwater marshes, where it flies back and forth, bill pointed down, until it spots a fish and dives. One study found that about a quarter of Forster's Tern dives were successful. Small fish are swallowed almost immediately, but larger ones may be dropped a few times first. Unlike gulls, Forster's Terns virtually never swim. Despite their fragile appearance, they have harsh, raspy voices. They normally nest once per season, producing two to three eggs in a clutch, but renest if they lose their eggs or young to flooding or predation.

Breeding adult

Black Skimmer

Rynchops niger L 18 in (46 cm)

The Black Skimmer has a distinctive, knife-thin bill; its lower mandible is longer than the upper.

KEY FACTS

Large, black-and-white seabird with very long wings, a remarkable red-and-black bill, and short, red legs. Winter birds have a white collar. Juvenile's upperparts are fringed with white.

+ voice: Nasal *ip* or *yep*

+ habitat: Marine areas with quiet water. Nests on sandy beaches.

+ food: Small fish caught by skimming

America's only skimmer is extremely sociable, nesting in colonies and associating in large flocks the rest of the year. When loafing on coastal beaches, skimmers, like other birds, invariably all face directly into the wind to keep air currents from ruffling their feathers. They usually forage singly, flying low, the knife-like tip of the lower bill skimming the water's surface. When it detects a prey item, the upper bill snaps shut. Detecting food by touch allows them to hunt at night under moonlight. When fishing by day, their slit-like pupils can be mostly closed to protect the eyes from harsh sunlight above and the glare of the water's surface below.

Breeding adult

Atlantic Puffin

Fratercula arctica L 12.5 in (32 cm)

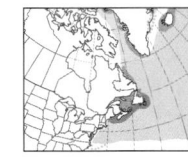

The Atlantic Puffin, a penguin-like flying seabird, lives on the open ocean except when breeding.

KEY FACTS

Compact seabird with a huge, multicolored bill, orange feet, and black-and-white plumage. Winter bird has dusky face and less colorful bill; juvenile's bill is much smaller.

+ voice: Growling notes at colonies

+ habitat: Breeds in the North Atlantic, from Maine northward; winters at sea, a few as far south as Virginia.

+ food: Fish, pursued underwater

Every spring, these "clowns of the sea" return to the island where they were hatched to raise their own young. They dig a nest burrow using their bill and shovel-like webbed feet. After the single egg hatches, they feed the chick small fish, bringing about ten per foraging trip. To carry so many, they brace each one as it's caught against the spiny upper palate with their raspy tongue. After nesting, adults shed their bill's bright outer plates and the horny decorations above and below their eyes. Puffins were extirpated from many Atlantic islands; reintroduction efforts are bringing them back.

Breeding adult

Rock Pigeon

Columba livia L 14.5 in (37 cm)

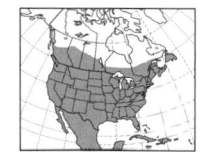

"Dove" and "pigeon" are interchangeable. In common usage, "dove" refers to smaller species.

KEY FACTS

Pigeons come in many color varieties: pure white, checkered, tan, red-brown, and more. Typical birds are gray with a white rump and black wing bars.

+ **voice:** Throaty cooing

+ **habitat:** Abundant in many cities, also around farmland

+ **food:** Seeds, fruits, bread crumbs, and littered food

Pigeons were first domesticated for food, sport, and companionship at least 5,000 years ago. Their powerful homing instinct has been exploited to send messages in times of war and peace and for racing. Their intelligence and trainability make them valuable research subjects for studying navigation and orientation, physiology, and how animals learn. City pigeons in the United States are feral descendants of domesticated birds, including racing pigeons and even war heroes. Unlike most invasive species, pigeons don't compete with native birds, and provide food for urban Peregrine Falcons.

Color variations

Eurasian Collared-Dove

Streptopelia decaocto L 12.5 in (32 cm)

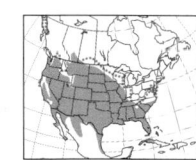

Released in the Bahamas in the mid-1970s, the Eurasian Collared-Dove soon reached the Florida peninsula.

KEY FACTS

Heftier and paler than Mourning Dove (next page), with a black half collar. Broad, squared-off tail is white with a black base. Perches in trees, overhead wires, rooftops.

+ **voice:** Three-syllable *coo-coo-cup*

+ **habitat:** Starting from Florida, this nonnative species has colonized much of the country.

+ **food:** Seeds and grain

These handsome doves thrive in urban and suburban habitats and in rural places where grain is available. They often feed side by side with other ground-feeding birds, share roosts with native doves, and may nest near or even in the same trees as House Finches, House Sparrows, and, surprisingly, predatory Loggerhead Shrikes. Eurasian Collared-Doves are prolific, the female often laying new eggs as soon as still-dependent chicks fledge. Clutches usually have two eggs, and the first to be laid is significantly larger than the second. Pairs stay together for a full breeding season, and some remain together in winter and even through the following year.

Mourning Dove

Zenaida macroura L 12 in (31 cm)

The Mourning Dove was named for its mournful song, given most persistently by unmated males.

KEY FACTS

Slender dove with long, pointed tail. Rich tan above with black spots, paler below. When taking flight, its wings make a loud whirring noise and spread tail shows large white tips.

+ **voice:** Slow, owl-like *oowoo-woo-woo-woo*

+ **habitat:** Abundant and widespread, except in dense forest

+ **food:** Seeds and grain

About 1 million hunters shoot about 20 million Mourning Doves every year, more than all other game birds combined. Despite the harvest, Mourning Doves are surprisingly long-lived; some banded ones survived in the wild over 10 years and at least one over 31 years. They're monogamous during the breeding season. The female produces two eggs in a rather flimsy nest. Both parents incubate, and both feed the young "pigeon milk" produced in their crops. Due to their dry diet of seeds, Mourning Doves need a lot of water, especially when feeding young. Pioneers knew water was nearby when they spotted one.

Juvenile

White-winged Dove

Zenaida asiatica L 11.5 in (29 cm)

Some doves and pigeons have "orbital rings"—colorful patches of bare skin around the eyes.

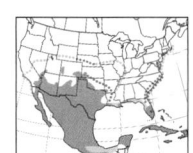

KEY FACTS

Midsize, tropical dove with a prominent white wing patch. Orange-red eyes surrounded by blue skin.

+ **voice:** Drawn-out cooing *who-cooks-for-you*

+ **habitat:** Found in woodlands, brush-lands, and desert towns in Texas and the Southwest.

+ **food:** Seeds, grain; also nectar, pollen, and cactus fruit

The White-winged Dove, more easily heard than seen, has a call that is reminiscent of an owl. Enormous colonies, limited to dense woodlands in the Rio Grande Valley and Mexico, declined as natural habitat was developed for agriculture, grazing, and housing. It slowly adapted to more fragmented nesting habitat, shifting to backyard feeders and ornamental plantings, and now the population and range are expanding. An introduced population has also been established in Florida. All doves pick up grit on the ground to help digest seeds. When they ingest shot along with tiny pebbles on the ground, their bodies may accumulate toxic levels of lead.

Monk Parakeet

Myiopsitta monachus L 11.5 in (29 cm)

The Monk Parakeet is native to Argentina. North America's only *native* parrot, the Carolina Parakeet, became extinct in 1918.

KEY FACTS

Midsize parakeet with bright green upperparts, pale gray face, bluish wing tips, and a long tail

+ voice: Loud, grating squawks and screams

+ habitat: Widespread, but decreasing, in Florida; also in scattered northern cities

+ food: Fruits, seeds; also visits bird feeders.

Monk Parakeets, commonly called "Quakers" by aviculturists, are the most widespread parrot that escaped captivity to become established in America. Native to temperate and subtropical Argentina, they have adapted to winters as far north as New York City, Chicago, and Portland, Oregon. In South America, they are considered agricultural pests. Fears of ravaged crops and fruit trees in America haven't materialized, and because so many people are fond of these charismatic birds, some eradication projects have been abandoned. This is the only parrot that builds a stick nest. Colonies build huge structures with individual apartments for nesting pairs.

Yellow-billed Cuckoo

Coccyzus americanus L 12 in (31 cm)

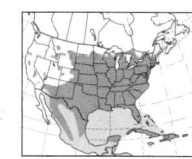

Although it is nicknamed the "rain crow," the Yellow-billed Cuckoo's calls aren't correlated with precipitation.

KEY FACTS

Gray-brown above, white below, long blackish tail with large white spots. Yellow bill with black on tip and upper edge

+ voice: Loud, hollow *kowlp-kowlp-kowlp*, staccato *kuk-kuk-kuk*, and series of *coo* notes

+ habitat: Open woods, orchards, streamside groves. Rare in West

+ food: Mostly caterpillars and large insects

These furtive birds are most easily seen near shrubs with webworms. American cuckoo calls include *coo-coo* notes, but none sound like cuckoo clocks, which were invented in the 18th century to mimic the Common Cuckoo of Germany's Black Forest. European cuckoos are brood parasites that lay their eggs in the nests of other species, leaving it to them to raise the young. American cuckoos raise their own young, though when food is abundant, Yellow-billed Cuckoos sometimes lay eggs in the nests of other cuckoos and species with blue-green eggs like theirs. In California, cuckoos sometimes breed cooperatively.

Adult

Greater Roadrunner

Geococcyx californianus L 23 in (58 cm)

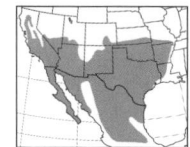

The Greater Roadrunner makes a series of slurred *coo-coo-coo* notes, but never says *me-beep.*

New Mexico's state bird is perfectly adapted for desert life. It gets enough water from its diet. When water is available, it drinks with gusto but never bathes, preferring dust baths. It lowers its body temperature at night, and raises it by sunbathing. It can fly short distances but usually runs, maintaining speeds of 20 mph (32 kph), using its tail as a rudder. It swallows horned lizards headfirst with the backside up so the horns won't jab vital organs; juveniles sometimes die when they swallow one before learning this. When alarmed or territorial, roadrunners expose orange skin behind their eye—one captive bird fled when its keeper wore orange.

Barn Owl

Tyto alba L 16 in (41 cm)

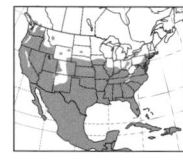

The only American bird in the family Tytonidae is nicknamed the "monkey-faced owl."

Spooky hisses and raspy screams, ghostly white wings, and nests in steeples above churchyard graves probably contributed to folklore associating owls with death. Barn Owls also nest in barns, nest boxes, and natural cavities, and some nests can be observed via live Internet streaming. They're territorial around their nest but may share foraging territories with others. They usually mate for life. Some males roost near the nest; others use separate roosts. Broods average about five chicks; the more food available, the more survive to fledging. The Barn Owl's ability to locate prey by sound in absolute darkness is the most accurate of any animal tested.

Eastern Screech-Owl

Megascops asio L 8.5 in (22 cm)

In eastern woodlands, this small owl is often the most common avian predator.

KEY FACTS

Small, robin-size owl with a big head, ear tufts, and yellow eyes. Two color morphs: gray and rufous. Cryptic plumage blends in with tree bark of roost site (see photo).

+ **voice:** Two typical calls: a long trill and a quavering whistle

+ **habitat:** Woodlands, suburbs, even large city parks

+ **food:** Small animals, including birds, rodents, insects

Screech-Owls spend their days in tree cavities and birdhouses year-round. If an intruder approaches one at the entrance to its cavity, it simply retreats. If the owl is on a branch, it stretches up and erects its feather tufts. If mobbed on a branch, it fluffs out, opens its eyes wide, and pulls back its tufts like an angry cat. Screech-owls usually select a mate of the same age, and pairs remain together for life. They often preen one another and their young, which reinforces their family bonds. They're very nurturing—in one case, a male Eastern Screech-Owl brooded flicker nestlings and tried to feed them bits of mouse.

Juvenile

Rufous morph

Gray morph

Great Horned Owl

Bubo virginianus L 22 in (56 cm)

The "feathered tiger" has the widest range and preys on more species than any American owl.

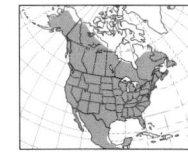

KEY FACTS

Very large owl with wide-set ear tufts. Overall color varies geographically from blackish brown (Pacific Northwest) to pale grayish (interior West) to reddish brown (East).

+ **voice:** Deep hooting *hoo hoo HOO-hoooo hoo*

+ **habitat:** Widespread; prefers woodlands with open edges.

+ **food:** Mammals, birds, snakes

The provincial bird of Alberta has a mellow hoot. Paired Great Horned Owls hoot back and forth, the pitch making it easy to tell which is which. Females are larger than males, but males have a deeper voice due to their larger syrinx (a bird's vocal apparatus). The heaviest ones weigh less than 4 pounds (1.8 kg), yet have been documented killing prey as large as Great Blue Herons and Sandhill Cranes. There is little evidence that they can carry prey that big, but they often consume just the heads of larger prey, and eat where they made the kill, if left undisturbed. Pairs may remain together for life.

Barred Owl
Strix varia L 21 in (53 cm)

The brown-eyed Barred Owl, one of the most vocal owls in North America, often uses nest boxes.

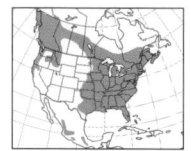

KEY FACTS

Large, chunky owl with rounded head and dark eyes. Mostly brown with barred breast and streaked underparts

+ **voice:** Deep hooting *who-cooks-for-you, who-cooks-for-YOU-ALL*

+ **habitat:** Mature forest; common in river bottoms and southern swamps

+ **food:** Small mammals, birds, reptiles, fish, and insects

Barred Owls are flourishing, expanding their range north and west, and even displacing endangered Spotted Owls. They mate for life and defend territories year-round, and may respond to imitations of their calls almost any time of year. Dueting Barred Owls often start with the familiar *who-cooks-for-you, who-cooks-for-YOU-ALL,* then move on to a series of ascending hoots ending with a loud *hoo-aw,* and finish off with a raucous jumble of calls sounding like maniacal laughter. Young ones sometimes make screechy begging calls all night long. Barred Owls live in swamps and wet forests, where they are known to pluck fish from ponds and streams.

Burrowing Owl
Athene cunicularia L 9.5 in (24 cm)

Unlike other owl species, the male Burrowing Owl averages slightly larger than the female.

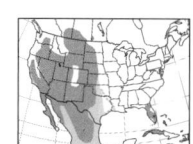

KEY FACTS

Brown with pale spots on back and brown bars on front. Glaring yellow eyes and long legs. Juvenile has buffy chest, no bars.

+ **voice:** Soft *coo-cooo* and chattering notes

+ **habitat:** Open areas with short grass, prairie dog towns, airports, golf courses

+ **food:** Insects, small mammals, birds, reptiles

In the West, this popular little ground-nesting owl usually nests and roosts in burrows dug by prairie dogs, skunks, tortoises, and so on. Some use culverts. The isolated Florida population is more likely to excavate their own burrows. Burrowing Owls tolerate higher levels of carbon dioxide than do other owls, an adaptation for living in long burrows with little air exchange. Females incubate the eggs and brood and feed the chicks; males hunt. They place animal dung around the entrance and within the nest. This may attract dung beetles, which they eat, but also may help control climate and carbon dioxide levels within the burrow.

Adult

Juvenile

Common Nighthawk

Chordeiles minor L 9.5 in (24 cm)

The Common Nighthawk has such an erratic, bat-like flight that it is nicknamed the "bullbat."

KEY FACTS

Floppy flight style on long, swept-back wings that have a white bar across the wing tip. Cryptic plumage makes it hard to see when perched.

+ **voice:** Nasal *peent* call is unique.

+ **habitat:** Woodlands, suburbs, and towns; declining in the East. Migrates to South America.

+ **food:** Insects caught in flight

Unrelated to raptors, nighthawks feed only on the wing, darting about in the evening sky, capacious mouths wide open to swallow moths and other flying insects whole. They fly low over water to drink. The tiny bill is loosely attached and the tongue a vestigial flap in the back of the mouth; grounded nighthawks cannot pick up food. They used rock-ballasted flat roofs for nesting, and their *peent* calls and booming breeding displays were once common evening sounds in both cities and wild open habitat. Changes in roof construction and increases in urban gulls and crows contributed to their decline over much of their range.

Juvenile

Eastern Whip-poor-will

Caprimulgus vociferus L 9.75 in (25 cm)

In 2011, the Whip-poor-will was split into two species: the Eastern Whip-poor-will and the Mexican Whip-poor-will.

KEY FACTS

Nocturnal and rarely observed, but often heard. Cryptic plumage makes it hard to see when perched. Large head and tiny bill, but very large mouth. Male's tail has large white tips, buffy in female.

+ **voice:** Loud, clear *whip-poor-will*

+ **habitat:** Open woodland

+ **food:** Flying insects; active at dusk, dawn, and on moonlit nights

Like nighthawks, whip-poor-wills belong to the nightjar family; unlike nighthawks, their mouths are bordered by long, stiff feathers called rictal bristles. Males produce their famous call from a variety of perches in their territory. The call attracts females but doesn't seem to elicit aggression from nearby territorial birds. They don't build a nest but, rather, lay their two eggs directly on leaf litter on the forest floor. Both parents incubate the eggs, brood the chicks, and feed them regurgitated insects. One study found that hatching occurs about ten days before a full moon, allowing maximum nighttime brightness for hunting while the chicks are most rapidly growing.

Chimney Swift

Chaetura pelagica L 5.25 in (13 cm)

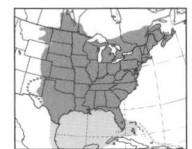

Strong claws and stiff tail spikes allow the Chimney Swift to roost on vertical structures.

KEY FACTS

Only swift in the East. Sooty brown, cigar-shaped bird, almost always seen in flight. Compared to a swallow, the swift has longer, swept-back wings and flies faster with stiffer wing beats.

+ **voice:** High-pitched, *chip* notes given in flight

+ **habitat:** Most common in areas with chimneys for nesting

+ **food:** Flying insects

On summer evenings, Chimney Swifts flutter in the sky, making chittering calls as they swarm and funnel into communal roosts. Before chimneys were available, they used hollow trees and large woodpecker holes. Now, chimneys are often capped and lined with metal; some people provide roosting towers as a substitute. Only one pair of swifts nests in a single structure, though others may roost in it with them. They break off sticks with their feet and carry them in their bill, using gluey saliva to hold the nest together and affix it to the substrate (see photo). The nest of an Asian species—the nest used in bird's nest soup—is built entirely from viscous saliva.

Ruby-throated Hummingbird

Archilochus colubris L 3.75 in (10 cm)

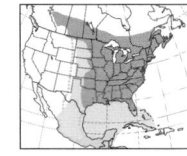

Hummingbirds, requiring protein as well as carbohydrates, are skilled at capturing tiny insects.

KEY FACTS

Only hummingbird in the East, Ruby-throat has glittering green upperparts. Male has red throat that looks black in some light; female's throat is white.

+ **voice:** Twangy *chips, tchew,* or *chih*

+ **habitat:** Forest edges, flowering gardens, sugar-water feeders

+ **food:** Flower nectar, small insects, spiders

Weighing less than a nickel, many Ruby-throated Hummingbirds fly nonstop over the Gulf of Mexico to the Yucatán Peninsula, a journey of at least 500 miles (805 km); in fall, they make this flight during hurricane season. They dive-bomb birds as large as Bald Eagles. Hummers arrive in spring before flowers open; they can feed on sap and insects at sapsucker drill holes. Nests are tiny and well insulated to hold the mother's belly tight against her pea-size eggs without letting her warmth escape. To stretch as chicks grow, the nests are made of bits of lichen woven with spider silk, which also holds the nest to the branch.

Adult ♂

♀

Anna's Hummingbird

Calypte anna L 4 in (10 cm)

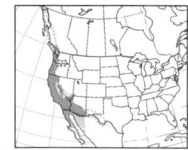

The only nonmigratory U.S. hummingbird, the Anna's Hummingbird wanders widely after nesting season.

KEY FACTS

Metallic green above and dingy grayish white and green below. Male has rose red throat and crown; female's throat is white with spotting.

+ **voice:** Male's song is a rhythmic series of scratchy notes.

+ **habitat:** Common around backyard gardens and flowering ornamental trees. Year-round resident

+ **food:** Flower nectar, small insects, spiders

The entire head of a male Anna's Hummingbird glitters like a sparkling amethyst, and he sings a complex, learned song as well as performing a flight display to attract females. At the bottom of his display dive, he makes a complex sound produced in part by the wings and in part by the spread outer tail feathers. Flight muscles comprise 28 percent of this hummer's body mass. Anna's Hummingbirds begin nesting at the onset of winter rains except in the more northern reaches of their range. The species has been expanding its range eastward, taking advantage of feeding stations and garden flowers.

Adult ♂

Adult ♀

Broad-tailed Hummingbird

Selasphorus platycercus L 4 in (10 cm)

The tiny Broad-tailed Hummingbird's wings produce a loud trilling sound rather than a hum.

KEY FACTS

Tail is impressively large. Male has a rose red throat; female has a white throat with tiny green spots and buffy underparts.

+ **voice:** Metallic *chip* notes; adult male's wings make a cricket-like trill in flight.

+ **habitat:** Summer resident of foothills and mountain meadows. Winters in Mexico.

+ **food:** Flower nectar, small insects, spiders

Broad-tails nest at high elevations in the central and southern Rocky Mountains. They can allow their body temperature to drop from a normal active temperature of 100° to 109°F (38°–43°C) down to as low as 54°F (12°C) when it's cold. In forward flight, their wings beat about 38 times per second; about 50 times per second while hovering to feed. As with other hummers, they don't form pair bonds—one male was reported mating with at least six different females. This species holds the North American hummingbird longevity record: A female caught and banded as an adult in Colorado in 1976 was retrapped, alive, more than 12 years later.

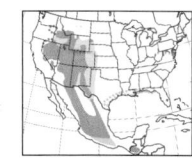

Adult ♂

♀

Rufous Hummingbird

Selasphorus rufus L 3.75 in (10 cm)

This bird has the shortest breeding season and sees the longest day lengths of any hummer.

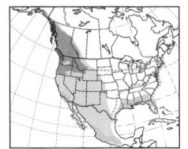

KEY FACTS

Adult male is copper-colored above with a reddish orange throat; the female has a white throat with green and red spots and buffy flanks.

+ voice: Hard, sharp call, *tewk*

+ habitat: Summer resident of open forests and streamside groves in the West. Winters in Mexico.

+ food: Flower nectar, small insects, spiders

The Rufous Hummingbird is extremely aggressive even by hummingbird standards, with an individual sometimes fending off dozens of other hummers to monopolize a feeder, chasing females from flowers and feeders even during breeding season. It can chase off chipmunks approaching its nest, although a tiny Least Chipmunk—weighing 1.6 ounces (44 g)—is about ten times heavier. Like all hummers, females feed young by regurgitation. Rufous Hummingbirds follow an elliptical migration route, heading north in spring along the coast; after the breeding season, they move south through the mountains. They also wander east more than any western hummer.

Belted Kingfisher

Megaceryle alcyon L 13 in (33 cm)

Unlike in most birds, the female Belted Kingfisher is more colorful than the male.

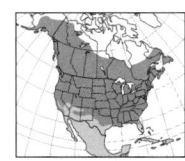

KEY FACTS

Stocky bird with big head, spiky crest, and large bill. Slate blue above. White below with single blue breast band in male; female has additional rusty band. Often perches over water.

+ voice: Loud rattle

+ habitat: Common and conspicuous along streams, rivers, and coastal estuaries. Nests in streamside burrow.

+ food: Fish and other aquatic life

A loud dry rattle alerts us to a Belted Kingfisher flying past or hovering in midair while staring at the water and sometimes plunging in to grab prey. Kingfishers carry their catch in their bill to a perch. They pound it to stun it and break off any long spines before eating it. Both sexes excavate the tunnel-like burrow that extends several feet into the side of a riverbank, gravel pit, or similar area. It can take three weeks or longer to build before the female lays five to eight eggs. Chick survival is very high in these protected nests except in years of flooding.

Red-bellied Woodpecker

Melanerpes carolinus L 9.25 in (24 cm)

This abundant woodpecker of the Southeast frequents bird feeders and backyard trees.

KEY FACTS

Familiar, zebra-striped woodpecker of the East. Male has a vibrant red crown; female has red only on her nape. White wing patch and rump visible in flight.

+ voice: Loud volley of churring notes

+ habitat: Year-round resident of open woodlands and suburban trees

+ food: Insects, nuts, fruits

Red-bellied Woodpeckers have extended their range north and west due to a combination of factors that may include habitat change, bird-feeding, and warming trends. At feeders, they take a wider variety of foods than most woodpeckers. European Starlings often take over their nest cavities; in turn, they take nest holes from Red-cockaded Woodpeckers, sometimes injuring or even killing the endangered birds. Males incubate the three to five eggs all night; both parents share duties during the day Except when incubating, males and females sleep alone in roost cavities. Fledglings often spend a few nights sleeping on branches before appropriating a roost cavity.

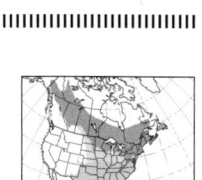

Yellow-bellied Sapsucker

Sphyrapicus varius L 8.5 in (22 cm)

Sapsuckers could more accurately be called sap-lappers; they lap sap with a brushy tongue.

KEY FACTS

Midsize woodpecker with striped face and large white wing patch. Male has red throat and forehead; female's throat is white. Juvenile is browner, lacks red.

+ voice: Nasal squeal *weeah* and catlike *meeww*

+ habitat: Northern forests; migrates south for winter.

+ food: Eats sap and insects attracted to sap wells.

If forest birds have a favorite avian neighbor, it is likely to be a sapsucker. Phoebes, kinglets, warblers, and other small birds visit their sap wells. Ruby-throated Hummingbirds associate closely with them before flowers open, feeding on sap and the insects drawn to it, and cavity nesters use their nest and roost holes. The slow, arrhythmic sapsucker drumming sound is easy to recognize. When a pair of sapsuckers spends time digging a new nest cavity, they produce fewer eggs than when they reuse an old nest, yet on average they fledge more chicks from new nests, perhaps because more parasites are present in old nests.

Adult ♂

Juvenile

Downy Woodpecker

Picoides pubescens L 6.75 in (17 cm)

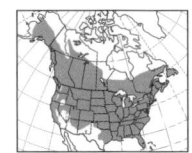

This diminutive woodpecker is named for the soft downy feathers on its lower back.

KEY FACTS

Black and white with a white stripe up the back and a short bill. Male has red spot on hind crown, lacking on female. Very similar Hairy Woodpecker is larger with a longer bill.

+ **voice:** Call is a sharp, high-pitched *pik!*

+ **habitat:** Year-round resident of woodlands

+ **food:** Probes and drills into wood for insects; comes to bird feeders.

Charming Downies visit feeding stations for suet and seeds, and sometimes even sugar water. In winter, they're drawn to mixed flocks of chickadees and other small songbirds. Foraging in the company of other wary birds allows each to focus more on feeding and less on searching for predators. The Downy also sometimes follows a foraging Pileated Woodpecker to probe for leftovers after the huge bird finishes up. Woodpecker chicks hatch at a less developed stage than most birds, probably because carbon dioxide levels build up in deep cavities while the parents are incubating. Flying in and out to feed young raises the oxygen level.

Northern Flicker

Colaptes auratus L 12.5 in (32 cm)

Civil War soldiers from Alabama often wore feathers of the "Yellowhammer" into battle.

KEY FACTS

Eastern birds have yellow underwings; western birds are pinkish red there. Males have a black (East) or red (West) whisker mark, absent on females.

+ **voice:** Loud *klee-yer!*

+ **habitat:** Woodlands, forest edges, suburbs

+ **food:** Specializes on ants, even digging into the soil for them; more berries in fall.

The state bird of Alabama often forages on the ground, hopping or running short distances. It also feeds in trees and roosts in cavities as other woodpeckers do. Eating ants gives the flicker a strong formic acid taste. John James Audubon said its meat was "not very savory," but predators don't object—piles of flicker feathers are often found along beaches on migration pathways. In one study, raptors killed 9 percent of adults with radio tags. Despite predation, wild flickers have lived longer than eight years. On its very first flight, a flicker chick may fly farther than 150 feet (46 m).

"Yellow-shafted" ♂

"Red-shafted" ♂

Least Flycatcher

Empidonax minimus L 5.25 in (13 cm)

Flycatchers belonging to the genus *Empidonax* are similar except for voice and habitat.

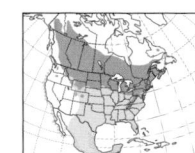

KEY FACTS

One of the 11 small flycatchers of the genus *Empidonax*, known as "empids" to birders. The Least is common in the East, and like other empids is olive green above with two wing bars and a white eye ring. Best identified by its song

+ voice: Harsh *che-BEK*, usually in a series

+ habitat: Summer resident of deciduous woods, orchards, parks

+ food: Mostly insects

The moment most birders hear *che-BEK,* they tick off Least Flycatcher and move on, often not realizing how fascinating and unique this unassuming little sprite is. Least Flycatchers nest in dense clusters rather than spreading out evenly throughout usable habitat. Each pair defends only a small territory from the others, but all aggressively chase other species from the area, perhaps why they are so seldom parasitized by cowbirds. They establish winter territories in Central America. To get the jump on the best spots, adults migrate as soon as their young are independent, not molting until they arrive in their wintering area.

Worn fall adult

Spring

Black Phoebe

Sayornis nigricans L 6.75 in (17 cm)

Phoebes catch prey on the wing. They seldom walk, hop, or even pivot by foot on a perch.

KEY FACTS

Small flycatcher, mostly black with a white belly. Constantly pumps tail up and down. Stays low, often alights on streamside rocks.

+ voice: Thin, whistled *pi-tsee* or *pi-tsew*

+ habitat: Permanent resident of open areas near water

+ food: Mostly insects; known to catch minnows

Calling Black Phoebes aren't as loud or insistent as their eastern relatives, but their rhythm lives up to the phoebe name. This handsome flycatcher is seldom far from water. Phoebes construct a muddy base against a structure such as a streamside boulder or open cavity and build the nest from mud and plant fibers. Natural sites are limited; nesting on man-made structures has allowed their numbers to increase. Established pairs often winter near each other and reuse nests year after year, getting a jump on breeding more quickly than those starting from scratch. Males roost on the nest rim or within a few feet of their mate.

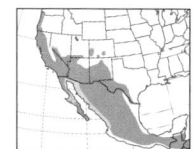

Juvenile

Eastern Phoebe

Sayornis phoebe L 7 in (18 cm)

All three of the North American phoebes have a distinctive habit of
bobbing their tails.

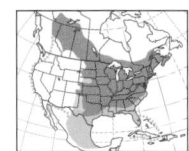

KEY FACTS

Drab eastern fly-
catcher, dark brown
above with pale
underparts; no wing
bars. Constantly
pumps its tail—rapidly
down and slowly up.

+ voice: Male sings
a harsh *fee-bee* that
gives the species its
name.

+ habitat: Wooded
areas, often near
buildings. Early spring
migrant, arriving in
March in many places

+ food: Flying insects

The Eastern Phoebe has been a popular research
subject for two centuries. It was the first bird to be
banded in the United States when, in 1804, Audubon
attached threaded leg bands to phoebe nestlings and
tracked them in following years. Observers have long
noted that phoebes are monogamous and seemingly
devoted mates; new DNA studies reveal that there is
considerable extra-pair paternity, particularly in the second brood each season. Banding and DNA studies con-
firm that some males pair with two females. Phoebes
prefer nest sites close to overhead
cover, building taller nests
to be closer to the
above surface. They
often use phoebe
nest platforms.

Worn summer
adult

Fresh fall

Eastern Kingbird

Tyrannus tyrannus L 8.5 in (21 cm)

The tiny red crown, found in both sexes of the Eastern Kingbird,
is revealed only in aggressive encounters.

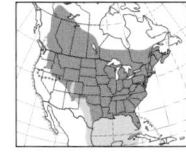

KEY FACTS

Midsize flycatcher with
slate gray upperparts,
snowy white belly, and
white-tipped tail. Sits
on open perches and
flies out after insects.

+ voice: High-pitched
sputtering notes

+ habitat: Open
areas with scattered
trees for nesting.
Despite its name, its
range extends far to
the west.

+ food: Flying insects

*A Kingbird sits atop
a soaring eagle.*

For half the year, *Tyrannus tyrannus* lives up to its scientific name—Ben Franklin even noted when dis-
paraging the Bald Eagle that "the little King Bird not
bigger than a Sparrow attacks him boldly and drives
him out of the District" (see photo). Predation on
adult kingbirds is very rare, yet ironically, an American
Kestrel once was observed grabbing a kingbird that
was distracted because it was attacking a Red-tailed
Hawk! These seasonal tyrants lead entirely different
lives on their wintering grounds in
the Amazon Basin. In the north,
they are extremely pugna-
cious, territorial flycatchers.
In the tropics, they become
sociable vegetarians, wandering
in large flocks to feed on fruits.

Western Kingbird

Tyrannus verticalis L 8.75 in (22 cm)

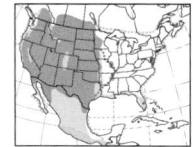

Some Western Kingbirds winter in South Florida, and a few wander to the East Coast during migration.

KEY FACTS

Midsize flycatcher with pale gray head and back, yellow belly, and white-edged tail. Like Eastern Kingbird, chooses conspicuous perches and is highly territorial.

+ voice: Fussy sputtering and *kip* notes

+ habitat: Summer resident of dry, open areas with scattered trees for nesting, often around farms and ranches.

+ food: Flying insects

These eye-catching birds sit on fences and other exposed perches, and conspicuously dive-bomb hawks, ravens, and other large birds. They expanded their original range as people planted trees in the Great Plains and cleared trees from heavily forested areas. Most nests are built in trees and shrubs, but many are set on such structures as utility poles, windmills, antennas, and even backyard basketball hoops, braced against the backboard. The nest is bulky, constructed of various fibers. Long strings incorporated into nests have sometimes entangled and killed nestlings; people should not set out yarns and strings longer than six inches for nesting birds.

Adult

Scissor-tailed Flycatcher

Tyrannus forficatus L 13 in (33 cm)

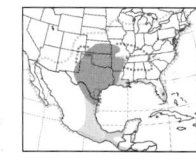

The Scissor-tailed Flycatcher and a tropical relative, the Fork-tailed Flycatcher, often wander far out of range.

KEY FACTS

Astonishingly long black-and-white tail, and pinkish belly and underwings. Pale gray head and back. Juvenile has much shorter tail and is paler overall.

+ voice: Fussy sputtering and *pup* notes

+ habitat: Summer resident of grasslands and prairies with scattered trees

+ food: Insects, especially grasshoppers, crickets, and beetles

A sitting Scissor-tailed Flycatcher is splendid enough. Sallying out to capture a flying insect, it exposes salmon pink wing linings and, as it hovers or turns abruptly, scissors its spectacular tail. In spring, males wheel and dip in ethereal courtship flights that leave even the most gushing superlatives wanting. Oklahoma's state bird was well chosen—its small breeding range is centered on the state. From spring through fall, it's hard to miss these extraordinary birds sitting conspicuously on fences and other obvious perches.

Adult ♂

Adult ♂

Juvenile

Loggerhead Shrike

Lanius ludovicianus L 9 in (23 cm)

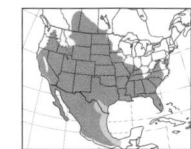

Nicknamed "butcher birds," shrikes are predatory songbirds that impale prey on thorns.

KEY FACTS

Small gray, black, and white songbird with the hooked bill of a predator. Similar in coloration to Northern Mockingbird, but note the shrike's black mask.

+ **voice:** Harsh *shack-shack*

+ **habitat:** Open country with scattered bushes. Declining in most areas.

+ **food:** Large insects, lizards, mice, and birds

Lacking talons, the shrike uses its sharp bill to grab a victim, and dispatches it with a bite to the nape. Using its bill again, the shrike carries off the carcass and impales it on a barb or thorn; the food may be eaten on the spot or saved for later. This 2-ounce (57 g) songbird has a high metabolic rate and must eat frequently. When hunting is good, food accumulates to be used when hunting is poor. Mockingbirds, caracaras, and Burrowing Owls sometimes raid the shrike's food stores, as do neighboring shrikes. Mated birds that don't migrate seem to stay close to each other year-round.

Adult

Red-eyed Vireo

Vireo olivaceus L 6 in (15 cm)

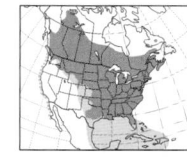

This unassuming bird is among the top ten most abundant land birds in North America.

KEY FACTS

Small, olive forest bird with striped face and gray crown bordered with black; the red eye is hard to see.

+ **voice:** Series of singsong phrases (*Here I am, over here, see me, where are you?*)

+ **habitat:** Summer resident of mature woodlands

+ **food:** Insects, particularly caterpillars

These often-overlooked birds sing what seems to be an endless monologue of robin-like phrases of two or three notes, pausing between each phrase as if for dramatic emphasis. They continue long after most songsters have quit for the day. The song is pleasing and the birds ubiquitous in summer, found in virtually every forest and small woodlot, yet many people live out their lives without ever knowing they exist. They quickly notice jays, cats, and other predators and draw the attention of other birds by scolding with querulous, harsh mews. Their compact cup nest suspended from a forked branch usually holds three to four young.

Blue Jay

Cyanocitta cristata L 11 in (28 cm)

Jays bury more acorns than they eat and have helped replant forests after glaciers melted.

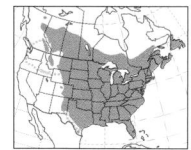

The provincial bird of Prince Edward Island, one of the most handsome, charismatic birds of America, attracts our notice with its loud squawks and colorful plumage. Feathers have no blue pigments; specialized cells in the barbs scatter light to produce blue. Adult jays feed primarily on seeds and fruits, along with insects during summer. Their four to five chicks require a high-protein diet, and so are fed insects supplemented with nestling birds of other species. When jays investigate a nest, many songbirds join together to drive them away. The rest of the time, jays are valued neighborhood watch guards, alerting everyone to danger.

Steller's Jay

Cyanocitta stelleri L 11.5 in (29 cm)

Corvids—crows, ravens, magpies, and jays—are among the most intelligent birds in the world.

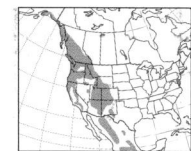

America's darkest jay, the provincial bird of British Columbia, is much more likely to visit campers for handouts than the Blue Jay, but is not as bold as its relatives, the Gray Jay and Clark's Nutcracker. It raids caches from them and from Acorn Woodpeckers. It has complex social interactions and vocalizations, including harsh notes, imitations of other species, and a quiet warbling song. Steller's Jays from a wide area join together to mob predators. In one case, a small group struck a perched Cooper's Hawk on the back, knocking it to the ground. Mated pairs remain together for life and are socially dominant over unmated individuals.

Western Scrub-Jay
Aphelocoma californica L 11 in (28 cm)

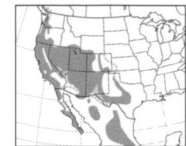

The Western Scrub-Jay has a curious habit of bobbing in several directions when it alights on a perch.

KEY FACTS

Familiar "blue jay" of the West—unlike Blue Jay of the East—has no crest or prominent white spots. Coastal scrub-jays are conspicuous; interior birds are shy and quiet.

+ **voice:** Harsh, upslurred *jay?* or *jreee?*

+ **habitat:** Year-round resident of chaparral, open woods, and backyards

+ **food:** Omnivorous. Nuts, fruit, insects; visits bird feeders.

Unlike its close relative, the Florida Scrub-Jay, this species is extremely vocal, sometimes following people while making harsh calls that warn other birds away. It preys on eggs and young of other species more often than Blue Jays do, and also hunts a variety of rodents and lizards. Scrub-jays wedge acorns and other hard nuts into tree crevices to hammer them open. Like other jays, they place a leaf or two over each cached food item they hide on the ground. Those in the habit of raiding other birds' caches are the most suspicious about being observed while hiding their own stores. Birds living in the interior West may be a separate species.

Black-billed Magpie
Pica hudsonia L 19 in (48 cm)

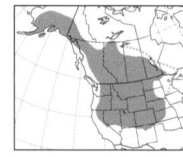

Magpie nests are huge masses of sticks with two entrances to an inner nest cup.

KEY FACTS

Striking black-and-white plumage; long, glossy tail and bluish wings. Magpies sometimes gather in large flocks. White wing patches visible in flight.

+ **voice:** Various harsh calls and raspy chatter

+ **habitat:** Year-round resident of western rangelands and foothills

+ **food:** Omnivorous. Fruit, grain, insects, small animals, carrion

Magpies raided Lewis and Clark's tents and stole food from the expedition. The explorers shipped four magpies to President Jefferson from Fort Mandan in 1845; only one survived the journey. This spectacular corvid scavenges on dead animals and sits on the backs of large mammals, picking off and eating ticks. It also may kill or injure small farm animals, and sits on large animals pecking flesh from open sores and wounds. Some states once offered a bounty on them. This extremely intelligent species can recognize its reflection in a mirror. It was considered the same species as the European magpie until 2000.

American Crow

Corvus brachyrhynchos L 17.5 in (45 cm)

Crows remember individual people perceived as threats and teach other crows to fear them.

KEY FACTS

All-black plumage. Larger than grackles and blackbirds; smaller than ravens. In the South, where the Fish Crow (*uh-uhh* call) also lives, the two species are best told apart by voice.

+ **voice:** Familiar call is harsh *caw, caw, caw.*

+ **habitat:** Various, especially open areas with scattered trees

+ **food:** Grain, insects, small animals, carrion, garbage

Henry Ward Beecher wrote, "If men had wings and bore black feathers, few of them would be clever enough to be crows." These fascinating birds maintain family and neighborhood ties that last generations. When West Nile virus decimated crows in Ithaca, New York, where many crows are individually marked for study, one widowed female adopted her dead neighbors' orphans; the orphans later helped her raise her own young. In fall and winter, crows join enormous flocks. When one detects a Great Horned Owl, it makes a loud "assembly call," bringing in dozens more that noisily and aggressively harass the owl. Many crows fly over highways searching for fresh roadkill.

Common Raven

Corvus corax L 24 in (61 cm)

The Common Raven, the largest songbird in the world, is the provincial bird of Yukon Territory.

KEY FACTS

All-black plumage and heavy bill. Often soars, unlike much smaller crows. Ravens have shaggy neck feathers and a wedge-shaped tail.

+ **voice:** Common calls are a croaking *kraaah* and a hollow *brooonk.*

+ **habitat:** Widespread in the West; uncommon in the East

+ **food:** Grain, insects, small animals, carrion, garbage

Many authorities consider the raven one of the most intelligent animals on Earth. It features as a benevolent trickster in native folklore of the Pacific Northwest. Its close associations with people, scavenging habits, and dark plumage contributed to its role as Poe's "ghastly grim and ancient" bird. Ravens have been associated with the Tower of London for centuries; they may have arrived to scavenge on executed corpses or victims of the 1666 Great Fire of London. A flock of seven captive ravens is still maintained in the Tower because of a superstition that if they disappeared, "the Crown will fall and Britain with it."

Horned Lark

Eremophila alpestris L 7.25 in (18 cm)

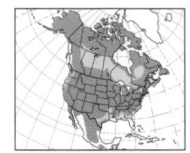

America's only native member of the lark family, the small Horned Lark is not related to meadowlarks.

KEY FACTS

Small songbird with black mask and forehead that extends upward as two "horns." Pinkish brown above with variable amount of yellow on face and a black bib. Forms large flocks in winter.

✦ **voice:** Song is series of tinkling notes.

✦ **habitat:** Open country; prefers bare ground.

✦ **food:** Seeds; feeds insects to its young.

A lovely yet unobtrusive song rings in the skies over farm country, airports, and other open areas. America's closest relative of Percy Bysshe Shelley's blithe spirit has a spectacular sky dance, flying straight up as high as 800 feet (244 m) where it breaks into its high-pitched, tinkling flight song while circling for many seconds. When finished, it drops headlong back to earth, wings closed until the last moment. Horned Larks blend in so well that they're usually overlooked on the plowed fields and rocky bare ground where they live, even when they gather in large flocks, but a close-up view through a scope or binoculars is well worth the effort.

Purple Martin

Progne subis L 8 in (20 cm)

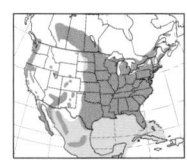

The Purple Martin, North America's largest swallow, spends the winters in the Amazon Basin.

KEY FACTS

Male is dark purple overall; female and juvenile are dingy grayish white below. In flight, rapid flapping alternates with short glides.

✦ **voice:** Loud, rich gurgling and whistles

✦ **habitat:** Open fields. Eastern birds nest exclusively in human-provided gourds and apartment houses. Migrates to South America.

✦ **food:** Flying insects

When a Purple Martin's tiny bill opens, revealing a truly capacious mouth, it looks as if the face split wide open. During nesting season, they dine on flying insects and make rich, liquid burblings and twitterings. Martins in the West and in Mexico nest almost exclusively in natural cavities. East of the Rockies, there is a dramatic switch. There, virtually all martins nest in birdhouses and hollowed-out gourds, and there are very few records of them nesting in natural cavities during the entire 20th century. Because so many martin houses are set out to accommodate several pairs, most martins in the East nest colonially.

Eastern ♀

Adult ♂

Western ♀

Tree Swallow

Tachycineta bicolor L 5.75 in (15 cm)

The Tree Swallow can digest berries when temperatures are too low for flying insects.

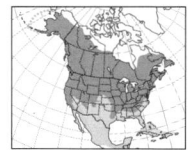

KEY FACTS

Adult has dark blue (or greenish), glossy upperparts and pure white belly. Tail is slightly forked. Juvenile is grayish brown above with diffuse band across chest.

+ voice: Liquid twittering and chirping

+ habitat: Needs open habitat and prefers to nest near water.

+ food: Flying insects; small fruits in winter

The first swallow to arrive north in spring, it nests in birdhouses, natural cavities, and woodpecker holes. It builds a nest of grasses and straws lined with white feathers. Nesting success is directly correlated with the number of feathers, which insulate eggs and chicks from extreme temperatures. Prairie grouse get into skirmishes while displaying—when they leave, Tree Swallows pick up the feathers the big birds lose. People can set out clean poultry feathers to help swallows. They complete breeding in early July and form enormous flocks. By late fall in the South, these flocks can number in the hundreds of thousands.

Spring adult

Cliff Swallow

Petrochelidon pyrrhonota L 5.5 in (14 cm)

Cliff and Barn Swallows gather at puddles to pick up mouth-size plops of mud for nests.

KEY FACTS

Compact body with short, square-tipped tail and buffy rump patch. Adult has white forehead and chestnut cheeks and throat. Juvenile is browner with less evident head pattern.

+ voice: Squeaking twitters and grating notes

+ habitat: Open areas with overhanging cliffs or structures to attach their mud nests to

+ food: Flying insects

Cliff Swallows build their adobe houses in large groups; a Wisconsin farmer once counted 2,015 nests on his barn. Females may start laying eggs before the roof of the nest is completed. The swallows nesting at San Juan Capistrano used to return every March when weather was warm enough to sustain flying insects. For people who carefully kept their eyes averted until March 19, the swallows indeed appeared on St. Joseph's feast day. They disappeared from the mission in recent years, probably in response to major landscape changes. Historically, the population spread from the West across the plains to the East, and numbers overall are fairly stable.

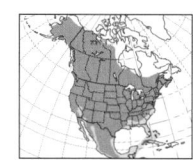

Barn Swallow

Hirundo rustica L 6.75 in (17 cm)

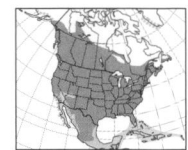

The Barn Swallow builds a large cup nest; the Cliff Swallow builds a globe with an entrance hole.

KEY FACTS

Cobalt blue above and buffy below with a chestnut throat. Long, graceful tail of adult is unlike any other North American swallow; juvenile's tail is shorter.

+ **voice:** Series of scratchy, warbling notes and grating rattles

+ **habitat:** Summer resident of open areas; shelters its mud nest on man-made structure.

+ **food:** Flying insects

Aristotle may have been thinking of this bird when he said, "One swallow does not a summer make." The most widespread swallow in the world nests under bridges and eaves, and in culverts, barns, and outbuildings. In Minnesota, a pair nested for several years in the lumber section of a home improvement store, hovering in front of the sensor to trigger the door to open so it could pass in and out. New World Barn Swallows once bred only in North America, spending the winter in South America as nonbreeders. In 1980, six pairs nested in Argentina. Now, a South American breeding population is well established.

Juvenile

Black-capped Chickadee

Poecile atricapillus L 5.25 in (13 cm)

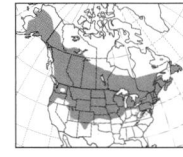

In late summer, sibling Black-capped Chickadees disperse, each joining a different winter flock.

KEY FACTS

Small, energetic songbird with oversize head patterned in black and white. The very similar Carolina Chickadee is found in the South.

+ **voice:** Call is a slow *chick-a-dee-dee-dee.*

+ **habitat:** Common resident of woodlands, wooded edges, suburbs, towns

+ **food:** Insects, seeds, and berries; often visits bird feeders.

The chickadee's inquisitive ways and acrobatic habits endear it to people, from hunters in deer stands to the housebound at their window. Chickadees are often the first birds to discover a new feeder and sometimes take seeds or mealworms from the hand. They nest in cavities they excavate themselves or in woodpecker holes and birdhouses. They prefer birdhouses filled with wood shavings for them to remove. Their song is a pure whistled *hey, sweetie.* When scolding, the more *dee* notes in their *chick-a-dee-dee-dee* call, the more dangerous the threat. Chickadees maintain strict hierarchies in winter flocks, which often include other songbirds.

Worn summer

Fresh fall

Tufted Titmouse

Baeolophus bicolor L 6.25 in (16 cm)

The handsome Tufted Titmouse, a crested relative of the chickadee, sings a whistle-like tune year-round.

KEY FACTS

Larger than a chickadee, with an obvious crest and a small black forehead. Juvenile lacks the black forehead.

+ voice: Song is whistled *peter-peter-peter*; call is chickadee-like *tsicka-dee-dee*.

+ habitat: Permanent resident of eastern forests and suburbs

+ food: Insects, seeds, and berries; visits bird feeders

Titmice nest and roost in cavities and birdhouses. To line their nests, they find soft fur, even plucking tail hairs from road-killed squirrels and sleeping raccoons. Some titmouse young disperse to other areas in late summer as chickadees do, but some remain with their parents all winter, and a few help their parents raise the following year's brood. Pairs are territorial year-round. Winter groups include a mated pair, their young, and young that dispersed from other areas; sometimes other small songbirds join the group. The Black-crested Titmouse of Texas and Mexico was until recently considered the same species.

Juvenile

Adult

Bushtit

Psaltriparus minimus L 4.5 in (11 cm)

The Bushtit, a sociable little bird, is famous for its enormous, dangling pouch nest.

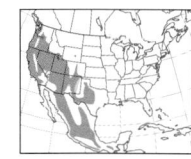

KEY FACTS

Tiny gray-brown songbird with a plump body and long tail. Male has dark eyes; female has pale eyes. Flocks of 10 to 40 birds forage together. Builds unusual hanging nest.

+ voice: Flocks twitter constantly.

+ habitat: Permanent resident of western thickets, scrublands, and backyards

+ food: Small insects and spiders

Active little Bushtits have two color variations that were once classified as different species. Researchers discovered that "black-eared" birds are juvenile males that are most common in Mexico, not a separate species. Bushtits huddle together at night or when resting by day, sometimes even when it isn't cold. Mated pairs take about a month to build their nest, which is huge for a bird weighing little more than two pennies. The nest maintains a more constant internal temperature, allowing the birds to spend more time foraging instead of incubating. In some places, the parents get help raising their young from unattached Bushtits.

Red-breasted Nuthatch

Sitta canadensis L 4.5 in (11 cm)

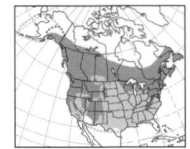

The Red-breasted Nuthatch is the only nuthatch that migrates
regularly, sometimes as early as July.

KEY FACTS

Chickadee-size song-
bird with black-and-
white head stripes and
cinnamon underparts.
Moves with a jerky,
toylike action. Often
very tame

+ **voice:** Series of
ehhnk ehhnk nasal
calls

+ **habitat:** Coniferous
woods and mountain
forests; some wander
south in winter.

+ **food:** Insects, spi-
ders, seeds

These handsome birds of fir and spruce forests
excavate their own nest cavities. The female
chooses a dead branch and starts digging. As she
enlarges the interior nesting space, she tosses
sawdust out the entrance—it accumulates at the
base of the tree. Both the male and female smear
globules of sticky conifer resin around
the entrance, which probably
deters predators. The parents
dive into the nest with precision to avoid touching the resin,
though at least one female was
found dead, firmly stuck. When conifer
seeds are scarce, large numbers
head south of their breeding
range in "irruptions."

White-breasted Nuthatch

Sitta carolinensis L 5.75 in (15 cm)

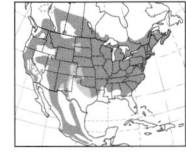

Nuthatches forage in bark crevices while they are upside down, thus finding
insects that other birds miss.

KEY FACTS

Larger than a chicka-
dee, with a stocky
body, short tail, and
dagger-like bill. Male
has black cap, blue-
gray upperparts;
female has gray cap,
dull gray back.

+ **voice:** Low-pitched
nasal *yank;* higher-
pitched in the West

+ **habitat:** Forests,
wooded suburbs,
parks

+ **food:** Insects,
spiders, seeds

America's only nuthatch associated with deciduous forests doesn't have stiff tail feathers
to brace itself against a trunk. Unlike woodpeckers and creepers, nuthatches have very short
tails and a long back claw. The name "nuthatch"
comes from the habit of wedging seeds and nuts
into crevices of bark and hacking them
open with the bill. White-breasted Nuthatches are territorial year-round,
though they may leave their territory to visit feeders when
food is scarce. After a pair is
established, they remain together as
long as they both live. They don't try
to nest more than once a year, even if
their first attempt fails.

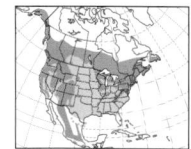

Brown Creeper

Certhia americana L 5.25 in (13 cm)

The Brown Creeper has forward-facing eyes that can examine bark closely with binocular vision.

KEY FACTS

Brown upperparts are mottled with pale streaks and dots that closely match the furrowed bark it clings to. Note white eyebrow; long, spiky tail; and cinnamon rump

+ voice: Call is a very high-pitched *seee.*

+ habitat: Mature forests. Birds in Southwest mountains may be a separate species.

+ food: Insects, spiders

The Brown Creeper's quiet habits, high-frequency songs and calls, and cryptic plumage make it one of the most inconspicuous of songbirds. To feed, a creeper begins at the base of a tree trunk and climbs upward, sometimes following a straight path, sometimes spiraling, until it reaches obstructing limbs or the top, when it drops like a leaf to the base of a nearby tree to start the process again. When it detects an insect or spider in the bark, it snaps it up with its delicate, curved bill. Creeper nests are so well hidden behind loose flaps of bark on decaying trunks that ornithologists didn't discover one until 1879.

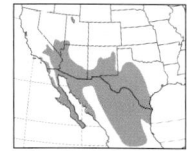

Cactus Wren

Campylorhynchus brunneicapillus L 8.5 in (22 cm)

The Cactus Wren gets all the water it needs from its food, and only rarely drinks at birdbaths.

KEY FACTS

Very large wren with brown crown, streaked back, heavily barred wings and tail, broad white eyebrow. Breast is densely spotted with black, less so in coastal California.

+ voice: Rapid series of harsh notes *cha cha cha,* heard all year

+ habitat: Conspicuous denizens of cactus country, arid scrub, and desert suburbs

+ food: Mostly insects and spiders

The well-chosen state bird of Arizona is perfectly adapted to extreme desert conditions. Males sing a rapid series of grating notes from conspicuous perches on and off all day, even in the heat of the afternoon. Females more rarely sing, their notes softer and higher pitched. Each pair builds several bulky, globular nests within spiny plants, and uses them for roosting year-round. They often take dust baths before retiring at dusk. Parents continue to feed their chicks for weeks after they fledge. Young birds may start nest-building activities just 12 days after fledging, using the structures for nighttime roosting.

Carolina Wren
Thryothorus ludovicianus L 5.5 in (14 cm)

The tail of South Carolina's state bird is usually cocked up, but it's lowered when the wren is singing.

KEY FACTS

Chunky, rufous-brown wren with buffy underparts, bold white eyebrow, and long twitchy tail. Pairs stay together year-round.

+ voice: Vocal throughout the year; song is a loud, rolling *tea-kettle tea-kettle tea-kettle.*

+ habitat: Eastern woods, brushy ravines, and backyards

+ food: Pokes into crevices for insects and spiders.

This backyard bird nests in crevices and cavities including birdhouses, glove compartments in abandoned cars, old shoes, mailboxes, flowerpots, and pockets of coats on clotheslines. They are much more often heard than seen. Their song is loud, but pleasing and seldom obtrusive. Many tropical wrens sing complex duets, but the Carolina Wrens' duets are more primitive. The male sings the familiar song, and occasionally the female chimes in with a buzzy phrase. Although nonmigratory, Carolina Wrens wander, and the species expanded its range northward in the 20th century, due in part to milder winters, reforestation of some areas, and bird feeders.

House Wren
Troglodytes aedon L 4.75 in (12 cm)

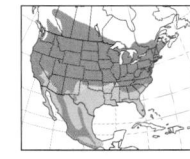

House Wrens may add spider egg cases to nest material; hatching spiders eat nest parasites.

KEY FACTS

Small, nondescript wren with finely barred brown plumage and medium-length tail, often held cocked up. Very assertive; will drive larger birds from nest hole it wants.

+ voice: Cascade of whistled notes

+ habitat: Summer resident across North America; found in open woods, thickets, and backyards

+ food: Mostly insects and spiders

House Wrens sing a cheerful, bubbly song virtually all day, are easily attracted to birdhouses, and eat a wide variety of insect pests, making them treasured backyard birds for all but those maintaining bluebird houses. House Wrens often peck or remove eggs and young nestlings of other nearby cavity nesters. Yet, unexpectedly, they've also been recorded feeding chicks of other species. While his mate was incubating, one male started feeding nestling flickers in the same tree. Several days later, when his own chicks hatched, he continued feeding the large woodpeckers as well as his own tiny young. House Wrens often take new mates for second broods within a season.

Golden-crowned Kinglet

Regulus satrapa L 4 in (10 cm)

The songs and calls of this tiny kinglet species are too high pitched for many people to hear.

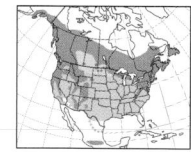

KEY FACTS

Tiny and very active. Male has black crown with yellow and red center; female's crown has a yellow center. Pale gray below.

+ **voice:** Call is a very high-pitched, often overlooked *tsii tsii tsii.*

+ **habitat:** Breeds in coniferous forest. Some birds move south in winter and are found in various wooded habitats.

+ **food:** Insects, spiders

A flame-colored crown is appropriate on a bird exuding as much heat as a Golden-crowned Kinglet. While the female is actively incubating, her eight to nine tiny eggs are kept at about 104°F (40°C). If that seems improbably warm, beneath her feathers, her body is about 111° (44°C). Many weigh less than a nickel, yet can survive winter in northern areas until temperatures drop to about -40°F (-40°C). Even in the dead of winter, they fuel their metabolic furnace on insects. In spring, they construct a tiny, thick-walled nest that will stretch out as their chicks grow. They nest twice a year but don't reuse their nests.

Ruby-crowned Kinglet

Regulus calendula L 4.25 in (11 cm)

The Ruby-crowned Kinglet's loud, rich warble includes notes sounding like *liberty, liberty, liberty.*

KEY FACTS

Tiny and very active. Slightly larger than Golden-crowned with more uniform olive plumage. Blank face with white eye ring. Male has red crown.

+ **voice:** Call is a husky, scolding *je-dit.*

+ **habitat:** Breeds in coniferous forest; migrates south and spends the winter in various wooded and brushy habitats.

+ **food:** Insects, spiders

Kinglets flit about while feeding, seldom remaining in any spot for more than a second or two, and constantly flicking their wings. The male Ruby-crown's song is exceptionally loud for his size; he often sings a muted version immediately before returning to the nest when eggs or chicks are present. Ruby-crowns produce the largest clutch of tiny North American songbirds—up to 12 eggs, which must be layered in the narrow nest. Females probably use their legs as well as the patch of bare skin on their belly (the brood patch) to keep all the eggs evenly warm. In winter, both kinglets join flocks, often with chickadees.

Blue-gray Gnatcatcher

Polioptila caerulea L 4.25 in (11 cm)

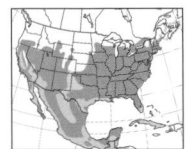

Other gnatcatchers are nonmigratory, but the Blue-gray Gnatcatcher retreats south in winter.

KEY FACTS

Small, slender song-bird; gray above and white below, with a long, twitchy tail. Breeding male has a thin black eyebrow, absent in winter.

+ **voice:** Call is a querulous *speeeee.*

+ **habitat:** Wide-spread in woodland settings; winters in southern states.

+ **food:** Insects, spiders

People seldom notice this tiny bird with its soft, high-frequency sounds, unless they're specifically looking for it. But a glimpse at a gnatcatcher richly rewards a search through the foliage. It is pugnacious, often attacking predators such as Cooper's Hawk and Northern Pygmy-Owl, diving at and hovering around their head and persistently following them. The tiny nest is built with lichens and spider silk, expanding as the four to six nestlings grow. This is the tiniest species regularly parasitized by Brown-headed Cowbirds. Cowbird eggs are too large for gnatcatchers to pierce or toss out, and survival of gnatcatcher chicks in a parasitized nest is very low.

Breeding ♂

American Dipper

Cinclus mexicanus L 7.5 in (19 cm)

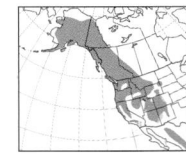

The American Dipper, or "water ouzel," lives along rushing western streams with clear, unpolluted water.

KEY FACTS

Aquatic songbird with a heavy, rounded body. Adult is uniformly sooty gray with a short tail and dark bill. Juvenile is paler overall with faint barring on underparts.

+ **voice:** Ringing, wren-like song of musical notes; audible over the din of rushing water

+ **habitat:** Mountain streams in the West; lower in winter

+ **food:** Aquatic insects and their larvae

The drab dipper elicits gasps of admiration from the graceful way it bobs up and down while standing on wet rocks in an icy river and then jumps in, walking underwater to search for food. Dippers nest on horizontal ledges or crevices, under overhanging dirt banks, or other noisy streamside sites, sometimes behind waterfalls. About half the nests are constantly sprayed, but there isn't a clear advantage for either wet or dry nests. To survive freezing temperatures in and out of frigid waters, dippers have a low metabolic rate, high oxygen-carrying capacity in their blood, and very thick plumage.

Juvenile

Eastern Bluebird

Sialia sialis L 7 in (18 cm)

Missouri and New York's state bird is beloved for its beautiful plumage and its soft, lovely song.

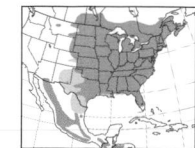

KEY FACTS

Male has flashy blue upperparts; female's plumage is more subdued. Rusty color on breast wraps up and forms a partial collar; belly is white. Juvenile is heavily spotted.

+ **voice:** Rich, musical warble *chur chur-lee chur-lee*

+ **habitat:** Open, rural areas. Most birds do not migrate.

+ **food:** Insects from spring to fall; small fruits in winter

Many people buy birdhouses in hopes of attracting bluebirds. But these red, white, and blue birds that inspired the "bluebird of happiness" are not backyard birds—they live in expansive fields, pastures, and orchards. Providing a few perches from which they can scan the ground for insects increases the likelihood of attracting them. Nesting males and females can be ferocious toward competitors. When pairs succeed in raising young, they often pair together the following year. Bluebirds increased when settlers cleared forests, and declined in the mid-20th century. Restoration projects helped bring back their numbers.

Mountain Bluebird

Sialia currucoides L 7.25 in (18 cm)

The state bird of Idaho and Nevada sometimes wanders east as far as Long Island in the winter.

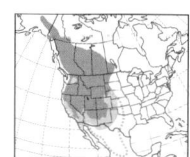

KEY FACTS

Male is sky blue above, paler blue below. Female is brownish gray overall with pale blue wings and tail, and flanks tinged with chestnut. Juvenile is spotted below.

+ **voice:** Warbled *tru-lee*; call is a thin *few*.

+ **habitat:** Western montane habitats; migrates to lowlands in winter.

+ **food:** Insects, small fruits

The Mountain Bluebird eats fewer berries and other plant food than other thrushes, and searches the ground for insect prey by hovering, more in the manner of a kestrel than a bluebird. As soon as a female accepts a mate, he starts following her closely wherever she goes until the chicks hatch. He does this "mate guarding" as she gathers nesting materials, even though he doesn't contribute to nest building. Most of the day, Mountain Bluebirds are quieter than other bluebirds, singing their soft, rich song most frequently before first light. Their alternate song, a soft, repetitious warble, can be sung at any time, but is often overlooked.

Hermit Thrush

Catharus guttatus L 6.75 in (17 cm)

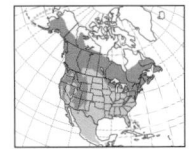

Vermont's state bird, the Hermit Thrush, only rarely sings its beautiful carol in winter or during migration.

KEY FACTS

Medium-size thrush—smaller than a robin—with a bright rufous tail. Upperparts are rich brown and the breast is buffy and spotted. Western birds are grayer brown. Juvenile has spotted upperparts.

+ **voice:** Series of clear, flutelike notes; call is a harsh *chup*.

+ **habitat:** Boreal and high-elevation forests; winters farther south.

+ **food:** Mostly insects and small fruits

The Hermit Thrush's down-to-earth feeding habits, nest placement on the ground or low in trees, and understated earth-tone plumage belie the heavenly images conjured by its song. Rather than a simple larynx, birds have an intricately designed syrinx where the trachea splits into the bronchial tubes, allowing them to create multiple sounds simultaneously. Ornithologists have chosen religious mnemonics to describe this thrush's ethereal song, such as *oh, holy holy, ah, purity purity, eeh, sweetly sweetly*. In spring, Hermit Thrushes arrive weeks earlier than other forest thrushes. This is the only forest thrush whose population has increased or remained stable over the past two decades.

Juvenile

American Robin

Turdus migratorius L 10 in (25 cm)

This state bird of three states—Connecticut, Michigan, and Wisconsin—spends fall and winter in sociable flocks feeding on fruits.

KEY FACTS

Fairly large, potbellied songbird with brick red breast, gray back, blackish head. Juvenile has paler breast with spots.

+ **voice:** Loud, musical song *cheerily cheer-up cheerio*

+ **habitat:** Common in many places, including lawns and parks, also wilder locations. Forms flocks in winter.

+ **food:** Earthworms and insects; more berries and fruit in winter

Named by homesick settlers for the familiar European "robin redbreast" (unrelated to thrushes), our robin is a thrush belonging to the same genus as the Eurasian Blackbird. Some robins spend all or part of the winter in northern areas where fruits remain. When temperatures average about 37°F (3°C), males return to territories, sing, and switch to high-protein food. Females arrive a few days later. Both parents feed nestlings. When the young fledge, the female renests as the male attends the fledglings. When the new clutch hatches, the older chicks are ready to be on their own.

Juvenile

Gray Catbird

Dumetella carolinensis L 8.5 in (22 cm)

Catbirds are best known for their mewing calls; males sing a long jumble of mixed sounds.

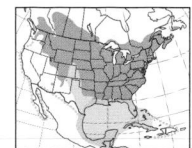

KEY FACTS

Slender, dark gray bird with a long tail, black cap, and chestnut undertail. Stays in underbrush and makes short, low flights.

+ voice: Mix of melodious and squeaky notes; call is a catlike *mew.*

+ habitat: Thickets, brushy forest, vine tangles; withdraws to the south in winter.

+ food: Insects; more small fruits in winter

Catbirds are easier to hear than see. They skulk in dense underbrush, and when one emerges to sing on an exposed branch, he often perches with his tail depressed, his slender, almost branch-like form keeping him hidden in full view. Often when we think catbirds must be nesting nearby because of their sounds, it isn't until after leaves fall that the nest is revealed in a conspicuous shrub. Catbirds eject cowbird eggs. They often destroy eggs and nestlings of other nearby birds, yet have been reported caring for orphaned young of other species. They may visit feeders for fruits, jelly, or mealworms. At least one wild catbird lived more than 17 years.

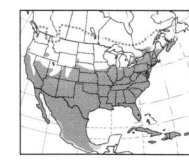

Northern Mockingbird

Mimus polyglottos L 10 in (25 cm)

Mockingbirds sing a long string of repeated phrases, adding new ones throughout their life.

KEY FACTS

Slender, gray-and-white songbird with a long tail and yellow eyes. Large patches of white in the wings and outer tail feathers. Juvenile has spotted breast and dark eyes.

+ voice: Mix of original and mimicked phrases, each repeated two to six times

+ habitat: Thickets and brushy areas; most birds are resident.

+ food: Insects; more small fruits in winter

The state bird of five states was one of Thomas Jefferson's favorite birds; he kept one as a pet in the White House. A male's repertoire may consist of more than 150 distinct song types; the number increases with age. Imitations may include mechanical sounds, vocalizations of other animals, and human voices, laughter, and screams. Unattached males may sing throughout the night. Pairs rear as many as four broods in a season. On the ground, mockingbirds frequently raise their wings to flash the white wing patches, perhaps to startle insects or predators or serve as a territorial display.

Juvenile

Brown Thrasher

Toxostoma rufum L 11.5 in (29 cm)

Thrashers are ground-foraging mimids; this species sings most imitations twice.

KEY FACTS

Rich rufous upper-parts, long tail, two whitish wing bars, and yellow eyes. Under-parts are extensively streaked. Skulks in the underbrush.

+ **voice:** Series of melodious phrases, each repeated two or three times

+ **habitat:** Eastern species favors dense cover; winters in the South.

+ **food:** Insects; more small fruits in winter

Georgia's state bird may have even more song types than mockingbirds—in 1981, a Brown Thrasher featured in *Ripley's Believe It or Not!* was credited with having a repertoire of 2,400 distinctly different songs. A popular mnemonic for their song is *Drop it drop it, cover it cover it, pull it up pull it up.* Birds nesting in shrubs or on the ground often lose young to predation; young Brown Thrashers leave, fully feathered, just nine days after hatching, minimizing their vulnerability. Thrashers take frequent dust baths. They often pick up ants from the ground and smear or place them in their breast feathers.

Curve-billed Thrasher

Toxostoma curvirostre L 11 in (28 cm)

A double whistle given by both sexes in many contexts is the Curve-billed Thrasher's most common call.

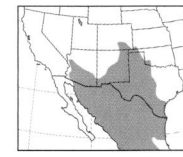

KEY FACTS

Most common thrasher in the Southwest. Robust bird with orange eyes and heavy, decurved bill. Dingy gray-brown plumage is spotted or mottled below. Long tail is tipped with white.

+ **voice:** Melodic series of low trills and warbles. Call is a sharp *whit-WHEET.*

+ **habitat:** Arid brush and cactus-rich desert

+ **food:** Insects, also seeds, cactus fruit

When William Swainson examined and named a specimen of this thrasher sent from Mexico in 1827, other thrashers with curved bills hadn't yet been described for science. Two populations of this species, one in the Sonoran Desert of Arizona and one in southern Texas and New Mexico brushland, may represent different species. The song doesn't include as many imitations or repetitions as songs of mockingbirds or Brown Thrashers. Pairs defend their territory year-round. They construct their loosely woven cup nest from thorny twigs. The chicks remain in the nest for 14 to 18 days after hatching.

European Starling

Sturnus vulgaris L 8.5 in (22 cm)

Hundreds of European Starlings can fly in tight units, maneuvering with breathtaking precision.

KEY FACTS

Stocky, short-tailed bird with pointed bill. Plumage is glossy black in summer, speckled with white in fall and winter. Juvenile is grayIsh brown.

+ **voice:** Elaborate song full of rattles, buzzes, and squeals

+ **habitat:** Very common around human structures; avoids dense forest and unbroken desert.

+ **food:** Insects, berries, seeds

Starlings didn't live in America until 1890, when a group trying to introduce to America every bird mentioned by Shakespeare released 60 in New York's Central Park. In 1891, they released another 40. From that small beginning, more than 200 million starlings now cover the continent. Their success came directly at the expense of several declining birds, because starlings compete aggressively for nest cavities. They also are a nuisance in cities and agricultural areas. These relatives of mynas are excellent mimics, and their songs also include several sounds innate to their species. Starlings are strong fliers, moving along at about 35 to 50 mph (56–81 kph).

Juvenile

Breeding ♂

Cedar Waxwing

Bombycilla cedrorum L 7.25 in (18 cm)

Waxwings are extremely sociable year-round, often nesting in synchrony in clustered groups.

KEY FACTS

Sleek plumage of warm browns and yellows. Brown crest and black "bandit's mask." White undertail coverts and waxy red tips to some wing feathers

+ **voice:** Soft, high-pitched whistle *zeee*

+ **habitat:** Widespread in areas with fruiting trees; flocks are somewhat nomadic.

+ **food:** Small fruits and berries; some insects

Waxwings have a charming habit of passing petals and berries down a long line of birds until one finally eats it. Wildlife rehabbers have observed that some tough-skinned berries pass through their digestive tract intact unless the rehabber manipulates the berries to soften them; passing food among a flock may make berries more digestible and may reinforce group bonds. Many pair off in April, but nesting is delayed until fruit is plentiful. The red tips on wing feathers that give waxwings their name are more numerous on older birds. Waxwings may use them as a signal of the reproductive value of potential mates.

Juvenile

Snow Bunting

Plectrophenax nivalis L 6.75 in (17 cm)

Snow Buntings and related longspurs were once classified with sparrows, but aren't related.

KEY FACTS

Brown, black, and white plumage during winter. By spring, brown fringes have worn off to reveal a starkly black-and-white bird that soon migrates far to the north.

+ **voice:** Whistled *tew* and a musical rattle

+ **habitat:** Winters in fields, dunes, lakeshores, often in large flocks. Breeds in the Arctic.

+ **food:** Seeds, insects

In late winter, these hardy birds snow-bathe on hard-crusted snow to abrade their dark feather tips, revealing their striking breeding colors. Males begin returning to the high Arctic in early April, competing for territories when harsh weather is still likely. Females return weeks later. Snow Buntings nest in skulls, cracks in large rocks, barrels, construction rubble, and so on, and the best sites are very limited. After the young are independent, they start forming flocks. Adults and young molt and build up fat before migrating. Birds in the back of a flying flock constantly overtake the ones in front, producing a rolling effect.

Winter ♂

Black-and-white Warbler

Mniotilta varia L 5.25 in (13 cm)

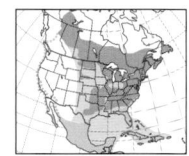

The Black-and-white Warbler probes bark for insects, sharing many habits with creepers and nuthatches.

KEY FACTS

Entire bird is striped with alternating lines of black and white. Breeding male has black throat and cheeks; these areas are white in female and immature in fall.

+ **voice:** Series of high, thin *wee-see* notes

+ **habitat:** Eastern warbler; common summer resident in mixed woodlands

+ **food:** Insects, spiders

This aptly named bird is one of the first warblers to arrive on its breeding grounds in spring. Like many warblers, it feeds and sings in trees but nests on or near the ground, at the base of a tree or fallen log. Its high-pitched song sounds a bit like a squeaky gate and is fairly easy for bird-watchers to recognize. During migration, it associates in feeding flocks with chickadees and other songbirds by day, making its long migration flights at night. Some fly nonstop over the Gulf of Mexico to reach the Yucatán Peninsula, but some winter in the Southeast and in Texas.

♀

Breeding adult ♂

Common Yellowthroat

Geothlypis trichas L 5 in (13 cm)

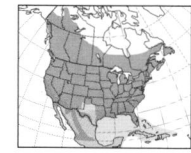

The Common Yellowthroat varies in appearance over its range, and the songster includes several subspecies.

KEY FACTS

Adult male has a broad black mask bordered above by gray or white, below by bright yellow throat and breast. Female lacks black mask and is duller yellow below.

+ **voice:** Loud, rolling song *wichity wichity wichity wich*

+ **habitat:** Common summer resident in marshes, shrubs, grassy fields

+ **food:** Insects, spiders

A loud *wichity wichity wichity wich*, punctuated with harsh call notes, is a characteristic sound of marshes and wet meadows. The songster, much smaller than his voice, can stay frustratingly hidden, but patience is rewarded when he hops to a conspicuous perch. Male yellowthroats sometimes feed their mate, and both parents equally share the feeding of nestlings. When the chicks fledge, the female may start a second nest, leaving care of the first brood to her mate. Yellowthroats seldom associate with other warblers. Like other nocturnal migrants, they call while flying in the dark, probably helping maintain a safe distance from others.

Adult ♂

♀

American Redstart

Setophaga ruticilla L 5.25 in (13 cm)

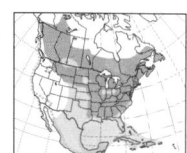

The popular American Redstart is known as *la candelita* (the little flame) in the West Indies.

KEY FACTS

Often fans tail and spreads wings. Male is glossy black and orange; female has similar pattern in olive and yellow.

+ **voice:** High, thin notes, often accented at end *tsee tsee tsee tsway*; call is rich *chip*.

+ **habitat:** Common summer resident in second-growth woodlands

+ **food:** Insects, spiders, some small fruits

The redstart has bristly feathers on the sides of the mouth that may help funnel in insects caught on the wing. Males don't assume their black-and-orange plumage until late in their second summer, after breeding season. These yearling males claim territories and sing, but females only consider them as mates when no adult males are available. After chicks fledge, parents may divide broods and go their separate ways, each feeding their half for up to a month longer. Ornithologists placed this distinctive species in its own genus until recent studies found it to be closely related to many other warblers.

Adult ♂

♀

Yellow Warbler

Setophaga petechia L 5 in (13 cm)

Bird-watchers always delight in seeing this warbler with brilliant, butter-yellow plumage.

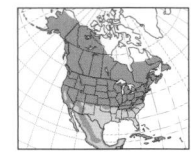

KEY FACTS

Plump warbler with a short tail and a prominent dark eye. Adult males and females are bright yellow below; male has prominent, reddish breast streaks. Immatures are duller.

+ voice: Song is rapid *sweet sweet sweet I'm so sweet.*

+ habitat: Widespread summer resident. Common in wet habitats

+ food: Insects

This species is one of the most common cowbird hosts. Cowbird eggs are too large for their small bills to pierce or eject. When one is detected, the warbler may desert, or may build a new nest floor above the eggs; scientists discovered one six-tiered nest containing 11 cowbird eggs. This approach essentially throws the baby out with the bathwater, because the warbler's own eggs are lost along with the cowbird egg. If a cowbird egg doesn't appear until the warbler has two or more eggs in the nest, she's more likely to care for it. Yellow Warblers have lived more than 11 years.

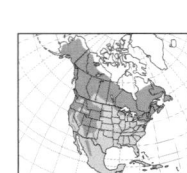

Adult ♂

Yellow-rumped Warbler

Setophaga coronata L 5.5 in (14 cm)

Large flocks of the Yellow-rumped Warbler may be joined by other species during winter and migration.

KEY FACTS

All have yellow rumps and yellow side patches. "Myrtle Warblers" from the North and East have white throats; "Audubon's Warblers" from the West have yellow throats. Birds are duller and browner in fall.

+ voice: Song is a simple warble.

+ habitat: Breeds in coniferous and mixed woods.

+ food: Insects; many berries in winter

This abundant warbler is a generalist, wintering from the central states to Panama, and exploiting more food sources in more habitats than other warblers. It gleans insects from leaves and the ground, and sometimes hover-feeds. Its digestive system is specially adapted to hold hard-to-digest items like seeds in the gut longer than more easily digested foods. The Yellow-rumped Warbler can also digest the wax in bayberries, allowing it to arrive earlier in the spring, depart later in the fall, and spend the winter farther north than other warblers.

Breeding ♂

Fall ♀

Eastern Towhee
Pipilo erythrophthalmus L 7.5 in (19 cm)

Towhees share their ground-foraging techniques with their junco and sparrow relatives.

KEY FACTS

Male has black hood and upperparts, rufous sides; female is chocolate brown where the male is black. Long tail. In the West, very similar Spotted Towhee has white spots above.

+ voice: Song is ringing *drink-your-tea!* Call is up-slurred *towhee.*

+ habitat: Found in tangles, thickets, overgrown fields

+ food: Insects, seeds, berries

In 1585, John White painted a male and female towhee during a visit to the short-lived settlement on Roanoke Island, bringing the bird to the attention of European ornithologists. Towhees eat, sing, and roost at eye level or below, often noisily scratching at the ground, moving leaves aside with a two-footed backward hop. They are difficult to see in the dense foliage where they spend most of their time. Light-eyed towhees live in Florida and often skulk behind Florida Scrub-Jays, especially when people are feeding the jays, perhaps to steal food from the jays' caches. This once abundant species has steadily declined in the Northeast.

California Towhee
Melozone crissalis L 9 in (23 cm)

The California Towhee allows close study; it's so sedentary that marked birds are easy to track.

KEY FACTS

Plain brown plumage with some orange-buff color around face and under tail. Scratches in leaf litter or on bare ground for food.

+ voice: Series of simple *chink* notes

+ habitat: Resident in chaparral and scrubland from Baja California to Oregon

+ food: Mostly seeds, some insects

This common and well-known bird is tolerant of urbanization and not particularly shy around human observers. Mated pairs remain together for life, both chasing any other towhees off their defended territory, and the two are usually found in very close proximity. During the breeding season, mated pairs produce squealing duets about three times every hour—these duets coordinate their breeding behaviors. Unexpectedly, over 40 percent of all nests contain "extra-pair" young—those with a different father or mother than the mated pair. This and the similar-looking Canyon Towhee found east of the Colorado River were once considered the same species, the "Brown Towhee."

Juvenile

Chipping Sparrow

Spizella passerina L 5.5 in (14 cm)

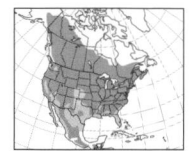

The "Chippy," named for its fast and slow trills, is a summer breeder in much of North America.

KEY FACTS

Slender sparrow with long tail and dark line through eye in all plumages. Breeding adult has chestnut cap and gray face; winter adult has streaked chestnut crown; first-winter bird has buff breast.

+ voice: Long trill of dry *chip* notes

+ habitat: Mix of trees and grassy openings. Northern breeders migrate.

+ food: Insects, seeds

This handsome backyard sparrow has adapted to many human-modified habitats. In spring, males arrive in many areas just as juncos are moving on. Females arrive a week or two later. The female builds the compact nest, usually in a conifer and often at eye level, as soon as weather patterns are favorable. She lines the nest with horsehairs and fur plucked from resting horses, sleeping dogs, and other animals. Males sometimes feed their incubating mate, and share in feeding the young. An Ontario study revealed that after a male has attracted a mate, he may wander through neighboring territories, mating with other females.

Breeding adult

1st winter

Field Sparrow

Spizella pusilla L 5.75 in (15 cm)

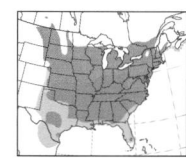

In spring, the male Field Sparrow sings with gusto, but after finding a mate, he sings much less often.

KEY FACTS

Long-tailed sparrow that is closely related to Chipping Sparrow, but has pale head with prominent white eye ring and pink bill.

+ voice: Series of clear, plaintive whistles that slowly accelerate into a trill

+ habitat: Fairly common in overgrown fields and open, brushy woodlands; found east of the Rockies

+ food: Insects, seeds

Field Sparrows live in brushy pastures and second growth, but shy away from similar habitat in suburbs and other settled areas. They seldom associate in flocks. The female builds the first nest of the season using grasses, on or near the ground at the base of woody vegetation. Later nests, as ground cover grows taller, are built in small saplings and shrubs. White-tailed deer eat the eggs and nestlings of ground-nesting sparrows far more than was suspected. Field Sparrows were probably most abundant in the 1800s, after the eastern forest was cleared but before widespread development. They have declined significantly since the Breeding Bird Survey began in 1966.

Song Sparrow

Melospiza melodia L 4.75–6.75 in (13–17 cm)

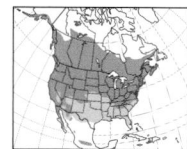

Thoreau described the song of the Song Sparrow as *Maids! Maids! Maids! Hang up your teakettle-ettle-ettle.*

KEY FACTS

Song Sparrows vary in overall coloration and size—small and pale in the desert to large and dark in Alaska—but all are russet and gray birds with bold streaks on the chest that often form a central spot.

+ **voice:** Series of notes with a trill in the middle or as a final flourish

+ **habitat:** Widespread and common, especially in brushy areas

+ **food:** Insects, seeds

This handsome, streaked sparrow nests in many suburban and even urban neighborhoods but is often overlooked, its pretty but quiet songs blending into background noise. Males communicate with neighboring males by song, and sing most persistently throughout the day while trying to attract a mate. One May day in her Ohio backyard, Margaret Morse Nice followed an 8-year-old marked male who had just lost his mate, counting his songs. Beginning at 4:44 a.m. (36 minutes before sunrise), he sang 2,305 songs that day. At least one Song Sparrow banded as an adult in Colorado was still alive when re-trapped in Colorado over 11 years later.

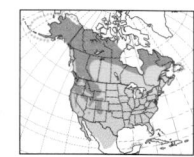

White-crowned Sparrow

Zonotrichia leucophrys L 7 in (18 cm)

The sparrows in the genus *Zonotrichia* are big, with bold markings and distinctive songs.

KEY FACTS

Adult has thick black-and-white crown stripes; immature has brown and light tan stripes. Back is striped with rusty brown and pale gray, and underparts are mostly pale gray.

+ **voice:** Mournful whistles followed by a trill

+ **habitat:** Breeds in mountains or far to the north; favors brushy edges in winter.

+ **food:** Seeds, insects

This handsome sparrow has a pleasing song, heard occasionally in winter and more often during spring migration. Throughout the year and throughout its range, this sparrow is found near grass, bare ground, and dense shrubs or conifers; it feeds mostly on the ground. Easy to maintain in captivity and observe in the wild, the White-crown is one of the most thoroughly studied of all birds. Thanks to it, scientists have learned more about evolutionary biology, how birds learn their songs, how songs vary geographically, migration, physiology, and many other subjects.

Immature

Adult

White-throated Sparrow

Zonotrichia albicollis L 6.75 in (17 cm)

Like avian chipmunks, these stripe-headed sparrows eat seeds on the ground.

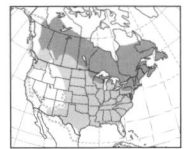

KEY FACTS

Medium-size sparrow with white throat and bright yellow spot near eye. Two color morphs—white-striped and tan-striped—describe the color of the stripe above the eye.

+ **voice:** Mournful, whistled song is heard year-round.

+ **habitat:** Breeds in the North; very common in East in winter; frequents bird feeders, brush, woodland edges.

+ **food:** Seeds, insects

This sparrow's clear, whistled *Old Sam Peabody, Peabody, Peabody* is heard in backyards as well as in northern forests where it breeds. Half of both males and females have white head stripes, the other half tan stripes; each prefers the opposite pattern in a mate, as if blond humans always selected a dark-haired mate and vice versa. About 96 percent of all pairs include one of each color. White-striped birds of both sexes sing and are territorially aggressive. Tan-striped birds of both sexes provide excellent parental care. Though tan-striped males are less aggressive than white-striped, their territories in an area tend to be the same size.

Tan-striped morph

White-striped morph

Dark-eyed Junco

Junco hyemalis L 6.25 in (16 cm)

This distinctive sparrow produces a shorter, more musical trill than the Chipping Sparrow.

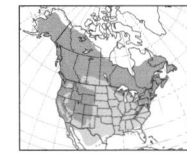

KEY FACTS

Juncos show a bewildering amount of variation. Eastern "Slate-colored" is mostly gray with a white belly; western "Oregon" has a blackish hood and rusty back. All have white outer tail feathers.

+ **voice:** Song is a simple trill.

+ **habitat:** Breeds in mountains or in the north; winter flocks favor fields and edges.

+ **food:** Seeds, insects

Our familiar "snowbird" has been a popular backyard bird since colonial days. When juncos fly away, white outer tail feathers alert others of danger. A high-protein diet before molt increases the amount of white, so the color also serves as a signal to potential mates that the bird may be a good provider. Juncos are territorial during the breeding season. In winter, they gather in huge flocks that mill about on the ground like "little gray-robed monks and nuns," as ornithologist Florence Merriam Bailey wrote in 1899.

"Slate-colored" ♂

"Gray-headed"

Scarlet Tanager

Piranga olivacea L 7 in (18 cm)

The tanagers of North America, in the cardinal family, aren't related to the tropical tanagers.

KEY FACTS

Breeding male has red body and glossy black wings; female is olive above and yellowish below. In fall, male resembles female.

+ voice: Robin-like, raspy song. Call is a harsh *chip-burr*.

+ habitat: Eastern hardwood forests; winters in South America.

+ food: Insects, fruits

This exceptionally vivid bird hides in plain sight in eastern deciduous forests due to its shy, secretive ways. In South America in winter and during migration, tanagers associate in flocks, but when they arrive on their breeding grounds, each male immediately establishes and defends a territory. The song consists of long, raspy but musical phrases often likened to a robin with a sore throat. Females return about a week after males; nesting begins soon after. After the young fledge, the parents stay with them for a couple of weeks before the family disperses. Males undergo a dramatic molt in late summer, but are seldom seen during this time.

Breeding adult ♂

Western Tanager

Piranga ludoviciana L 7.25 in (18 cm)

The Western Tanager, a medium-size songbird, was first collected for science by Lewis and Clark in Idaho.

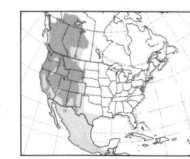

KEY FACTS

Breeding adult male has a red head; mostly yellow body; and black wings, tail, and back. Female is olive-gray above and yellowish below. Both sexes have conspicuous wing bars.

+ voice: Robin-like, raspy song. Call is a *pit-er-ick*.

+ habitat: Western pine forests; winters in Mexico, Central America.

+ food: Insects, fruits

One of the most strikingly beautiful of all North American birds, this inhabitant of coniferous forests ranges farther north than any other tanager. Its red color is produced from a pigment found in very few birds. Rhodoxanthin is found in conifer needles, and tanagers probably get it from eating insects that feed on the needles. Mate selection may take place during winter or on migration. Even during the breeding season, Western Tanagers aren't very aggressive, but do escort intruders off their territory. Rare sightings in the East are usually at feeders.

Breeding adult ♂

Northern Cardinal

Cardinalis cardinalis L 8.75 in (22 cm)

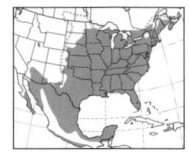

Many sports teams are named for the Northern Cardinal, the most popular state bird of all.

KEY FACTS

Male is instantly recognizable by red color, long crest, black face, and triangular red bill. Female is tawny brown with red accents; similar juvenile has black bill.

+ **voice:** Call is a sharp *chip*. Songs include a loud *cheer cheer cheer*.

+ **habitat:** Resident in the East

+ **food:** Insects, seeds; common at bird feeders

Which is more beautiful: a cardinal's plumage or voice? Bright color indicates the quality of a cardinal's diet. Redder males and females with brighter underwings provide more food to their chicks; those males defend higher quality territories, too. Unexpectedly, captive females don't show a preference for brighter mates. Both sexes sing. Females incubate the eggs, and males provide more food for the nestlings. Cardinals are nonmigratory, but many disperse in fall. The range has crept northward since the early 1800s. The seven states that named the cardinal state bird form an interconnected block.

Rose-breasted Grosbeak

Pheucticus ludovicianus L 8 in (20 cm)

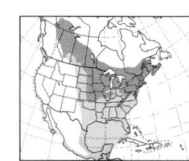

This songbird is nicknamed "cutthroat" for its plumage, and "potato bug bird" for its diet.

KEY FACTS

Adult male is black and white, with a rose-colored breast and huge pale bill. Female looks like a big finch—brown above, streaked below, with two wing bars.

+ **voice:** Rich warbling song. Call is a very sharp *eek!*

+ **habitat:** Eastern forests and edges; winters in Central America, Cuba.

+ **food:** Insects, fruit, seeds, buds

From the time grosbeaks arrive in spring, their striking appearance and song make them very welcome at feeders. Females have a subtle beauty; spring males lack any trace of subtlety. Gaudy as they are, males incubate the eggs for hours each day, often singing while on the nest. Both parents feed the young. Like cardinal chicks, fledglings hop about but cannot fly for a few days, usually staying hidden in foliage. Families remain together until migration. They hybridize with the Black-headed Grosbeak where their ranges overlap in the Great Plains. Hybrid females produce smaller clutches, and more of their eggs don't hatch.

Breeding adult ♂

Indigo Bunting
Passerina cyanea L 5.5 in (14 cm)

Based on DNA studies, buntings belong to the same colorful family as the cardinal.

During the breeding season, male Indigo Buntings sing from conspicuous perches, continuing later in the afternoon and longer into summer than most songbirds. They don't learn their father's song; rather, when they're on their own first territory, they learn their song by interacting with neighboring males. Females, which raise the chicks with little help, are quiet and secretive, and seem to have no preference between males who sing the same song as nearby birds and those singing a different tune. They may raise three or four broods in summer before retreating to the tropics for winter. A few winter in Florida.

Breeding adult ♂

Breeding adult ♀

Painted Bunting
Passerina ciris L 5.5 in (14 cm)

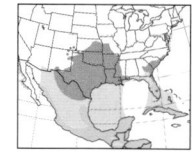

When skulking in vegetation, even the gaudiest male Painted Bunting can be surprisingly hard to see.

This "most wanted" species on many birders' lists is called the *nonpareil* (without equal) in French. There are two separate populations. Western birds make a short postbreeding migration to Arizona and parts of Mexico to molt, and then continue on to Mexico and beyond. Eastern birds molt on their breeding grounds before migrating to the Caribbean. These nocturnal migrants sometimes alight in exhaustion on ships in the Gulf of Mexico. Young males resemble females until their second autumn, when their brilliant adult feathers grow in. Their beauty entices people in the winter range to capture many for the pet trade, a factor in the species' decline since the 1960s.

Adult ♂

♀

Western Meadowlark
Sturnella neglecta L 9.5 in (24 cm)

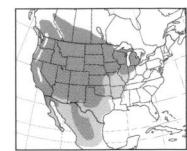

Many members of the blackbird family, such as meadowlarks and orioles, have brightly colored plumage.

KEY FACTS

Chunky, robin-size songbird with flat head and long, slender bill. Bright yellow breast is crossed with black "V," and upperparts are cryptic mix of brown and tan. In flight, note the white outer tail feathers.

+ voice: Series of bubbling, flutelike notes; sharp *chuck* note

+ habitat: Grasslands and agricultural fields

+ food: Probes soil for insects, grain, seeds.

In 1805, Lewis and Clark collected "a kind of larke" in Montana and described differences in tail and song from the Eastern Meadowlark. Audubon, taken with its song and mystified about why it had been overlooked since the expedition, gave it the scientific name *neglecta*, but it wasn't accepted as a separate species until 1910. The Eastern Meadowlark's simple whistle lacks the universal appeal of the Western's rich, bubbling song—the Western is state bird of six states, the Eastern of none. Meadowlarks sing on perches around their territory's border, often along roadsides. They expose prey by inserting their bill into soil and forcing it open.

Red-winged Blackbird
Agelaius phoeniceus L 8.75 in (22 cm)

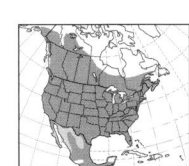

When America's most abundant, familiar blackbird first appears in marshes and meadows, spring is here.

KEY FACTS

Stocky, short-tailed bird that forages on the ground. Adult male is glossy black with vivid red shoulders; female is dark brown above and heavily streaked below.

+ voice: Song is a gurgling *konk-la-ree*; call is a flat *chack*.

+ habitat: Marshes, meadows; huge winter flocks congregate in agricultural areas.

+ food: Seeds, grain, insects

Males arrive on breeding marshes before females and immediately start displaying, singing exuberantly while exposing their red epaulets to perfection. Those with the best displays and territories may attract as many as 15 females. DNA tests of nestlings indicate that females often mate with more than one male as well. Among the most abundant species on the continent, Red-wings form tremendous flocks during migration and winter, often damaging crops. Millions are killed each year via shooting, trapping, pesticides, and spraying roosts with wetting agents during cold weather. Humans provide both their primary sources of food and causes of mortality.

Adult ♀

Adult ♂

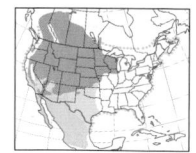

Yellow-headed Blackbird

Xanthocephalus xanthocephalus L 9.5 in (24 cm)

The Yellow-headed Blackbird and many other blackbirds nest colonially and flock together the rest of the year.

KEY FACTS

Brilliant yellow head and black body of adult male are unmistakable. Female has buffy yellow breast; head and upperparts are dusky brown.

+ voice: Song ends in a long, strangled buzz; call is a rich *croak*.

+ habitat: Prairie wetlands and western marshes; flocks winter in agricultural areas.

+ food: Insects, seeds, grain

This bird's breeding range is centered on prairie wetlands, often in the same marshes as Red-wings. Yellow-heads dominate, claiming territories over the deepest water with cattails and bulrushes. They nest in association with some terns, cooperatively attacking predators. People liken the Yellow-head's song to the grating noise of a rusty hinge and may call it ugly, bizarre, weird, or hilarious; females find it irresistible. Experiments reveal that if a Red-wing's epaulets are blackened, he loses his territory, but if a Yellow-head's head is blackened, he can retain his territory, usurp another's territory, and attract mates.

Spring adult ♂

Brewer's Blackbird

Euphagus cyanocephalus L 9 in (23 cm)

The Brewer's Blackbird, a "miniature grackle," has such lustrous plumage that it's been called the "satin bird."

KEY FACTS

Male is black with purple gloss on head, green gloss on body. Female is flat, brownish gray with dark eyes; male's eyes are yellow.

+ voice: Wheezy, unmusical *que-ee*; call is a harsh *check*.

+ habitat: Open areas of the West, from grasslands to city sidewalks

+ food: Insects, seeds, grain

Formerly only found in the West, Brewer's Blackbird was not recorded nesting east of the Great Plains until 1914, when it spread east in a dramatic range expansion. Newly arriving blackbirds supplanted Common Grackles in wild and rural areas, but the grackles won the competition in cities and towns. In turn, as grackles spread west, they started replacing Brewer's Blackbirds in cities and towns but not grasslands. Some birds breeding in Canadian prairies migrate west, wintering in the coastal regions of British Columbia and Washington. Where abundant, they are an important food source for falcons, yet at least one banded bird in California survived over 12 years.

♀ ♂

Common Grackle

Quiscalus quiscula L 12.5 in (32 cm)

Grackles, like many birds, smear ants into their feathers, possibly to control parasites.

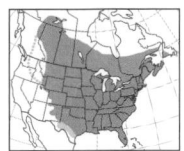

KEY FACTS

Blackbird with flared tail and pale eyes. Iridescent plumage varies from purplish in South and most of East Coast to bronzy with blue head. Female has nonglossy, brown body.

+ **voice:** Short, creaky *readle-eak;* call is a deep *chuck.*

+ **habitat:** Open woods, farmland, marshes, suburbs

+ **food:** Insects, seeds, grain

Right when the first robins arrive, grackles suddenly appear, males puffed up in territorial displays, strutting about, bills pointed skyward. When they fly, they lower and extend the central tail feathers to form an impressive keel. They forage on lawns, devouring slugs, cutworms, and other pests. During nesting, they become secretive. About half the males remain with their mates during incubation, and often help feed the noisy young. Grackles raid nests and attack small birds—perhaps what led Ogden Nash to "deem the grackle / an ornithological debacle."

"Bronzed Grackle"   "Purple Grackle"

Great-tailed Grackle

Quiscalus mexicanus ♂ L 18 in (46 cm) ♀ L 15 in (38 cm)

Until 1983, the Great-tailed Grackle and the coastal Boat-tailed Grackle were considered the same species.

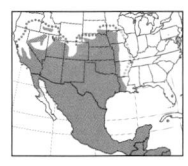

KEY FACTS

Very large grackle with pale eyes. Purple-glossed male has extremely long keel-shaped tail; brownish female is smaller and shorter-tailed, with buffy throat.

+ **voice:** Loud series of whistles, trills, and ratchet-like sounds

+ **habitat:** Farms, feedlots, wetlands, lawns, and urban areas

+ **food:** Insects, small animals, grain, garbage

It's impossible to ignore breeding colonies of these flamboyant birds with their whistles, squawks, screeches, bugle-like calls, and soft tinkling notes, which can continue all night at roosts near artificial lighting. Displaying males may engage in fierce battles. Only the largest ones with the longest tails hold territories, each containing one to several trees. Several females may construct nests and raise young in the same tree. At all ages more females than males survive. A rapid range expansion northward during the last century, coinciding with irrigation and urbanization, has brought this formerly tropical and subtropical species as far north as Minnesota.

Brown-headed Cowbird

Molothrus ater L 7.5 in (19 cm)

The parasitic Brown-headed Cowbird occurs in open or semi-open country and often travels in flocks.

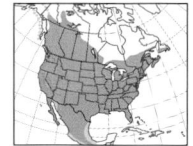

KEY FACTS

Small, compact black-bird with short tail and dark eyes. Male is glossy black with rich brown head. Female is plain, gray-brown; juvenile is also brown, but streaked below and scaly above.

+ **voice:** Male song is a liquid gurgling; calls include rattles and chattering.

+ **habitat:** Woodlands, farmlands, suburbs

+ **food:** Seeds, grain, some insects

"Buffalo birds" were once limited to short-grass plains following bison, feeding on insects where the heavy-hoofed giants cut into prairie sod. Widespread extirpation of bison coincided with plowing the prairies and cattle ranching; rather than disappearing, cowbirds increased and spread. Females search for nests of small songbirds in which to lay their eggs—they produce as many as 40 eggs each year. Females check on these nests, often trashing one if their egg has been removed. Some parasitized species raise a cowbird or two and their own young successfully; others have declined significantly since cowbirds arrived.

Hooded Oriole

Icterus cucullatus L 8 in (20 cm)

Orioles belong to the genus *Icterus*, which means jaundiced, for the birds' yellow color.

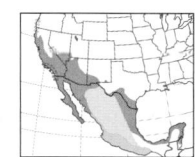

KEY FACTS

Slender, long-tailed oriole. Male is deep yellow or orange with a black bib, back, and tail. Female is olive above, yellow below; first-spring male is similar, but has black bib.

+ **voice:** Calls include loud, whistled *wheet* and rapid chattering.

+ **habitat:** Summer resident, usually nests in shade trees and palms; winters in Mexico.

+ **food:** Insects, spiders, fruits, nectar

Quieter than other orioles, this beautiful bird expanded its range as ornamental plants and feeders became popular. Those breeding in northern Mexico and the United States were once entirely migratory, but now many remain in southern California and Texas year-round. In California, they're sometimes called the "palm leaf oriole" for building their suspended nests in fan palm leaves. Rio Grande Valley birds once depended on Spanish moss for nest construction, but today, nests are more often woven of palm leaf fibers. They are a primary host for Bronzed Cowbirds.

Breeding adult ♂

Baltimore Oriole

Icterus galbula L 8.25 in (21 cm)

The state bird of Maryland was named for the colors on Lord Baltimore's family crest.

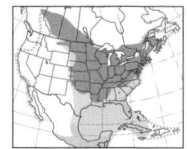

KEY FACTS

Stocky, short-tailed oriole. Brilliant orange and black male can be hard to see in the treetops; female is paler orange below, brownish olive above; head can be blackish.

+ **voice:** Series of rich, fluting whistles; staccato chatter call

+ **habitat:** Summer resident of eastern woodlands

+ **food:** Insects, fruit, nectar; jelly feeders, cut fruit

This striking bird visits feeding stations for oranges and sugar water. It nests in tall shade trees, especially near water. The female anchors the pouch-like nest at the fork of twigs too slender to support most predators. Sometimes, adults and nestlings become entangled in long nest fibers, so people are cautioned never to provide string or yarn longer than 6 to 8 inches (15–20 cm). Baltimore Orioles molt into fresh plumage before migrating to the tropics in fall. Scattered individuals may winter in northern states. From 1973 to 1994, this and the western Bullock's Oriole were considered to be the same species.

1st spring ♀   Breeding adult ♂

American Goldfinch

Spinus tristis L 5 in (13 cm)

With coast-to-coast popularity, the American Goldfinch is the state bird of Washington, Iowa, and New Jersey.

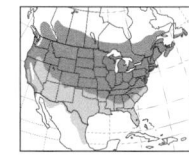

KEY FACTS

Small finch with short tail. Breeding male is bright yellow with black cap; breeding female is olive-brown above, yellow below. Winter male is grayish brown below with yellow face, tan upperparts.

+ **voice:** Twitters, trills, *su-wee* notes; flight call *per-chik-o-ree*

+ **habitat:** Widespread in overgrown fields, pastures, suburbs

+ **food:** Mostly seeds; common feeder visitor

The "wild canary" is one of the few vegans among birds, even feeding nestlings regurgitated seeds. It undergoes a complete molt of body feathers both in early spring and in fall. It waits longer than other birds to begin nesting, both because it takes time to recover from the spring molt and because goldfinches use thistle seeds for nest construction and food. Cowbird nestlings don't survive long when raised by goldfinches because of the low-protein diet. Some goldfinches winter quite far north; others move long distances. One banded in March in Ontario was recovered, dead, in Louisiana, 1,000 miles (1,600 km) away, eight months later.

Breeding ♀

Breeding ♂

House Finch

Carpodacus mexicanus L 6 in (15 cm)

In backyards and apartment balconies, the popular House Finch often nests in hanging baskets.

KEY FACTS

Sparrow-size finch with a stubby bill, curved on top. Male has red eyebrow, breast, and rump (sometimes yellow). Female and juvenile are grayish brown with streaked underparts.

+ **voice:** Lively warble call is whistled *wheat*.

+ **habitat:** Widespread in city parks, farms, backyards, and forest edges

+ **food:** Seeds, buds, fruit; visits feeders.

Originally a western species, House Finches were once trapped and sold as "Hollywood Finches." In 1940, to avoid prosecution under the Migratory Bird Act, unscrupulous caged-bird dealers released captive finches on Long Island. The birds established a population that spread along the coast and westward. In 1994, an epidemic of a common poultry eye disease struck in the mid-Atlantic states, decimating local populations and rapidly spreading north and west. The disease was especially deadly in eastern populations, probably in part because eastern birds lacked genetic diversity, having descended from a handful of birds.

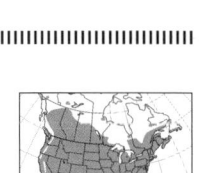

House Sparrow

Passer domesticus L 6.25 in (16 cm)

Unrelated to sparrows native to America, the House Sparrow is the sparrow mentioned in the Bible.

KEY FACTS

Big-headed sparrow with heavy bill and short tail. Male has black bib, gray crown, and chestnut nape; pattern less evident in winter. Female has buffy eyebrow and streaked back.

+ **voice:** Series of *chirrup* notes

+ **habitat:** Abundant in human-altered habitats, especially cities, towns, and farms

+ **food:** Seeds, grain, insects; visits feeders.

This artful dodger has attached itself to civilization throughout history. Immigrants who missed the familiar European bird released captives many times in the 1800s, most famously in New York City around 1850. People claimed the sparrows would devour insect pests, but they mostly eat grains. Their rapid spread sparked declines in native birds, agricultural losses, and long-standing debates among ornithologists. Currently, they are slowly decreasing in America and more rapidly in their native Europe. In winter, they roost in buildings, outside lights, and near other heat sources. Some fly underneath parked cars to warm themselves until the engine cools.

Breeding ♂

Fall ♂

A Texas Horned Lizard lends a prehistoric presence among arid-land flowers. The reptile's distinctive horns and flattened, toadlike body suggest the erroneous common name Horned Toad.

3 Reptiles & Amphibians
The Water–Land Connection

Reptiles and amphibians represent widely distributed groups of vertebrates that show many vestiges of their ancient lineages. From crocodilians that seem to lumber straight from the Triassic period some 200 million years ago, to snakes and lizards that slither and scurry, to salamanders and frogs that play out their double lives in water and on land, reptiles and amphibians offer an understanding of the past and present.

■ Reptiles & Amphibians

Reptiles and amphibians are traditionally grouped together for their biological and ecological similarities rather than close relationships based on evolution. Amphibians are the older of the two. They evolved from lobe-finned fishes about 360 million years ago and have retained a connection to watery or moist habitats, at least for the purposes of reproduction. The ancestors of reptiles appeared about 40 million years later and evolved into the creatures we know today. They have impermeable skin that minimizes water loss and other adaptations for life out of water. Despite this evolutionary sequence, reptiles often get first billing in field guides.

The study of reptiles and amphibians is the subject of the field of herpetology. The species represented in these two groups often are called herpetofauna, or merely "herps." These terms ultimately derive from the ancient Greek for "to creep." From a North American perspective, reptiles include crocodilians, turtles (and tortoises), lizards, snakes, and an amphisbaenian (worm lizard). Amphibians include salamanders, frogs, and toads.

■ What Is a Reptile?

Reptiles are ectothermic, which means that they cannot control body temperature by internal physiological processes. They rely on strategies such as basking in the sun to raise temperature and "rev up" metabolic systems, shuttling between sun and shade, and resting underground to avoid conditions that could fry or freeze them. Reptilian bodies are covered with scales, shields, or plates, and most reptiles that have toes have claws on them. Most species lay shelled eggs (all crocodilians and turtles do), but a few lizards and a fair number of snakes give live birth. Young reptiles are usually miniature versions of their parents at hatching or birth.

Reptiles occupy a wide variety of habitats in the United States and Canada, just about every type except alpine tundra. Many are secretive and elusive, even those that live in close proximity to humans. Many are fossorial (burrowing) and nocturnal, adding to their mystique.

■ What Is an Amphibian?

Amphibians are also ectothermic, but unlike reptiles,

they have nonscaly skin through which they breathe, and their need for moist skin limits potential habitats when they are active. As the name "amphibian"—meaning "both [kinds of] life"—suggests, most amphibians lead at least partially aquatic lives. Their eggs are laid in water and hatch into aquatic larval forms before transforming into land-based adults. But a number of amphibians lay eggs on land (with precautions to keep them from drying out) and skip the aquatic larval life stage entirely. And to complicate matters, some adult amphibians never move on, but retain their larval appearance and lifestyle when they become sexually mature.

Amphibians are found in diverse habitats, but their habitat requirements make their overall North American distribution more limited than that of reptiles. Nevertheless, they inhabit domains as diverse as desert arroyos and alpine meadows.

■ Identifying Reptiles & Amphibians

This chapter will help you to learn about reptiles and amphibians and how to identify and observe them. The entries in this chapter introduce 148 of the 700-plus species of reptiles and amphibians found in North America north of Mexico. It is a small but representative sample.

The chapter sections present the species in basic groups arranged according to our current understanding of their classification. The classifications are under constant revision as we learn to better

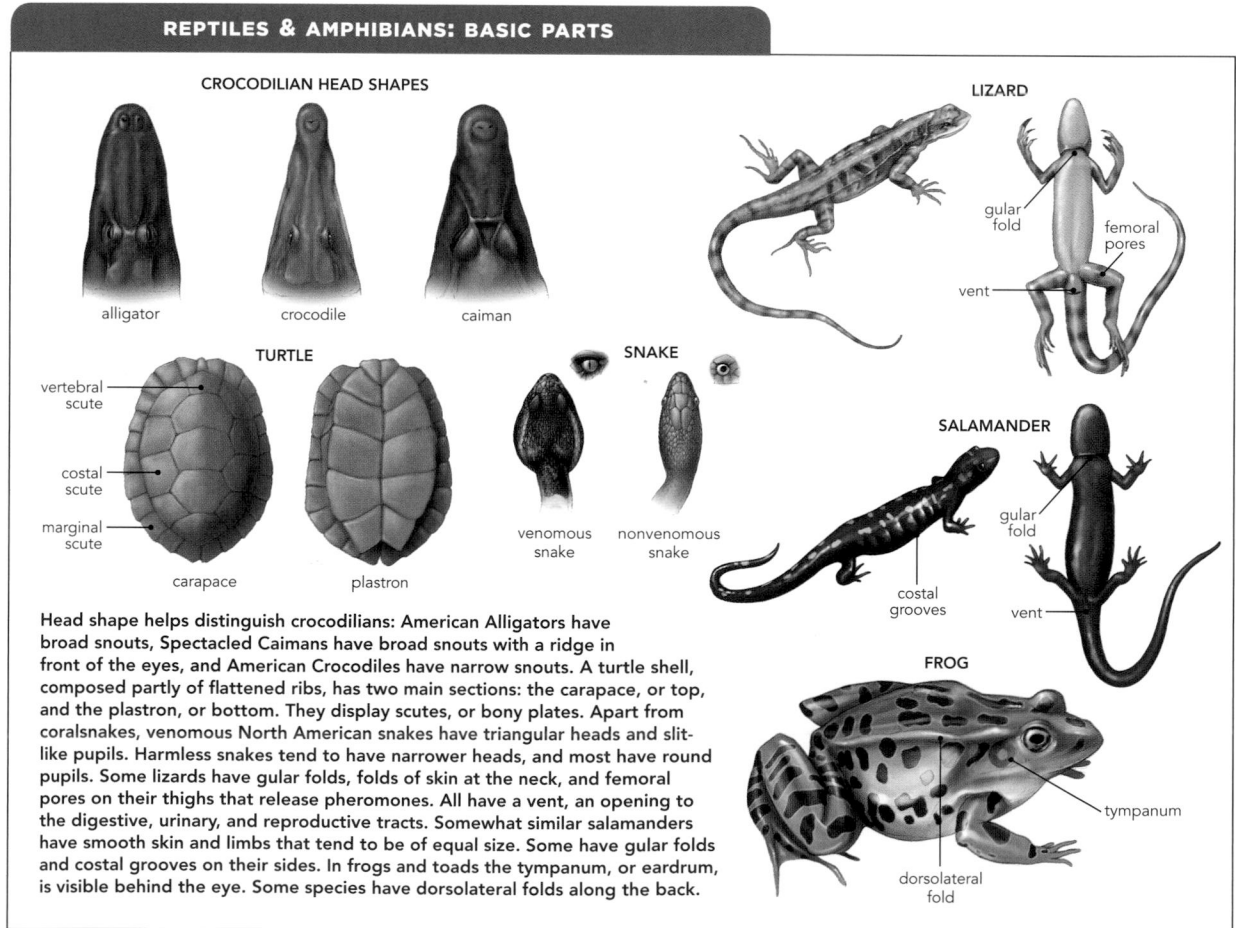

REPTILES & AMPHIBIANS: BASIC PARTS

CROCODILIAN HEAD SHAPES

alligator crocodile caiman

LIZARD

gular fold
femoral pores
vent

TURTLE

vertebral scute
costal scute
marginal scute

carapace plastron

SNAKE

venomous snake nonvenomous snake

SALAMANDER

gular fold
costal grooves
vent

FROG

tympanum
dorsolateral fold

Head shape helps distinguish crocodilians: American Alligators have broad snouts, Spectacled Caimans have broad snouts with a ridge in front of the eyes, and American Crocodiles have narrow snouts. A turtle shell, composed partly of flattened ribs, has two main sections: the carapace, or top, and the plastron, or bottom. They display scutes, or bony plates. Apart from coralsnakes, venomous North American snakes have triangular heads and slit-like pupils. Harmless snakes tend to have narrower heads, and most have round pupils. Some lizards have gular folds, folds of skin at the neck, and femoral pores on their thighs that release pheromones. All have a vent, an opening to the digestive, urinary, and reproductive tracts. Somewhat similar salamanders have smooth skin and limbs that tend to be of equal size. Some have gular folds and costal grooves on their sides. In frogs and toads the tympanum, or eardrum, is visible behind the eye. Some species have dorsolateral folds along the back.

A hatchling Loggerhead Sea Turtle heads for the open ocean, a perilous journey often made at night by moonlight.

understand relevant relationships—but this is not without controversies. Scientific and common names provided follow the *Scientific and Standard English Names of Amphibians and Reptiles of North America North of Mexico,* Edition 6.1, which is published by the Society for the Study of Amphibians and Reptiles and is endorsed by the professional herpetological societies in North America. Within each genus, the species are mainly listed in alphabetical order of their species names, as not all species have been investigated equally and their interrelationships are imperfectly known. The measurements provided give a length that includes the animal's head and body (carapace only for turtles), including the tail for crocodilians, lizards, and snakes. Each chapter also offers a set of characteristics that will help you identify a species and compare it with others, as well as additional details about habitat, range, and natural history.

■ Where to Find Them

The key to observing reptiles and amphibians in nature is to visit their regular haunts and understand

||||||||||||||||||||||||||||||||||||||||||||||||||||||||||||||||||

How doth the little crocodile
Improve his shining tail,
And pour the waters of the Nile
On every golden scale!
—LEWIS CARROLL

||||||||||||||||||||||||||||||||||||||||||||||||||||||||||||||||||

their habits. Depending on where you live, hiking a trail through the woods might bring you close to box turtles, ratsnakes, and salamanders, especially if the area is moist. You might have to roll a log or move a rock. To roll a log, stand parallel to the log and slowly roll it toward you, allowing any creatures beneath to escape in the opposite direction. You may have to do this a number of times. Don't roll too many logs or lift too many stones in the same area, though, and always replace them gently in their original positions—they are the homes for many species, after all.

To find turtles, frogs, and watersnakes, visit a pond or stream on a sunny day, being careful to approach stealthily so that the numerous basking animals don't disappear into the water. You may be able to see a stack of turtles perched perilously on a log, or a watersnake in a face-off with a frog. Going out at night with a flashlight or headlamp can be effective, especially on a hot summer night, as can driving slowly and carefully along country or desert roads, particularly after summer rains, when lizards, snakes, and toads are on the prowl.

Frogs often are heard but not seen. They make a number of different calls and vocal reactions, including males' distinctive "advertisement" calls that they produce singly or in chorus to attract potential mates to breeding sites. Learning to identify frogs by their calls is easier than ever thanks to the Internet. Search any species to find audio clips and YouTube videos, which show the frog in call mode. Cornell University's online Macaulay Library (*www.macaulaylibrary.org*), a treasure trove of more than 65,000 audio files that covers many animal groups, is a good place to start.

Caution should be exercised in encounters with venomous snakes (and the Gila Monster). Most are not looking for a fight, but some species are more skittish than others, so all should be given a wide berth. If you are planning to explore in an area with venomous snakes, educate yourself about precautions and the availability of medical help.

This chapter does not offer any tips on collecting reptiles and amphibians in the field. That knowledge can be found in field guides dedicated to these animals. If you want to get to know your local herpetofauna, a visit to a local nature center may allow you to see species up close and learn more about them. It is an excellent way to prepare for chance encounters in nature, when a fleeting glimpse may be all you get.

■ Environmental Implications

We can all take individual responsibility for preserving herpetofaunal abundance and diversity by stewardship of our own yards and gardens and through respect for their connectedness to the health of local watersheds and the greater environment. Reducing or eliminating the use of pesticides, herbicides, and fertilizers; planting native species that can serve as food sources directly or indirectly (as food for prey species); and never releasing nonnative pet species into natural settings are contributions we all can make. It is also wise not to release native reptile and amphibian species that have been in captivity. If you need to give up a captive animal of any kind, contact your local nature center or animal shelter for advice.

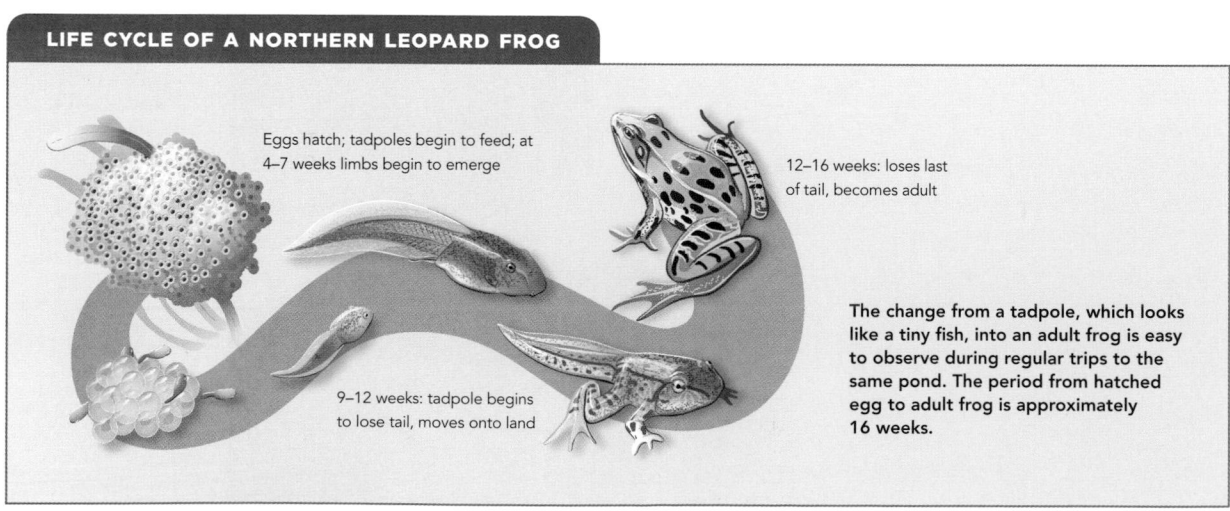

LIFE CYCLE OF A NORTHERN LEOPARD FROG

Eggs hatch; tadpoles begin to feed; at 4–7 weeks limbs begin to emerge

12–16 weeks: loses last of tail, becomes adult

9–12 weeks: tadpole begins to lose tail, moves onto land

The change from a tadpole, which looks like a tiny fish, into an adult frog is easy to observe during regular trips to the same pond. The period from hatched egg to adult frog is approximately 16 weeks.

REPTILES & AMPHIBIANS

American Crocodile
Crocodylus acutus L 8–14 ft (2–4 m)

The American Crocodile's longer and thinner V-shaped snout distinguishes it from the alligator. Living crocodilians have changed little since they first appeared some 200 million years ago.

KEY FACTS

Body is brownish, gray, or olive green, usually paler than alligator, often has dark bands that fade with age; young lack alligator's yellow bands.

+ **habitat:** Brackish and saltwater swamps

+ **range:** Southernmost Florida, including Everglades and Keys

+ **food:** Fish, birds, frogs, mammals, and invertebrates

The elusive crocodile is a far rarer sight in its limited U.S. range than is the alligator. In addition to its tapered snout, the crocodile displays the fourth tooth of its bottom jaw when its mouth is closed, another aid in identification. Like the alligator, the attentive crocodile mother carefully removes hatchlings from the nest mound and gingerly carries them to the water. The species' slow reptilian digestion allows it to survive for months without food. Although it was removed from the U.S. endangered species list in 2007, it is still listed as threatened.

Spectacled Caiman
Caiman crocodilus L 4–8 ft (1–2 m)

An introduced crocodilian, the Spectacled Caiman is native from southern Mexico southward. A curved, bony ridge on the top of its snout in front of the eyes gives it a bespectacled look.

KEY FACTS

Body is brown, olive, or light yellow; snout is somewhat tapered; back has dark crossbands that become obscured as color darkens.

+ **habitat:** Freshwater marshes, ponds, lakes, and canals

+ **range:** Southern Florida, including Everglades

+ **food:** Fish, amphibians, birds, mammals, and insects

The Spectacled Caiman has established itself in the canal systems of southern Florida, where it competes with the native American Alligator for space and food resources. The species became popular in the pet trade when the alligator was given protected status; much of its presence can probably be accounted for by pet release and escape, and the destruction of commercial exotic pet facilities by Hurricane Andrew in 1992. The smallish, aggressive species is sensitive to the cold and reproduces in a manner similar to alligators and crocodiles; clutches of hard-shelled eggs are laid in nest mounds and hatch in the late fall.

American Alligator

Alligator mississippiensis L 6–16 ft (2–5 m)

A broader and rounder snout helps distinguish the American Alligator from the crocodile. It is the longest native North American reptile and is found over a much larger range than the crocodile.

KEY FACTS

Adult is brown to black to green; underside is pale; juvenile body dark with yellow crossbands.

+ **habitat:** Freshwater and brackish marshes, swamps, bayous, and rivers

+ **range:** Coastal plain from extreme southeastern Virginia to the Florida Keys to Texas

+ **food:** Fish, turtles, birds, snakes, and mammals

Watchful waiting is the American Alligator's strategy. It may spend much of the day basking on land or floating in water with only ears, eyes, and snout protruding above water, until motivated to move—perhaps by a quick meal. Male alligators announce readiness to mate with resonant bellows; interested females respond less forcefully. Once mated, the female builds a mounded nest to receive the 20 to 60 eggs she will lay, which she then covers with vegetation. She checks on the nest until the hatchlings start to appear about two months later. The surviving young may stay with their mother for several years.

Snapping Turtle

Chelydra serpentina L 8–18 in (20–46 cm)

The large head, limbs, and tail of the Snapping Turtle keep it from being able to retreat very far into its shell, leading it to rely on its powerful snapping jaws for defense.

KEY FACTS

Carapace is tan to brown with 3 rows of keels; head is large; long tail has sawtoothed top.

+ **habitat:** Shallow fresh water and sometimes brackish bodies of water with mud bottoms

+ **range:** Central and eastern U.S. to southern Canada

+ **food:** Invertebrates, fish, small mammals, carrion, and plant matter

More likely to be found on land than the Alligator Snapping Turtle, the Snapping Turtle nonetheless is a highly aquatic species. It often buries itself in mud with only eyes and nostrils visible. A dedicated omnivore, it acquires a wide range of food by ambush and by foraging. It often picks up an extensive coating of algae, and as it ages its carapace becomes smoother, and the keels, or ridges, on it become less prominent. Snappers mate under water. An adult female on a scouting mission for a nest site on land can be particularly aggressive if confronted. She digs a nest and lays a clutch of 20 to 50 eggs, which hatch in 9 to 18 weeks.

Alligator Snapping Turtle

Macrochelys temminckii L 15–26 in (38–66 cm)

The well-camouflaged Alligator Snapping Turtle opens its mouth and wriggles its fleshy tongue, which resembles a worm, hoping a fish will take the bait. If it does, the mouth snaps shut.

KEY FACTS

Carapace has 3 distinct keels and extra row of small marginal scutes; huge head has hooked beak.

+ **habitat:** Deep lakes and streams

+ **range:** Southwestern Georgia and northern Florida west to Texas; north to Kansas and Mississippi Valley to Iowa

+ **food:** Worms, mollusks, other invertebrates, and plant matter

The largest freshwater turtle in North America, the Alligator Snapping Turtle can weigh up to 150 pounds (68 kg). A record-holding 50-year-old captive snapper weighed more than twice that. Its built-in fish bait allows the species to remain fairly sedentary on the muddy bottom in deep water by day. It is otherwise nocturnal, creeping along while foraging and surfacing frequently to breathe. This species has a short run of doubled marginal scutes on its carapace, distinguishing it from the Snapping Turtle. These turtles mate under water. The female digs a cavity in the soil on land and lays 20 to 50 spherical eggs.

Loggerhead Musk Turtle

Sternotherus minor L 3.5–5 in (9–13 cm)

The head of the Loggerhead Musk Turtle is quite large compared with the carapace; in old males it is huge. Younger Loggerheads have three well-developed carapace keels, or ridges.

KEY FACTS

Carapace is dark brown, often with 3 keels or 1 sharp keel; one-hinged plastron yellow to pink with whitish patches.

+ **habitat:** Springs, streams, rivers, swamps, and sinkholes

+ **range:** Northern Georgia to southern Alabama and northern Florida

+ **food:** Insects, snails, other invertebrates, and plant matter

Musk turtles can be distinguished from similar mud turtles by their smaller, single-hinged plastrons. The Loggerhead differs from the Eastern Musk by the appearance of barbels, or short filaments, only on its chin and by the lack of lines on the sides of its head. The Loggerhead's large head, maxed out in old males, accommodates strong jaws that are used to crush the shells of mollusks. Loggerheads crawl along muddy bottoms of shallow waters, emerging to bask—and sometimes climb a snag to do so—but only briefly, before returning to the water. Young can produce stinky musk from their glands before they hatch.

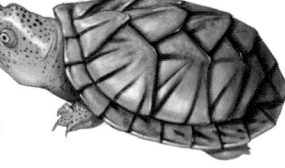

Eastern Musk Turtle

Sternotherus odoratus L 2–4.5 in (5–11 cm)

The Eastern Musk Turtle is also fittingly known as the Stinkpot for the nasty yellowish fluid it exudes from two pairs of glands just under its carapace when annoyed.

KEY FACTS

High-domed carapace is olive-brown to gray, sometimes with 3 keels; small single-hinged plastron.

+ habitat: Streams, ponds, lakes, canals, and ditches

+ range: Southern Ontario and much of eastern U.S. to Wisconsin and Texas

+ food: Insects, crayfish, other invertebrates, tadpoles, fish, and plant matter

The Eastern Musk Turtle shows two light lines on each side of its head, and barbels on both its chin and its throat, features that differentiate it from the Loggerhead Musk Turtle, which has no stripes and only chin barbels. Its domed carapace may be obscured by a heavy growth of green algae, which often gives it the appearance of a streambed stone. This species wanders the bottoms of both brackish and freshwater bodies, but may also climb snags or overhanging tree trunks to bask. It mates under water and the female lays one to nine hard-shelled, white oval eggs in a shallow nest under a stump or perhaps in the wall of a muskrat lodge.

Striped Mud Turtle

Kinosternon baurii L 3–4.75 in (8–12 cm)

Mud turtles have two hinges on the plastron, a feature that distinguishes them from the one-hinged musk turtles. The flexible hinges allow the turtle to tuck in its head and limbs securely.

KEY FACTS

Smooth carapace has 3 long, light stripes on top; plastron has 2 hinges; males have blunt spine at end of tail.

+ habitat: Swamps, ponds, drainage canals, and wet meadows

+ range: South Carolina to southern Florida, including Keys

+ food: Insects, snails, algae, carrion, and plant matter

The scutes of the Striped Mud Turtle's carapace may be almost translucent, offering a glimpse of the bony underlying structures. The carapace's diagnostic stripes may be interrupted and also may fade with time. Two light stripes on the side of the head help aid identification. This species typically roams farther afield and is more terrestrial in general than other North American mud turtles. It is an adventuresome scavenger, foraging even in manure, a habit that gives it the nickname "cow-dung cooter." Males have rough patches of clasping tubercles on the insides of their rear legs, which help with stability during mating.

Yellow Mud Turtle

Kinosternon flavescens L 3.5–6 in (9–15 cm)

The small Yellow Mud Turtle and its dime-size hatchlings are under threat from predation, habitat loss and alteration, and other factors. They are listed as endangered in a number of states.

KEY FACTS

Smooth carapace is olive to brown with dark-outlined plates; yellow to brown plastron has 2 hinges, 11 plates; chin and throat yellowish.

+ **habitat:** Streams, rivers, ponds, lakes

+ **range:** South-central U.S.; isolated populations in Illinois and Iowa

+ **food:** Insects, worms, snails, tadpoles, plant matter

The Yellow Mud Turtle favors bodies of water with muddy bottoms, not seeming to mind whether they are natural or artificial, such as cattle tanks or irrigation ditches. This shy species appears on land during rains or when it sets out to forage or wants to hunker down in cool weather under cover of leaves or brush. It also buries itself in mud to escape drought. Yellow Mud Turtles often travel surprisingly long distances from natural habitats to reach artificial bodies of water, especially in drier regions. The yellowish throat of a basking Yellow Mud Turtle often is visible. The male's plastron is concave to aid in mating.

Eastern Mud Turtle

Kinosternon subrubrum L 3–4 in (8–10 cm)

The Eastern Mud Turtle, despite its name, is capable of wandering far from mud-bottomed watery habitats. It will migrate overland to find another body of water if its own dries up.

KEY FACTS

The carapace is olive to dark brown; plastron has 2 hinges; males have blunt spine at end of tail.

+ **habitat:** Marshes, ponds, swamps, other wetlands

+ **range:** South-central and eastern U.S. south from Long Island

+ **food:** Worms, insect larvae, snails, other invertebrates, fish, algae, and aquatic plants

The omnivorous Eastern Mud Turtle prefers shallow-water habitats, readily accepting those with brackish water, and it is frequently found in coastal marshes and on islands. Like musk turtles, mud turtles have two pairs of musk glands just under their carapace that produce an odor a bit less noxious than that of the musk turtle. During breeding season, females lay up to three clutches of eggs. They dig nests in well-drained soil with vegetation for cover and occasionally use the lodges of beavers or Muskrats. They deposit one to six oval, hard-shelled, pinkish or bluish white eggs. The hatchlings emerge in about ten weeks.

Western Pond Turtle

Actinemys marmorata L 3.5–8.5 in (9–22 cm)

The Western Pond Turtle has a lower shell profile compared to other pond turtles. The species is in rapid decline throughout most of its range.

KEY FACTS

Carapace is olive to brown; may have pattern of dark, radiating lines or yellow spots; males often have few markings.

+ **habitat:** Marshes, swamps, ponds, lakes, and rivers

+ **range:** Washington to California; isolated Nevada population

+ **food:** Fish, frogs, aquatic plants, and carrion

This very aquatic species is usually found in freshwater habitats, but it does not avoid brackish or even saltwater situations. It prefers habitats with a thick growth of aquatic vegetation and a muddy or rocky bottom. A dedicated basker, the Western Pond Turtle will face down rivals with an open-mouthed display to protect a favored basking site. If otherwise disturbed, it will quickly retreat to the water. The female nests from April through August, depending upon latitude. She lays a clutch of 3 to 14 eggs in a nesting hole that she digs in a sunny location. The hatchlings emerge in about three months.

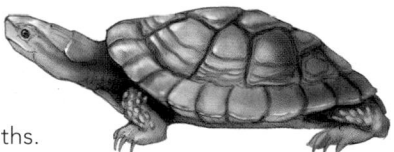

Spotted Turtle

Clemmys guttata L 3.5–4.5 in (9–11 cm)

The Spotted Turtle bears a pattern of yellow spots on its head, limbs, and carapace that look as if they were flicked from a paintbrush, giving it a nickname of polka-dot turtle.

KEY FACTS

Carapace has few to many yellow spots; head has orange or yellow spots.

+ **habitat:** Shallow streams, wet meadows, woodlands, bogs, and beaver ponds

+ **range:** Southern Maine west to Great Lakes, south to northern Florida; isolated populations elsewhere

+ **food:** Insects, other invertebrates, frogs, and carrion

The Spotted Turtle is best observed in spring, before vegetation takes over and makes it difficult to find. It occurs, often in groups, in areas with Bog Turtles (*C. muhlenbergii*), which have orangish blotches on the sides of the head, and Blanding's Turtles (*Emydoidea blandingii*), which have small yellow flecks overall and a yellow chin. The Spottie is a kind of laid-back species, taking to the water in an unhurried fashion if disturbed while basking. The adult female's eyes are orange and the male's eyes are brown. The female digs a shallow nest and deposits three to five oval eggs. Hatchlings emerge in several months, or may overwinter in the nest.

Wood Turtle

Glyptemys insculpta L 5–8 in (13–20 cm)

One of the most terrestrial turtle species, next to tortoises and box turtles, the Wood Turtle seems to carry an entire development of stepped pyramids on its back.

KEY FACTS

Carapace is brown; scutes have concentric ridges forming irregular pyramids; neck and forelegs often orange.

+ **habitat:** Woodland streams, swamps, marshes, farmlands

+ **range:** Nova Scotia to Minnesota and south to Virginia and West Virginia

+ **food:** Insects, worms, other invertebrates, tadpoles, fruit

The Wood Turtle puts in plenty of time in watery realms, but it also does walkabouts that take it far from water, where it hibernates in the muddy bottoms. It likes to forage in opportunity-rich environments such as recently plowed farm fields. This turtle is an adept climber and can scale barriers including chain-link fences. Also known as redleg, the species was once a major food source. The Wood Turtle differs from the Blanding's Turtle and the Eastern Box Turtle by its lack of a hinged plastron. The female lays clutches of six to eight eggs that hatch in three or four months or overwinter and hatch the following spring.

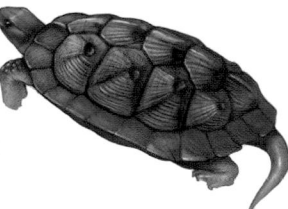

Eastern Box Turtle

Terrapene carolina L 4.5–6 in (11–15 cm)

The Eastern Box Turtle, beloved woodland icon, may be the most easily recognized turtle. It is noted for withdrawing its head and limbs tightly into its shell, thanks to a flexible, hinged plastron.

KEY FACTS

High-domed, keeled, tan, brown, or black carapace and plastron may have prominent or obscure yellow, orange, or olive marks.

+ **habitat:** Forests, meadows, pastures, other open areas

+ **range:** Eastern U.S. from southern Maine to eastern Kansas, south to Texas

+ **food:** Fruit, flowers, vegetables, insects, worms, slugs

This well-adapted land-based turtle may appear to have more in common with tortoises, but in fact it is more closely related to species of water turtles. It often takes to water or mud to have a good soak and comes out in numbers following a rain shower. The male's plastron is concave, and its rear claws are curved to facilitate mating. The female can store sperm from a single mating and use it to produce fertile eggs for a number of years. She digs a shallow nest and deposits three to eight eggs that hatch in about three months. A close-up look reveals that males often have red eyes and females' eyes are yellowish.

Ornate Box Turtle

Terrapene ornata L 4–5 in (10–13 cm)

The Great Plains counterpart of the Eastern, the Ornate Box Turtle is in general more aggressive and more inclined to carnivorous feeding habits.

KEY FACTS

High-domed brown to black carapace has pattern of radiating yellow lines; hinged plastron has similar patterning.

+ **habitat:** Prairies, plains, pastures, open woods, and sandy areas

+ **range:** Mid-central and south-central U.S. to Southwest

+ **food:** Insects, worms, berries, small vertebrates, and carrion

Ornate Box Turtles have a more uniform appearance in carapace and plastron color and patterning than other box turtles and usually show no sign of a central keel. They also are more tolerant of arid environments, basking quickly in the mornings and setting off to forage before the heat sets in. They will also burrow to escape the heat. In areas with cattle, they can be found rummaging in dung piles in search of beetles. This species overlaps with the Eastern Box Turtle and occasionally hybridizes with it. Females that nest early in the May–July season, laying two to eight eggs in a shallow cavity, may produce a second clutch.

Diamond-backed Terrapin

Malaclemys terrapin L males 4–5.5 in (10–14 cm) females 6–9 in (15–23 cm)

For a long time, the Diamond-backed was known primarily for its value as a gourmet food item. Culinary demand and coastal development seriously reduced its numbers.

KEY FACTS

Brown to black carapace is keeled; scutes have growth rings; plastron yellowish or greenish; head and limbs are spotted.

+ **habitat:** Salt marshes, tidal flats, and other brackish and salt waters

+ **range:** Atlantic and Gulf coasts from Cape Cod to Texas

+ **food:** Crustaceans, mollusks, other invertebrates, and fish

The female Diamond-backed is up to twice as long as the male, which has a thicker tail and a concave plastron that facilitates mating. The female digs a cavity above the high-tide line and deposits 4 to 18 leathery eggs, which hatch in two to four months. Diamond-backed appearance varies, and seven distinct subspecies distributed sequentially along the coasts are recognized. The central carapace keel is more prominent in the Gulf Coast races; the unhinged plastron also differs widely in color and pattern. "Terrapin" comes from an Algonquian word for "the turtle." In Europe, the term is used for other species of freshwater turtles.

Northern Map Turtle

Graptemys geographica L males 3.5–6.25 in (9–16 cm) females 7–10 in (18–26 cm)

The Northern Map Turtle shows fewer of the decorative "bells and whistles" of most other map turtles, which are so called because of the detailed patterns on their carapaces.

KEY FACTS

Low carapace has keel and elaborate markings; yellow spot behind eye preceded by 2–3 diagonal lines.

+ **habitat:** Rivers and lakes

+ **range:** Great Lakes watershed to Tennessee and Alabama; Arkansas and Missouri watersheds; scattered in New York and Pennsylvania

+ **food:** Insects, crayfish, snails, and mollusks

If the carapace patterns of map turtles look like waterway maps, their intricately yellow-lined head and limbs resemble geological contour maps. The female Northern Map Turtle is significantly larger than the male and her head and jaws are also disproportionately larger. This allows her to feed on shelled animals such as freshwater clams, which require more crushing power. The female also tends to lose her distinctive markings as she ages. The Northern Map likes rivers with logjams, which allow it to bask in company, sometimes stacked on top of one another. If disturbed, it will disappear rapidly into the water.

Black-knobbed Map Turtle

Graptemys nigrinoda L males 3–4.5 in (8–11 cm) females 4–7.5 in (10–19 cm)

The multiple embellishments of the Black-knobbed Map Turtle include unnatural-looking knoblike projections along the carapace ridge and heavily serrated carapace margins.

KEY FACTS

Olive-brown carapace has black, knoblike projections on vertebral scutes; prominent centers on marginal scutes; yellow mark behind eye is connected to neck stripe.

+ **habitat:** Streams and rivers

+ **range:** Waterways of Mobile Bay drainage system in Alabama

+ **food:** Insects and other invertebrates

The jagged serrations on the edges of its carapace give the Black-knobbed Map Turtle the alternate common name of Black-knobbed Sawback. This species and most other map turtles occupy narrow and localized ranges based on a river drainage system, most of them in the South. The ranges of different Black-knobbed subspecies are not exclusive, and intergradations between them often make identification difficult. The subspecies *G. n. nigrinoda* differs from its neighbor, *G. n. delticola,* the Delta Map Turtle, by the former's lighter overall appearance above and the smaller extent of the black markings on its plastron.

Red-eared Slider

Trachemys scripta elegans L 5–8 in (13–20 cm)

A prominent, wide, reddish—or sometimes yellow—line that extends behind the eye is a distinctive feature of the Red-eared Slider, but sometimes this stripe is missing.

KEY FACTS

Carapace has yellow streaks and bars; red stripe on head; dark smudge on plastron scutes.

+ **habitat:** Quiet rivers, streams, swamps, ponds, and lakes

+ **range:** West Virginia to New Mexico, south to the Gulf; introduced elsewhere

+ **food:** Aquatic plants, insects, other invertebrates, fish, and tadpoles

Pond sliders in general *(T. scripta)* are the most common North American turtle and have been widely studied as a species and in the context of ecological investigations. They favor shallow, quiet waters with muddy bottoms and much vegetation, and are among the species frequently seen in short, vertical stacks on basking logs in ponds. The Red-eared Slider is prone to melanism, in which its shell patterns, limbs, head, and tail become obscured to varying degrees with black blotches, even to the point of producing an all-dark turtle. The Red-eared Slider is commonly raised on farms and sold as a pet, which should never be released.

River Cooter

Pseudemys concinna L 7–13 in (18–33 cm)

The common name of this species likely comes from an African word, *kuta,* meaning "turtle." The River Cooter was a traditional food source in the South.

KEY FACTS

Brown carapace often has concentric yellow markings; red bar on marginal scutes; plastron reddish; males have elongated front nails.

+ **habitat:** Streams, rivers, lakes, swamps, and marshes

+ **range:** Coastal plain from Virginia to Georgia and Alabama

+ **food:** Aquatic plants, berries, insects, fish, and carrion

The River Cooter frequently occurs in mixed company with the Painted Turtle and sliders, often sharing a basking log. It tends to "jump ship" quickly when disturbed. An able swimmer, it does well in stronger streams and rivers and tolerates brackish water. A male courts a potential mate by stroking her face with his elongated front claws. If interested, she will meet him on the bottom for mating; she sometimes initiates mating. Between late May and July, the female will lay 8 to 20 pinkish white eggs. These will hatch in several months; hatchlings may overwinter in the nest. This species congregates in floodplains of river oxbows.

Northern Red-bellied Turtle

Pseudemys rubriventris L 10–12.5 in (25–32 cm)

The Northern Red-bellied Turtle, often called the Red-bellied Cooter, is generally the largest basking turtle in its range. An isolated population lives in southeastern Massachusetts.

KEY FACTS

Brown to black carapace has red bars on marginal scutes; plastron reddish; arrow shape on head.

+ habitat: Ponds, lakes, streams, rivers, and marshes

+ range: Southern New Jersey to northeastern North Carolina and along Potomac River

+ food: Aquatic plants, crayfish, other invertebrates, and tadpoles

The Northern Red-bellied is usually found in deeper water than many other basking turtles. The adult female shows a vertical red line on her side scutes, and the male has elongated front claws, as the River Cooter does. An adult can be distinguished from Painted Turtles, with whom they often bask, by size and the two yellow spots on the sides of the Painted's head. Northern Red-bellieds are skittish and quickly disappear into the water if disturbed. Females dig a nest and lay 8 to 20 eggs; hatchlings may overwinter. As with many turtles, hatchling survival rates are low due to predation, environmental factors, and human interference.

Painted Turtle

Chrysemys picta L 3.5–7 in (9–18 cm)

True to its name, the Painted Turtle has many colorful embellishments that distinguish it from other species. It has the widest range of all North American turtles.

KEY FACTS

Smooth, dark carapace has red pattern on marginal scutes; plastron is yellow, often decorated

+ habitat: Sluggish, shallow streams, rivers, and lakes

+ range: U.S.–Canada border to Nova Scotia, northern and south-central U.S.; isolated in Southwest

+ food: Aquatic plants, insects, and other invertebrates

In addition to the bright markings and patterns on its shell, this ubiquitous basking turtle shows red lines on the neck, which become yellow on the head, and has stripes on the legs and tail. Females average larger than males and lay multiple clutches—up to four a year in the South—of 2 to 20 oval eggs. Three subspecies have distinct variations in pattern on the carapace and plastron and are geographically distinct, but show intermediate characteristics where the ranges overlap. As with many other turtle species, the young are generally carnivorous and transition to a more herbivorous diet as they mature.

Chicken Turtle

Deirochelys reticularia L 4–6 in (10–15 cm)

When the Chicken Turtle sticks its neck out, it becomes obvious that the length of that striped neck measures about three-fourths the length of its shell.

KEY FACTS

Carapace has a network of fine lines; hind legs are striped; front legs have broad stripe.

+ **habitat:** Ponds, lakes, swamps, and marshes

+ **range:** Coastal plain from North Carolina to eastern Texas; isolated population in southeastern Virginia

+ **food:** Insects, crustaceans, tadpoles, frogs, and aquatic plants

The carapace of the Chicken Turtle lacks ridges and serrations in the rear, but it widens over the hind legs, giving a slight caboose effect. Males are somewhat smaller than females. These water turtles very rarely occur in streams and rivers, and often roam about on land. Despite differences in appearance, the Chicken Turtle and Blanding's Turtle (*Emydoidea blandingii*) share similarities such as long necks and narrow jaws, and are closely related. Their ranges do not overlap, but show a North–South distribution. Formerly a popular food in the South, this species gets its name because, well . . . it tastes like chicken.

Blanding's Turtle

Emydoidea blandingii L 5–7 in (13–18 cm)

As a northern species, the Blanding's Turtle can handle cold temperatures. A hinged plastron allows it to close up, but not as securely as a box turtle.

KEY FACTS

Smooth black carapace has yellowish spots and thin lines; neck very long; chin and throat bright yellow.

+ **habitat:** Ponds, lakes, marshes, bogs, and creeks

+ **range:** Discontinuous range from Nova Scotia to Nebraska

+ **food:** Insects, crayfish, other invertebrates, and aquatic plants

The Blanding's Turtle has a kind of bug-eyed look, and its slightly domed shell and ability to pull into it have led to the nickname "semi-box turtle." However, the Blanding's Turtle differs from the box turtle in a number of ways, including a notched instead of hooked beak. The somewhat similar Spotted Turtle has larger, more diffuse spots and no hinge on the plastron. The Blanding's Turtle often basks and moves about on land, but otherwise is very aquatic. The nesting season is short; 6 to 11 eggs are laid in June or July and hatch in August or September. Together, the Chicken Turtle and Blanding's Turtle are sometimes known as "American snakenecks."

Mojave Desert Tortoise
Gopherus agassizii L 8–15 in (20–38 cm)

The genus *Gopherus* strongly hints at a major characteristic of this North American tortoise: an ability to dig like a gopher with scaly, flattened forelimbs. The rear limbs are round and stumpy.

KEY FACTS

High-domed brown carapace has growth rings on each scute, often with a yellowish center; plastron extends under chin in males.

+ **habitat:** Canyons, slopes, dunes, washes, and oases

+ **range:** Southeastern California, southern Nevada, and southeastern Utah south to Mexico

+ **food:** Grasses, herbs, and cacti

Mojave Desert and Gopher Tortoises enjoy some similarities, but location fixes their identification, as their ranges don't overlap. Nevertheless, the former has a straight plastron extension, compared to the one bent up under the chin of the Gopher Tortoise. Males frequently face off and ram each other with these extensions. The Mojave Desert Tortoise digs burrows in compacted sand. Usually these are only a few yards long, but some stretch up to 30 feet (9 m). The female lays a clutch of usually four to six eggs in a nest that is often near the mouth of a burrow. This species may den communally in the winter.

Gopher Tortoise
Gopherus polyphemus L 6–9.5 in (15–24 cm)

Adept diggers, Gopher Tortoises excavate long sloping tunnels with terminal chambers. Many animals such as Burrowing Owls, snakes, and invertebrates frequently share these accommodations.

KEY FACTS

Domed carapace is brown to tan; scutes with concentric rings often have light centers; male's front plastron extension bends upward.

+ **habitat:** Sandy areas between grasslands and forests

+ **range:** Southwestern South Carolina to Florida and eastern Louisiana

+ **food:** Grasses, forbs, and berries

Locals refer to the Gopher Tortoise as "a cow with a shell" for its placid grazing habits. The animal emerges from its burrow in the cool of the morning to feed. After mating, females lay a small clutch of 2 to 14 spherical eggs with brittle shells in a cavity often at the mouth of a burrow. The eggs hatch in about 100 days. Temperature during incubation determines sex of the hatchlings; higher temperatures produce females. The Gopher Tortoise species is an ecological linchpin. Its tunnels provide shelter and protection for some 300 other species within a range that is severely impacted by development and dense human population.

Loggerhead Sea Turtle
Caretta caretta L 31–47 in (79–120 cm)

Like most other sea turtles, Loggerheads come onto land only to nest. Females haul themselves onto beaches at night, excavate holes, and lay large clutches of eggs at about two-week intervals.

KEY FACTS

Carapace is reddish brown and somewhat heart shaped; plastron is yellowish; head is very large.

+ habitat: Oceans, bays, estuaries, and tidal rivers

+ range: All oceans, nesting regularly on Atlantic coast as far north as North Carolina

+ food: Marine invertebrates, fish, and algae

The Loggerhead moves its flippered forelegs in a figure-eight stroke to swim; its webbed hind legs work as rudders. When young, it tends to stay in the open ocean for a number of years, and then moves closer to the continental shelf and coastline. It can take 25 years to reach reproductive maturity, with a female laying eggs at two- or three-year intervals. Hatchlings must escape their sandy nests and navigate a threat-filled expanse of beach to reach the sea. Despite much international protection, the future of sea turtles is threatened by commercial fishing practices, poaching, and development along beaches.

Green Sea Turtle
Chelonia mydas L 36–48 in (90–120 cm)

The Green Sea Turtle gets its common name not from the color of its shell (which is mostly brown), but from the color of its body fat. The species is highly prized as a food source.

KEY FACTS

The smooth carapace is brown, sometimes with an olive tinge, and has blotches or a radiating pattern.

+ habitat: Oceans, bays, and estuaries

+ range: Warm waters of Atlantic and Pacific; sometimes nests in southeastern Florida

+ food: Algae, sea grasses, sponges, and other marine invertebrates

The female Green Sea Turtle returns to the beaches where she hatched to lay clutches of about 100 eggs in the sand. This journey may carry her several thousand miles from the sea-grass "pastures" where she grazes. The species is prone to a disease that causes a proliferation of soft-tissue lesions, which affects the turtle by interfering with internal organs and makes it less agile when the lesions appear externally. Of unknown origin, the disease is prevalent near areas with dense human populations. Gelatinous substances known as "calipee" and "calipash," under the shell, are harvested for use as thickeners in turtle soups.

Leatherback Sea Turtle
Dermochelys coriacea L 50–70 in (120–170 cm)

Leatherbacks are the largest turtles worldwide. A record breaker measured more than 9 feet (2.7 m) and weighed more than 2,000 pounds (900 kg).

KEY FACTS

Carapace and plastron have smooth, leathery, blackish skin; carapace has 7 prominent keels; forelimbs are flippers, hindlimbs are webbed.

+ **habitat:** Oceans

+ **range:** Atlantic and Pacific, nesting along Atlantic as far north as North Carolina

+ **food:** Jellyfish and other marine invertebrates

The only member of its family, the Leatherback Sea Turtle is a pelagic species, swimming the open oceans on long migratory journeys, often in pursuit of large aggregations of prey, such as jellyfish. The giant turtles have unique, specialized physiological processes that allow them to maintain a body temperature well above ambient water temperature. They swallow slimy prey with the help of long spines in the mouth and an esophagus that points backward and also discharges the stinging cells of jellyfish. The female Leatherback returns to her own nesting beach every two or three years and lays multiple clutches of 50 to 170 eggs each.

Hawksbill Sea Turtle
Eretmochelys imbricata L 30–35 in (75–90 cm)

The Hawksbill Sea Turtle's hooked beak, which can inflict a serious bite, gives it its common name. Its species name, *imbricata,* refers to the overlapping scutes of its carapace.

KEY FACTS

Brown carapace is often yellow-streaked; head has 2 pairs of scutes between eyes.

+ **habitat:** Coastal waters, bays, and estuaries

+ **range:** Mainly tropical waters; seen on Atlantic coast to New England and on Pacific to southern California

+ **food:** Sponges, jellyfish, mollusks, crustaceans, fish, and algae

The Hawksbill female is mainly a tropical nester, but she sometimes nests on Florida Atlantic beaches. The species eats invertebrates that other sea turtles and other animals typically avoid, such as toxic sponges. Atlantic, Pacific, and Indian Ocean subspecies of the Hawksbill Sea Turtle are recognized. These turtles are taken for their meat, eggs, and the attractive, often translucent carapace, which is made into tortoiseshell jewelry and other luxury items (although most "tortoiseshell" sold today is plastic). In older individuals, the carapace scutes may no longer overlap, but lie adjacent to one another.

Kemp's Ridley Sea Turtle

Lepidochelys kempii L 23.5–27.5 in (60–70 cm)

Formerly known as the Atlantic Ridley Sea Turtle, this species is about as wide as it is long, giving it a nearly round carapace. It is the smallest North American sea turtle.

KEY FACTS

The carapace is gray or green with 3 keels and has 5 costal (side) scutes; the plastron is yellowish.

+ habitat: Shallow coastal waters

+ range: Gulf of Mexico; sometimes seen as far north as New England and Nova Scotia

+ food: Crabs, other marine invertebrates, and fish

Like other sea turtles, the Kemp's Ridley is highly endangered. In the past, tens of thousands of this species would gather for a mass mating event in the Gulf of Mexico, and thousands of females would subsequently come ashore to lay eggs. Greatly reduced numbers of these turtles still dig nests on beaches in daytime and deposit about 100 spherical eggs. The adult Olive Ridley Sea Turtle *(L. olivacea)* has an overall olive shell and six to eight costal scutes on its carapace. It is slightly larger than Kemp's turtles and is found along the Pacific coast, but eventually may end up in Florida due to small migrant populations in the Caribbean.

Smooth Softshell

Apalone mutica L males 4.5–7 in (11–18 cm) females 6.5–14 in (16–36 cm)

Softshell is somewhat of a misnomer. Softshell species have a bony shell, but it is overlain with a covering of soft, leathery skin, not the scutes—or hard plates—that most other turtles have.

KEY FACTS

Carapace is olive to brownish; adult male and juvenile carapace is marked with dark dots; adult female carapace has blotches.

+ habitat: Rivers and streams

+ range: Drainage areas of major rivers in the central U.S.

+ food: Crayfish, other invertebrates, frogs, and fish

This very aquatic turtle is known as much for its snorkelish snout as it is for its pancake-like shell. It moves quickly both on land and in the water. It may bask on land but also spends a good deal of time immersed in mud or sand in the shallows, its long neck extending the snout to the surface for air. Softshells in general come equipped with sharp beaks and claws that give them defensive protection that their shells do not. The female digs nests with her webbed hind feet and lays a clutch of 4 to 32 round, fairly hard-shelled eggs. These hatch after an incubation period of about two and a half months.

Spiny Softshell

Apalone spinifera L males 5–9 in (13–23 cm) females 6.5–18 in (16–45 cm)

The Spiny Softshell has small spines on the front edge of its carapace that distinguishes it from the Smooth Softshell. As in the Smooth species, female Spinys are much larger than males.

KEY FACTS

Carapace is olive-gray to brownish; adult males have black-bordered spots; adult females have dark splotches.

+ **habitat:** Creeks, ponds, lakes, and rivers

+ **range:** Central and southeastern U.S.; scattered populations elsewhere

+ **food:** Fish, frogs, crayfish, and other invertebrates

Examination of a Spiny Softshell's tubelike snout (which is a risky undertaking, because the turtle bites) reveals that the septum projects into both nostrils; these projections are not present in the Smooth Softshell. Like other softshell turtle species, the Spiny is very aquatic. It is usually within a quick plunge of water, and it uses its snout as a snorkel while submerged. During the May to August nesting season, the female digs a hole in a sand or gravel bank and deposits 4 to 32 round eggs. The eggs usually hatch in two months, but hatchlings may not emerge until the following spring.

Western Banded Gecko

Coleonyx variegatus L 4.5–6 in (11–15 cm)

Despite its delicate coloring and appearance, the Western Banded Gecko is adept at surviving the rigors of desert life. A nocturnal species, it waves its tail when tracking prey.

KEY FACTS

Soft body is pink to tan to yellow; brown bands on body and tail may become dis-jointed with age; pad-less, slender toes.

+ **habitat:** Arid rocky areas, canyons, and dunes

+ **range:** South-eastern California, southern Nevada, southwestern Utah into Arizona and New Mexico

+ **food:** Insects and spiders

Unlike geckos with fixed eyelids, the Banded Gecko has eyelids that open and shut. Its eyes bulge noticeably, and its pupils are vertical. The Western Banded Gecko is a vocal species and often chirps or squeaks when caught. The tail has a constriction at the base where it can break off neatly if grabbed by a predator. The male can be identified by the prominent spurs at the sides of the base of the tail; these are smaller or lacking in the female. A mainly terrestrial species, this gecko hides among rocks or under logs or brush during the hottest part of the day. Females lay a clutch of two eggs up to three times a year.

Mediterranean Gecko

Hemidactylus turcicus L 4–5 in (10–13 cm)

Mediterranean Geckos usually prefer to share buildings with human inhabitants. As the name suggests, this is an alien species, which now is established in about 20 U.S. states.

KEY FACTS

Body is pinkish to brownish with irregular dark spots and rows of white tubercles extending onto tail; enlarged toe pads; clawed toes; round, often banded tail.

+ **habitat:** In tree bark, rock crevices, and buildings

+ **range:** Scattered in Gulf Coast states; isolated colonies elsewhere

+ **food:** Insects

Unconcerned by human proximity, the nocturnal Mediterranean Gecko scurries up walls and across ceilings in pursuit of insects drawn to light sources. It emits faint, mouselike squeaks as it goes about its activities. The male is territorial and makes chirping sounds when approaching a potential mate. This lizard has large, bugged-out eyes with vertical pupils. The white underside is semitranslucent and often reveals the oval, hard-shelled eggs that can make up a quarter of a female's weight. Females nest communally, depositing a pair of eggs under tree bark or palm leaves or in building cracks and crevices.

Green Anole

Anolis carolinensis L 5–8 in (13–20 cm)

The only native *Anolis* species in the continental United States, the Green Anole shows the characteristic throat fan in males, a feature appearing in territorial disputes and courtship displays.

KEY FACTS

Body is green, brown or mottled, dewlap, or throat fan, usually pink; long, pointed snout; enlarged toe pads; rounded tail.

+ **habitat:** Trees, shrubs, vines, walls, and fence posts

+ **range:** Southern Virginia to Florida, including Keys; to eastern Texas

+ **food:** Insects, spiders, and other small invertebrates

Active during the day, the arboreal Green Anole can change colors depending on a change in mood or environmental factors. Color change and extension of the throat fan are accompanied by body movements, including head bobbing and push-ups. Unlike other lizards, many *Anolis* females, including the Green Anole, lay a single egg every two weeks during mating season, in this case March to September. The female deposits the egg in leaf litter, rock piles, or debris, and it hatches in five to seven weeks. Sometimes mistakenly called a chameleon (an Eastern Hemisphere lizard), this is the species commonly sold in pet stores.

Brown Anole
Anolis sagrei L 5–8 in (13–20 cm)

This species was introduced from the West Indies to Florida and Texas. In locations they share with Green Anoles, the Browns are usually found closer to the ground.

KEY FACTS

Color is tan to gray to brown, turning darker with change; dewlap white-bordered reddish orange; snout short; tail laterally compressed.

+ **habitat:** Close to ground in shrubs, trees, fences, walls, and rock piles

+ **range:** Peninsular Florida, southeastern Texas, and adjacent states

+ **food:** Insects and spiders

The Green Anole may turn brown, but the Brown Anole never turns green, a characteristic that helps differentiate the two, along with the Brown's laterally compressed tail and reddish orange dewlap edged in white. When folded up, the dewlap appears as a white line under the Brown male's chin. Some males have a crest along the tail. The females vary; some have a thin, central yellow stripe down the back that is flanked by brown half-moon shapes. The territorial male goes through a typical anole display with head bobs and push-ups. The species' coloration always stays within the gray to brown range, potentially becoming darker.

Eastern Collared Lizard
Crotaphytus collaris L 8–14 in (20–36 cm)

The large-headed male Eastern Collared Lizard in prime breeding condition exhibits bright coloration and a distinctive pattern that makes it one of North America's most attractive lizards.

KEY FACTS

Body is often green with black and white collars; crossbands and spots on back; spotted tail.

+ **habitat:** Rocky and hilly areas with shrubs, piñon-juniper woodlands, grasslands, and riparian areas

+ **range:** Southwestern Illinois to central Texas, west to Utah and Arizona

+ **food:** Insects, spiders, other lizards, and plant matter

The paler female Eastern Collared Lizard's physical attributes are more subdued than the male's, and she lacks the dorsal bands found on his back. However, she displays red or orange spots and bars along her sides when carrying fertilized eggs; these fade after the eggs are laid. Up to about a dozen eggs are deposited in a nest under a rock or in a rodent burrow in midsummer. This aggressive species seizes prey abruptly, often rushing at it bipedally on its hindlegs with the tail lifted off the ground like a little tyrannosaur. It seems to prefer habitats with large boulders for basking and many rock crevices for shelter and safety.

Long-nosed Leopard Lizard

Gambelia wislizenii L 8.5–15 in (22–38 cm)

The large-headed Long-nosed Leopard Lizards frequently wait out prey from atop a rock, hopping off and giving chase—often bipedally—at a fast speed. They often target smaller lizards.

KEY FACTS

Body is gray or brown with dark spots from head to tail; rounded tail.

+ **habitat:** Arid or semiarid areas with gravel or sand and low grasses and shrubs

+ **range:** Southern Oregon and Idaho to New Mexico, west to southern California

+ **food:** Insects, spiders, other lizards, snakes, rodents, and plant matter

This species shares many features and behaviors with the closely related collared lizards. Like the Eastern Collared Lizard, the female Long-nosed Leopard shows orange spots and bars on her sides when carrying fertile eggs. Females lay about three to eight eggs once or twice a mating season, from May to August. Older and large females typically lay larger clutches. In cool weather the body color of the Long-nosed Leopard Lizard noticeably darkens, which obscures the spotted pattern. Otherwise, the species blends into its environment, especially when it hides in the shade of a shrub, another common location for ambushing prey.

Desert Iguana

Dipsosaurus dorsalis L 10–16 in (25–40 cm)

When most desert animals have scattered to shady places to beat the high heat, Desert Iguanas often are just starting their day. They can tolerate temperatures up to 115°F (46°C).

KEY FACTS

Head is brown or pale; neck and trunk have netlike pattern and grayish spotting; dark-banded tail.

+ **habitat:** Arid brush and scrub with sand and rocks

+ **range:** Western Arizona, southern Nevada, and southeastern California

+ **food:** Leaves, buds, and flowers of creosote bush; other plants

When the Desert Iguana finally wants to escape the heat, it climbs into a creosote bush for shelter—and a built-in meal; the lizard's range approximates that of the arid-land shrub. To rest for the night or hibernate, these lizards choose rodent burrows, which they seal up at entrances with plugs of sand. Heat tolerance is one thing, but Desert Iguanas are a very skittish species and disappear at the mildest disturbance. They often stick close to a burrow retreat when basking. Breeding adults show a pink tinge on the sides of the belly. On closer inspection, the bands on the tail are made up of closely spaced dark spots.

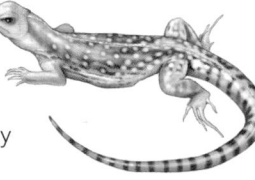

Green Iguana

Iguana iguana L 30–79 in (75–201 cm)

Introduced from Central and South America, Green Iguanas are larger than the largest U.S. native lizards, Gila Monsters. They can weigh up to 18 pounds (8 kg).

KEY FACTS

Body is usually green; head has large, round scales below ear; ridge of spines on back and front edge of dewlap; young are bright green.

+ **habitat:** Dense stands of trees, often near water

+ **range:** Southeastern Florida and several Florida Keys

+ **food:** Leaves, flowers, and fruit

The Green Iguana is frequently found basking in tree canopies that overhang water sources; if threatened, it jumps in and can stay submerged up to 30 minutes. The male Green Iguana is bigger than the female, with a larger head and longer back and dewlap spines. Very aggressive, it spars with other males for access to females. Multiple females will enter the territory of a particularly promising-looking male. After mating, the female deposits 28 to 40 eggs in a nest, which hatch in about 90 days. This is a popular pet species, but it tends to fare poorly because it seldom receives the kind of care and conditions it requires.

Common Chuckwalla

Sauromalus ater L 11–16.5 in (28–42 cm)

The Common Chuckwalla's former species name, *obesus*, aptly describes its appearance. Its bulk is exaggerated by the small size of the scales on its body.

KEY FACTS

Dark body may be black in male; has folds at throat; paler blunt-tipped tail very thick at base.

+ **habitat:** Shrublands with rock outcrops, rocky hillsides, and lava flows

+ **range:** Western Arizona, southern Nevada, and southern California

+ **food:** Leaves, buds, and flowers; some insects

Despite its large size, the Common Chuckwalla is a timid lizard. When threatened, it will often retreat to a rock crevice and gulp in air to inflate its body until it is well wedged in—not a difficult feat for such a bottom-heavy lizard. Like most reptiles, the Chuckwalla begins its day with a metabolism-awakening bask and then sets off to forage. It finds much of the sustenance it needs from the creosote bush and other arid-land vegetation. Young Chuckwallas show dark bands across the body and tail that fade with age in adult males; females tend to retain the bands. In some parts of the range, adult males have orange tails.

Zebra-tailed Lizard

Callisaurus draconoides L 6–9 in (15–23 cm)

Black bands across its long, tapering tail give this agile lizard its common name. Zebra-tails curl their tails to display the distracting pattern as they flee predators such as snakes, birds, and foxes.

KEY FACTS

Body is gray to brown to whitish, with blotches above; head has 2 throat folds; ear opening dark.

+ **habitat:** Dry areas of compact soil with sparse vegetation and rocks

+ **range:** Northern Nevada to southeastern Arizona and southern California

+ **food:** Insects, spiders, other lizards, vegetation, and carrion

The slim body and long legs of the Zebra-tailed Lizard help make it one of the fastest desert lizards. This species prefers open spaces and often is active in the heat of the day, although it forages mainly in the morning. It digs its own burrows for escape and shelter. The banded tail also plays a part in intraspecific communication. The Zebra-tailed Lizard wags its tail to signal a warning that a predator is in the vicinity; males also wag to advertise their general fitness. Like many lizards, male breeding-season displays include head bobbing and push-ups. Breeding males also develop metallic green-blue patches on their sides.

Common Lesser Earless Lizard

Holbrookia maculata L 4–5.5 in (10–14 cm)

Unlike most other lizards, earless species lack external ear openings. This feature is likely an adaptation to burrowing in sand. Small head and back scales and a nasal valve also shut out sand.

KEY FACTS

Color varies from tan to brown to gray to white; 4 rows of dark blotches on back; mid-back light; paired black marks on belly.

+ **habitat:** Exposed sand or gravel with short grass, piñon-juniper woodlands, sagebrush, and farmland

+ **range:** Great Plains to northern Texas and Southwest

+ **food:** Insects, spiders, and lizards

The small, slender-bodied Common Lesser Earless Lizard tends to blend into its environment with an overall background body color keyed to the predominant color of the terrain. In addition to the variable coloration, the male and the female show some differences: The male has blue-bordered belly markings, and the female shows red or orange on the throat during breeding season. The female lays one or two clutches of up to a dozen eggs between April and September, which hatch in a month or two. In addition to fleeing like the collared or the leopard lizard, it relies on headfirst burrowing to get out of harm's way.

Texas Horned Lizard
Phrynosoma cornutum L 2.5–7 in (6–18 cm)

The squat, flat, short-legged Texas Horned Lizard often goes by the erroneous names of horned toad or horny toad, especially in the pet trade.

KEY FACTS

Body is red to yellow to gray; spines on head with 2 long central spines; 2 dark lines below eye; 2 rows of pointed scales on sides; white line down back.

+ **habitat:** Arid areas with sparse grass and shrubs

+ **range:** Kansas to Texas, west to southeastern Arizona

+ **food:** Ants, other insects, and spiders

Although it tolerates heat better than many lizards, the Texas Horned Lizard may get out of the sun by shuffling sideways until it has dug itself into loose sand or soil, a typical feature of its preferred habitat. It also shelters in rock crevices, burrows of other animals, and under bushes. Color, pattern, and texture help camouflage the lizard when it is out and about. It often "parks" near an anthill, devouring passing ants one at a time. When confronted, it may shoot streams of blood from sinuses in the corner of its eyes to a distance of several feet. This is often enough to deter predators such as foxes and Coyotes.

Pygmy Short-horned Lizard
Phrynosoma douglasii L 2.5–5.5 in (6–14 cm)

The Pygmy Short-horned Lizard has all the typical horned lizard features, but on a smaller scale. Unlike most lizard species, which are oviparous, it bears live young.

KEY FACTS

Body is yellow to gray to reddish brown; head spines very short; one row of pointed scales on side; 2 bars on neck.

+ **habitat:** Plains, piñon-juniper woodlands, and mountains

+ **range:** U.S. Northwest into northern California, perhaps extinct in British Columbia

+ **food:** Ants and other insects

Well adapted to cold, the Pygmy Short-horned Lizard ranges farther north and to higher elevations than most lizards except for its close relative the Greater, or Mountain, Shorthorned Lizard. The Pygmy species occurs to about 6,000 feet (1,830 m). Both species bear live young; the Pygmy Shorthorn female bears 3 to 15 young between July and September. Hissing, biting, and shooting blood from the corners of its eyes are among horned lizard defenses. By day this species "parks" in the vicinity of anthills and other insect sources; at night it will bury itself in the soil with the side-shuffling motion common to horned lizards.

Common Sagebrush Lizard

Sceloporus graciosus L 5–6.25 in (13–16 cm)

Common Sagebrush Lizards are spiny lizards, with keeled (ridged) and pointed scales on their back. This species occupies habitats ranging to about 10,500 feet (3,200 m).

KEY FACTS

Body is gray or brown with irregular bands above; light stripes on sides; rust behind forelimbs.

+ **habitat:** Sagebrush, other brushlands, woodlands, and mountain forests

+ **range:** Washington to California, east to New Mexico, north to Montana

+ **food:** Insects, spiders, and other invertebrates

The Common Sagebrush Lizard can be found in open areas with basking rocks or logs, but it seldom strays far from the protection of rock crevices, rodent burrows, or brush. Mostly terrestrial, it sometimes climbs trees to forage or escape predators. The breeding male has blue belly patches and a blue throat. The neck and sides may show yellow or orange, which is emphasized in the breeding female. The female lays a single clutch of two to seven eggs in June or July that hatch in about two months. This species is similar to the Western Fence Lizard, but it is generally smaller; has smaller, more numerous scales; and occurs at higher elevations.

Desert Spiny Lizard

Sceloporus magister L 7–12 in (18–30 cm)

The sturdy Desert Spiny Lizard has large keeled-and-pointed scales. Like many lizards, its color darkens in cool weather to absorb more heat and lightens in warm weather to reflect heat.

KEY FACTS

Body is brownish with crossbanding; triangular patches on shoulders; males have blue on belly and throat.

+ **habitat:** Brush and scrublands in arid and semiarid areas at low elevations

+ **range:** Southern Nevada to California, east to western Texas

+ **food:** Insects, spiders, lizards, and plant matter

Though a desert species, this spiny lizard often is found in habitats that include nearby water sources. Rocks are also a requirement, along with easy access to the protective shelter of brush, rodent burrows, and wood rat nests. The male is territorial and seeks to exclude other males, which sometimes revert to more juvenile coloration to slip past. The female tends to remain in or near the areas where she hatched. The species becomes inactive in winter for a length of time determined by local climate. Mated pairs of males and females are commonly seen. The female lays one or more clutches of 4 to 19 eggs in spring or summer.

Western Fence Lizard
Sceloporus occidentalis L 6–9.25 in (15–23 cm)

This spiny lizard species is found in a wide variety of habitats, except outright desert. They range to more than 9,000 feet (2,700 m), preferring areas with rocks, woodpiles, and even old buildings.

KEY FACTS

Body is black, gray, or brown with blotches or crossbars; blue patches on lower sides; male has blue throat.

+ habitat: Grasslands, shrublands, woodlands, coniferous forests, and farmland

+ range: Washington to California, Idaho to western Utah

+ food: Insects and other small invertebrates

When a male Western Fence Lizard wants to impress, it flattens its body to show the deep blue coloring of its lower sides. This display gives the species the nickname "blue-belly." When courting a female or challenging other males, it also performs head bobs and push-ups like some other lizards. Studies show that a Western Fence Lizard's body contains a protein that neutralizes the Lyme disease bacterium. The protein clears vector ticks of the bacterium when they bite the lizard, so that they no longer contain the disease. The species becomes active on mild days in midwinter. Females lay one clutch a year containing 3 to 13 eggs.

Eastern Fence Lizard
Sceloporus undulatus L 3.5–7.5 in (9–19 cm)

The Eastern Fence Lizard is the only spiny lizard in most of its range. For its habit of running up pine trees it is known also as the Pine Lizard.

KEY FACTS

Body is gray to brown above; females have barred pattern across back; males have blue bellies, throats, and blue-green sides.

+ habitat: Woodlands, grasslands, and shrublands

+ range: New Jersey to Florida and west to the Mississippi River

+ food: Insects, spiders, and other invertebrates

The Eastern Fence Lizard eats a wide range of insects and other small arthropods. Routine behaviors change with habitat and location. The species is more arboreal in the eastern part of the range; it climbs trees (and walls) to escape from potential danger. It often makes the ascent in stages, remaining motionless for a while before ascending higher. In the prairie portion of the range, it is found more often on the ground or sheltering in rodent burrows. Female Eastern Fence Lizards also show blue patches on the throat, but they are less conspicuous than those on males.

Ornate Tree Lizard

Urosaurus ornatus L 4.5–6.25 in (11–16 cm)

This small climbing lizard is well camouflaged against the bark of trees and a jumble of rocks. It also frequently hangs out on fence posts.

KEY FACTS

Body is brown to gray with crossbands and 2 bands of larger scales along the back; male has blue belly patches, blue to orange throat patches.

+ **habitat:** Arid areas with shrubs and trees, often along streams

+ **range:** Southwestern Wyoming to west-central Texas, west to California

+ **food:** Insects and other small invertebrates

Quick and agile, the Ornate Tree Lizard runs when disturbed to a nearby rock or tree, climbing up quickly on the far side to stay out of view. It forages in daytime, including midday in cooler weather. Females are territorial, and their territories may overlap with those of several males. Males establish a dominance hierarchy, with the power—and most of the mating opportunities—going to the top-ranked lizard, usually the heaviest. Females lay up to six clutches of 2 to 13 eggs each year. The female will search to find the right location for digging a nest. This species comes together in groups in the winter to hibernate.

Common Side-blotched Lizard

Uta stansburiana L 4–6.75 in (10–17 cm)

The Common Side-blotched Lizard is just that: one of the most common lizards of the Southwest. It appears in a variety of arid and semiarid habitats up to 9,000 feet (2,700 m).

KEY FACTS

Body ranges from brown to gray to yellow to black; pattern varies; blue-black spot behind each front leg

+ **habitat:** Grasslands, shrublands, and other low-growing vegetation

+ **range:** Central Washington to western Texas, west to California, north to Oregon

+ **food:** Insects, spiders, and other small invertebrates

There is significant color and pattern variation among different populations of Common Side-blotched Lizards. The diagnostic side blotch—the mark behind the front legs—is more pronounced in males. These terrestrial and diurnal lizards are active year-round in the southern parts of their range, where breeding is extended and females lay up to seven clutches of one to eight eggs. Northern lizards become less active in winter, except on warm days, and they have fewer and smaller clutches within a shorter breeding season. Side-blotched females can retain sperm for several months, and they can fertilize more than one clutch.

Southern Alligator Lizard
Elgaria multicarinata L 10–16.5 in (25–42 cm)

Alligator lizards have lateral folds composed of smaller scales on the sides of their body that allow room for expansion to accommodate the ingestion of food or the development of eggs.

KEY FACTS

Body is reddish brown to yellowish gray; back and tail have dark crossbands; centers of belly scales have dark stripes.

+ **habitat:** Grasslands and open woods, especially oaks

+ **range:** Southern Washington to southwestern California

+ **food:** Insects, spiders, and other invertebrates

Like other alligator lizards, the Southern Alligator Lizard has a long body and tail and proportionately small limbs, so a speedy escape is not a sound defensive strategy. Instead, these lizards wiggle vigorously when grabbed and may bite and defecate to discourage predators. The tail is fragile and breaks off easily; a wiggling tail remnant allows the rest of the animal time to slip away. The species is mainly diurnal, although it may be active during the night in summer. It often climbs vegetation in search of insects and other small prey and eats a wide range of invertebrates, including scorpions and Black Widow spiders.

Texas Alligator Lizard
Gerrhonotus infernalis L 10–16 in (25–40 cm)

Armored scales on the head, body, and tail—in addition to strong, formidable jaws—give the Texas Alligator Lizard and others of this group their common names.

KEY FACTS

Body is yellow to reddish brown, with a lateral groove on side; the back and tail have light, irregular crossbands.

+ **habitat:** Rocky hillsides and wooded canyons

+ **range:** West-central Texas

+ **food:** Insects, spiders, and small snakes, lizards, and mammals

The Texas Alligator Lizard has a rigid body and short legs, and moves with plodding deliberation compared with more agile and speedy lizards. When threatened, it may puff up its body with air to appear much larger; a groove along both sides of the body stretches out to accommodate this significant expansion. This diurnal lizard uses its semiprehensile tail for stability when foraging in vegetation. The species breeds year-round; the female lays between 8 and 20 eggs in a nest in moist ground and incubates them for about two months. During that time, the female protects the nest with her eggs, but her care ends when the eggs hatch.

Slender Glass Lizard

Ophisaurus attenuatus L 22–41 in (56–105 cm)

The limbless Slender Glass Lizard has a tail that breaks easily into several pieces; few individuals make it through life with their original tail. The replacement tail is usually shorter and darker.

KEY FACTS

Body is light brown with groove along sides; dark lines below groove; dark line down middle of back.

+ **habitat:** Dry grasslands and open woodlands

+ **range:** Virginia to Florida, west to Texas, north to Wisconsin and Indiana

+ **food:** Grasshoppers, insects, other invertebrates, and small mammals

Often mistaken for a snake, the Slender Glass Lizard displays unsnakelike movable eyelids and ear openings. This diurnal species does not like to be handled and will whip its body back and forth, which often causes the tail to shatter even if it is not directly grabbed. The grooves on the sides of its body, shared with related alligator lizards, allow for some expansion to accommodate food and eggs. Unlike snakes, glass lizards do not have flexible jaws, so the size of their prey is more limited. Females lay a clutch of 4 to 19 eggs, which they brood by coiling around it. The species often shelters in abandoned rodent burrows.

Gila Monster

Heloderma suspectum L 18–24 in (45–60 cm)

The only venomous lizard in the United States, the Gila Monster delivers its venom from glands in its jaw while chewing on its victim. For small prey, the bite alone is often lethal.

KEY FACTS

Face and feet are black; body pattern is yellow, orange, or pink and black; thick tail; curved claws.

+ **habitat:** Desert shrubland and oak woodland

+ **range:** Southeastern California, southern Nevada, southwestern Utah into Arizona and New Mexico

+ **food:** Young birds, eggs, other lizards, insects, and carrion

Bulky, big-tailed, and with small, beadlike scales on its body, the Gila Monster looks formidable as it flicks its forked black tongue in the air. This carnivorous lizard locates food primarily by taste and often is seen licking the ground while it walks. The species is active during the day or at dusk and after dark in summer. It retreats to burrows that it digs or appropriates from burrowing mammals and also in thickets or under rocks. In times of scarcity, the Gila Monster lives off the fat of its tail. As its bite is rarely fatal to humans (and then usually due to secondary health factors), there is no reason to kill it on sight.

Nile Monitor

Varanus niloticus L 5–7 ft (1.5–2 m)

Introduced around 1990, mostly likely as dumped or escaped pets, the Nile Monitor is reptile non grata in southern Florida. These large African lizards threaten native wildlife.

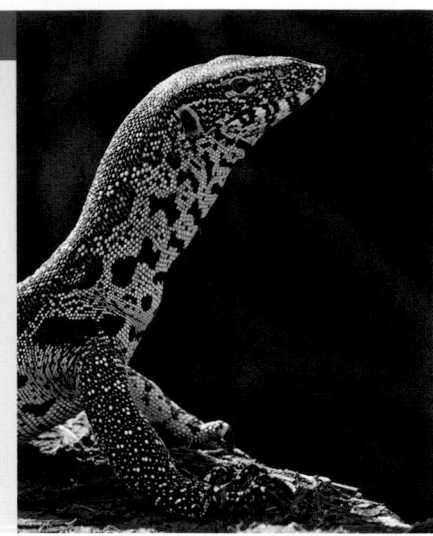

KEY FACTS

Body is gray to brown to black with beady scales and bands of yellow stripes.

+ **habitat:** Swamps, marshes, river, lake, canal banks, and suburban and urban areas

+ **range:** Southern Florida, breeding on the west coast

+ **food:** Crabs, fish, turtles, snakes, other lizards, frogs, birds, eggs, and mammals

The strong, feisty, and intelligent Nile Monitor has established a menacing presence in southern Florida. It finds the area's numerous watery habitats to be very suitable and has branched out to household visitations, entering through doggie doors and cat flaps. The species preys on many vulnerable native species, such as crocodilians, wading birds, Burrowing Owls, and sea turtles; it can wipe out an entire clutch of turtle eggs in one meal. The monitor is thought to breed only on the southwestern coast, including Sanibel Island. Residents are urged to report monitor presence by calling a hotline: 1-888-IVE-GOT1.

Six-lined Racerunner

Aspidoscelis sexlineata L 6–9.5 in (15–24 cm)

Racerunners rely on their speed to get them out of dangerous situations. These close relatives of whiptails prefer open spaces where they can run unimpeded.

KEY FACTS

Body has 6 or 7 light stripes bordered by dark bands; males have green or blue throat, females have white; brown tail striped on sides.

+ **habitat:** Grasslands and open woodlands

+ **range:** Southeastern and south-central United States

+ **food:** Insects

The Six-lined Racerunner is the only representative of the whiptail lizards that occurs widely in the eastern United States. This species is active by day—"fieldstreak" is a common nickname—especially in the morning, taking refuge or shelter when needed in vegetation, under rocks, or in burrows they dig or borrow from other animals. Subspecies include an eastern (*A. s. sexlineata*) and a prairie (*A. s. viridis*) race that differ by range, with some overlap where ranges meet, and by the latter's green color on the front half of the body. The somewhat similar skink species display shininess that racerunners lack.

Tiger Whiptail

Aspidoscelis tigris L 8–12 in (20–30 cm)

These whiptails are noted for their jerky movement with a side-to-side head snap reminiscent of *Jurassic Park* velociraptors. The many subspecies display a wide range of markings.

KEY FACTS

Back has varying light stripes, spots, bars, other dark markings; underside pale with dark spots, sometimes all black.

+ **habitat:** Desert and dry, open woodlands

+ **range:** Western Idaho and eastern Oregon to southern California and western Texas

+ **food:** Insects, scorpions, and other small invertebrates

Also known as the Western Whiptail, the Tiger Whiptail shows the characteristic long, thin tail of its named group. Like other whiptails, its body has small, granular scales above and larger, rectangular ones on the underside. A plethora of subspecies display differences in the number of dorsal stripes and the presence and colors of spots, bars, and chest spotting. The interesting Marbled Whiptail (*C. t. marmoratus*) shows a marbled pattern on back and sides that often overlays faint striping. From May to July, Tiger Whiptail females lay a clutch of one to four eggs in cooler climates and two clutches in warmer ones.

Plateau Striped Whiptail

Aspidoscelis velox L 8–10.75 in (20–27 cm)

All Plateau Striped Whiptails are female. Through parthenogenesis, these females lay unfertilized yet fertile eggs. The eggs hatch into females that are essentially clones of the mother.

KEY FACTS

Body has 7 light stripes on back; underside, especially chin, often has bluish cast; tail is pale blue.

+ **habitat:** Piñon-juniper and oak woodlands, pine and fir forests, and chaparral

+ **range:** Western Colorado and New Mexico to northeastern Arizona and southeastern Utah

+ **food:** Insects and spiders

The Plateau Striped Whiptail is less skittish in the face of disturbance than many other whiptails, and even when it flees a scene, it may not go far. It forages by day for spiders and insects such as grasshoppers and crickets in leaf litter at elevations up to 8,000 feet (2,400 m). Those living at lower elevations are typically found near streams. The Plateau Striped and other parthenogenetic species of whiptail (about half of those found in the United States) developed from the hybridization of bisexual species. Female Plateau Stripeds lay a clutch of three to five eggs in June to July that hatch in August.

Common Five-lined Skink

Plestiodon fasciatus L 5–8.5 in (13–22 cm)

Skinks are known for their high energy levels and the tendency to ditch their tails easily. Common Five-lined Skinks show off their signature pattern mainly when young.

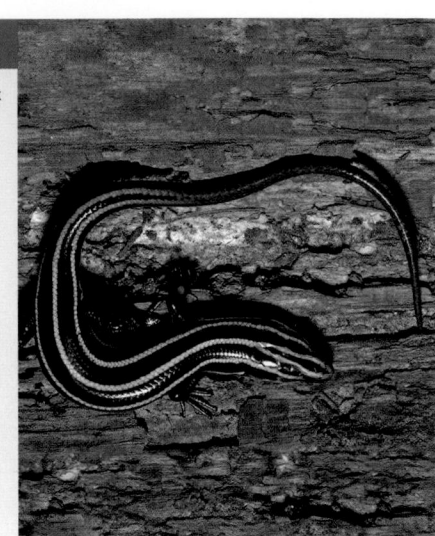

KEY FACTS

Body is brown to black with 5 light stripes; tail is blue to gray.

+ **habitat:** Humid woodlands with thick substrate and gardens

+ **range:** Eastern U.S. from southern New England to northern Florida, west to Texas, and north to Great Lakes

+ **food:** Insects, other invertebrates, other lizards, and mice

During its lifetime, the male's distinct lines often fade, giving the adult male a solid-brown appearance; females tend to retain their lines in a muted form. In the spring breeding season, the male's jaws turn a reddish orange. The female lays 4 to 15 eggs in nests that she excavates; she sticks close to the nest while tending the eggs. Juveniles are usually heavily striped and have bright blue tails, which serve to divert attention from their head and can quickly be shed at fracture planes. These skinks are diurnal and usually terrestrial, favoring habitats with plenty of stumps and fallen logs that offer a rich supply of insects.

Broad-headed Skink

Plestiodon laticeps L 6.5–12.75 in (16.5–32 cm)

The Broad-headed Skink male's large and distinctive reddish orange head in breeding season contributes to the species' common nickname "red-headed scorpion."

KEY FACTS

Body is brown with 5 light stripes; head is broad; tail is blue to brown.

+ **habitat:** Moist woodlands and open areas with debris cover

+ **range:** Southeastern Pennsylvania to central Florida and east to Kansas and eastern Texas

+ **food:** Insects, other invertebrates, berries, and fruit

As with other five-lined skink species, the juvenile Broad-headed's bright markings and colorful blue tail will disappear with age, most thoroughly in the adult male. The Broad-headed Skink differs from the Common Five-lined Skink in the wider width of its head. Scales above the mouth, behind the eye, and under the tail also show diagnostic differences. The Broad-headed Skink is diurnal and semi-arboreal, looking for insects and larvae high in trees and making use of tree cavities. It seeks shelter in leafy woodland substrate and even within human debris. The female lays 6 to 16 eggs in cavities and tends to them in characteristic skink fashion.

Great Plains Skink
Plestiodon obsoletus L 6.5–13.25 in (16.5–34 cm)

Although it is the continent's largest skink, the Great Plains Skink can be elusive. This diurnal species spends a lot of time under rocks and will inflict a painful bite on an interfering hand.

KEY FACTS

Body is beige or gray with yellowish sides; scales dark edged; scales between limbs angle upward.

+ habitat: Grasslands, woodlands, canyons, and deserts with rock deposits

+ range: Great Plains from Nebraska through Texas and west to central Arizona

+ food: Insects, spiders, and other invertebrates

The Great Plains Skink is rarely found far from its rock shelter. When it goes out to forage, it usually keeps the excursion brief. It also digs its own burrow and takes over burrows of other species. At breeding time in April and May, the male shows orange patches on the sides of the head. The female lays large numbers of eggs, up to two dozen per clutch. Like other skinks, the mothers tend the eggs until they hatch in about 40 days, adjusting the humidity by turning the soil. They also look after the hatchlings for several weeks. Juveniles have black bodies with white and orange dots along their lips and a bright blue tail.

Western Skink
Plestiodon skiltonianus L 6.5–9.25 in (16.5–23 cm)

These four-lined skinks lack the light medial line down the back that five-liners show; the line is replaced with a broad brown line. Western Skinks show a faded pattern as they age.

KEY FACTS

Shiny body has a brown central line edged in black and four light ones; tail is blue or gray.

+ habitat: Grasslands, shrublands, forests, and open woods with rocky areas

+ range: Southern British Columbia to California and northern Arizona

+ food: Insects, spiders, earthworms, and other invertebrates

As in other skink species, a breeding male Western Skink sports orangish red around the mouth and chin, but it doesn't stop with that. It also often displays the reddish coloring on its sides, and on the tip and underside of its tail. Typical of many *Plestiodon* species, Western Skink young are brightly marked with striking blue tails. When a predator grabs a tail, it breaks off cleanly and wiggles independently to distract the pursuer while the animal scurries away. This shy, secretive species really keeps a low profile and is sighted most often from the bright flash of blue tail in a younger individual.

Little Brown Skink

Scincella lateralis L 3–5.5 in (8–14 cm)

Also known as the Ground Skink, the Little Brown Skink lives up to both of its straightforward common names. The small, agile species wriggles laterally like a snake when it runs.

KEY FACTS

Body is brown with shiny, overlapping scales and a dark line on the sides; lower eyelid has a transparent disk.

+ **habitat:** Moist forests, forested grasslands, and urban gardens

+ **range:** New Jersey to Florida and west to central Texas and Nebraska

+ **food:** Insects, spiders, and other invertebrates

The clear "window" in the lower eyelids of the Little Brown Skink offers protection from irritants and allows the lizard to see when its eyes are closed. The species prefers habitats with deep leaf litter that provides insects and shelter from predators. In some parts of the range, this skink observes an eight-month breeding season; a female may lay up to five clutches of one to seven eggs on a nearly monthly basis between April and August. Unlike the *Plestiodon* skinks, the female Little Brown does not tend eggs in the nest; the eggs incubate and hatch without supervision. All skinks in the genus *Scincella* are terrestrial.

Florida Wormlizard

Rhineura floridana L 7–11 in (18–28 cm)

Looking like a giant pinkish purple earthworm, the Florida Wormlizard is an amphisbaenian, a group of reptiles that are mostly limbless and adapted to living underground.

KEY FACTS

Body is pink or fuchsia with scales formed in rings; the ears and eyes are not visible; there are no limbs; the tail is flat and bumpy.

+ **habitat:** Dry, sandy soil

+ **range:** Northern and central Florida

+ **food:** Earthworms, termites, and spiders

On closer inspection, the wormlizard reveals that—unlike an earthworm—it has a defined head and tail. The head is wedge shaped for efficient burrowing and lacks external eyes and ears; the short tail is flattened and somewhat tapered. Like the earthworm, the wormlizard is often flushed out after a heavy rain or when soil is overturned, at times fooling birds in search of tasty worms, which may receive a nasty bite instead. Not much is known about the Florida Wormlizard's breeding habits; the female lays one to three eggs in summer that hatch a few months later. This species is the only amphisbaenian present in the United States.

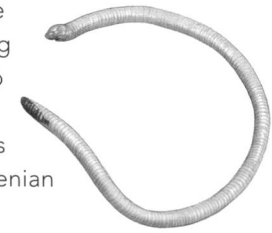

Western Threadsnake

Rena humilis L 7–16 in (18–40 cm)

The Western Threadsnake is a species of blind snake, a primitive group of burrowing snakes that cannot see, but have vestigial eyes that are visible as black dots beneath ocular scales.

KEY FACTS

The cylindrical body is brown, pinkish, or purplish with smooth, shiny scales; tail tip has small spine.

+ **habitat:** Deserts, grasslands, and canyons

+ **range:** Southeastern Utah, southern Nevada and California to southern Arizona and New Mexico to western Texas

+ **food:** Termites and ants

Also known as the Slender Blind Snake or Western Blind Snake and formerly placed in the genus *Leptotyphlops,* the Western Thread-snake is an elusive presence in the American Southwest. It lives underground or beneath rocks, emerging at night in search of termites and ants and their young. Its streamlined body and countersunk jaw allow it to burrow rapidly into the sandy and gravelly soil of its preferred habitats. After mating in the spring, the female lays a clutch of two to six slender eggs, sometimes in a communal nest, and tends the eggs during incubation. This species occurs at elevations up to 5,000 feet (1,500 m).

Rubber Boa

Charina bottae L 14–33 in (36–84 cm)

The appearance of the Rubber Boa strongly favors that of a fake rubber snake. Only two boas, a primitive group of snakes, are native to the United States and Canada.

KEY FACTS

Body is uniformly olive to tan to brown with small scales; tiny eyes have vertical pupils.

+ **habitat:** Damp woodlands, coniferous forests, chaparral, and grasslands

+ **range:** British Columbia to southern California, east to Utah, Wyoming, and Montana

+ **food:** Small mammals, birds, lizards, and salamanders

The top of the Rubber Boa's head is covered with a number of large scales. Males show small spurs behind the vent, or anal opening, which represent vestigial hind limbs. This snake is active at night and is an adept burrower, climber, and swimmer. In true boa fashion, it constricts its prey. When threatened, a Rubber Boa often forms a tight protective ball with the blunt tail protruding, perhaps mimicking the head, which is carefully protected in the center of the coils. Often, a number of these snakes will rest together under a warm rock. Female Rubber Boas bear two to eight live young, which can live more than 40 years.

Boa Constrictor
Boa constrictor L 5–7 ft (1.5–2 m)

A popular snake in the pet trade, the introduced Boa Constrictor poses a threat to native wildlife in southern Florida. Though confined to a small area, it has the potential to expand its range.

KEY FACTS

Body is tan to light brown with a dark brown saddle pattern across back, tapering to triangles; pattern is darker and redder toward tail.

+ **habitat:** Pine rocklands, hardwood hammocks, limestone walls, and along canals

+ **range:** Miami area of southern Florida

+ **food:** Mammals, birds, lizards, and fish

A native of Central and South America, the Boa Constrictor is a species in which the scientific name has taken hold as the common name. It is widely regarded as a very attractive snake—a reason for its popularity as a pet; however, the exact origin of the southern Florida population is not known. The Boa Constrictor is capable of a swift defensive strike if disturbed. Its bite is not venomous, but it can be severe and require careful attention. This snake uses its strong jaws to restrain prey, and it kills the prey by constriction. The female bears about 20 to 30 live young that are more than a foot long at birth.

Burmese Python
Python bivittatus L 7–12 ft (2–3.6 m)

An even bigger threat to ecological balance than the Boa Constrictor, the introduced Burmese Python has found southern Florida very much to its liking and is reproducing prolifically.

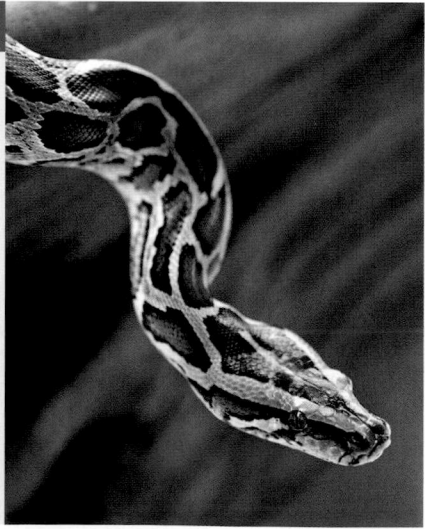

KEY FACTS

Large body is tan to brown; the back and sides have reddish brown blotches; a light V-shaped pattern extends backward from the eyes.

+ **habitat:** Swamps, marshes, canals, and forested and open areas

+ **range:** Southern Florida

+ **food:** Mammals, birds, and even alligators

For about 20 years, intentional releases and accidental escapes of pet Burmese Pythons have created a Florida population that shows no signs of decreasing, despite organized snake roundups and other eradication measures. This formidable Southeast Asian predator goes after many kinds of prey, including some endangered species, and it competes with native predators. A record-length female python was captured in the Miami area in 2013 that measured 18 feet, 8 inches (5.5 m). She carried no eggs, but the previous record holder held a whopping 87 eggs—an indication of the difficulty of eradicating this invasive species.

North American Racer

Coluber constrictor L 3–5 ft (0.9–1.5 m)

The North American Racer skims along the ground almost like a hovercraft, with its head held up. It often is viewed in rapid transit across roads or through the grass.

KEY FACTS

Body is smooth, satiny black, brown, blue, or greenish; may be white, yellow, or gray below.

+ habitat: Fields, grasslands, brush, meadows, open woodlands, and rocky hillsides

+ range: Much of central and eastern U.S. and southernmost Ontario

+ food: Insects, lizards, snakes, birds, and small rodents

When the North American Racer feels threatened, it vibrates its slender, pointed tail, creating a rattler-like buzzing sound in dry leaf litter. It also can escape upward into bushes or trees. If grabbed, it will flail back and forth and bite repeatedly to defend itself. This snake bears the scientific name *constrictor*, but it does not constrict to kill its prey. Instead, the diurnal species subdues prey by pinning it down with a loop of its body, then swallowing it whole. Females may lay up to two dozen granular eggs, often in communal nests. These snakes may also hibernate communally in the company of other species.

Coachwhip

Masticophis flagellum L 3.5–5 ft (1–1.5 m)

A long tail resembling a braided whip distinguishes Coachwhips and related snakes. Contrary to folk belief, this species does not pursue a prey animal and whip it to death.

KEY FACTS

In West, body is tan to brown to yellow; In East, dark brown to black near the head and lighter to the rear; long, thin tail.

+ habitat: Grasslands, sandy pastures, scrublands, and open woodlands

+ range: Southern half of the U.S.

+ food: Frogs, lizards, snakes, baby turtles, and birds

Slender and sleek, the Coachwhip is one of North America's fastest snakes. Two-toned coloration in the eastern snake is distributed in different proportions among individuals; there is also a black form with a reddish tinge near the tail and a pink southwestern form. Coachwhips are often seen on the ground but also climb trees. They scout prey by raising their head above surrounding vegetation as they travel. This nonvenomous snake may advance on a subject menacingly; the bite does not puncture the skin, but tears it. It may also feign death. The female Coachwhip lays 5 to 12 grainy eggs under debris or in mammal burrows.

California Kingsnake
Lampropeltis californiae L 30–41 in (75–105 cm)

Other snake species form a portion of the California Kingsnake's menu, including venomous species such as rattlers. The kingsnake is resistant to the venom's effects.

KEY FACTS

Body is black or brown with white or yellow crossbands or a white striped mid-back; face is mostly white.

+ **habitat:** Chaparral, grasslands, woodlands, and marshes

+ **range:** Southern Oregon to southern California, east to western Arizona

+ **food:** Small mammals, lizards, snakes, amphibians, birds, and eggs

Despite the hissing and tail-rattling defense that the California Kingsnake mounts when threatened, it is a tameable snake and has become a popular species in the pet trade. It has been selectively bred to produce varied colors and patterns, including an albino version. Escaped or released pet California Kingsnakes are sometimes spotted in the eastern and central United States, but they are not known to have established breeding populations there. These snakes are active during the day in cooler weather or at dawn, dusk, and during the night in hot weather. This kingsnake kills prey with powerful constrictions.

Yellow-bellied Kingsnake
Lampropeltis calligaster L 30–41 in (75–105 cm)

Storms and low-pressure systems often bring the otherwise secretive Yellow-bellied Kingsnake out into the open. It frequently is seen crossing roads after rains.

KEY FACTS

Body is brownish gray to tan with brownish, black-edged back markings and smaller side markings; blotched yellowish belly.

+ **habitat:** Grasslands, open woodlands, and savannah

+ **range:** Maryland to northern Florida, west to eastern Nebraska and Texas

+ **food:** Rodents, birds, lizards, frogs, and other snakes

The Yellow-bellied Kingsnake spends much of its time burrowed in loose soil or under rocks. In the open, it is well camouflaged by its pattern, which in older individuals may become obscure. This species' coloration sometimes causes it to be mistaken for a Copperhead and killed. Like all kingsnakes, the Yellow-bellied is oviparous. The female lays between 3 and 21 eggs in June or July in an underground cavity; the eggs hatch in about 7 to 11 weeks. There are three Yellow-bellied subspecies: the Prairie Kingsnake, the Mole Kingsnake, and the South Florida Mole Kingsnake, which is found in disjunct colonies in central Florida.

Eastern Kingsnake

Lampropeltis getula L 36–47 in (90–120 cm)

The Eastern Kingsnake shows regularly spaced white crossbands with irregular borders that give the appearance of chain wrapped around the body. It is also known as the Chain Kingsnake.

KEY FACTS

Body is black or dark brown with white, cream, or yellow crossbands forming a chain pattern.

+ **habitat:** Pine woods, swamps, marshes, stream banks, fields, old houses, and barns

+ **range:** New Jersey to northern Florida, west to Appalachians and southern Alabama

+ **food:** Rodents, lizards, birds, eggs, and snakes

The smooth and shiny Eastern Kingsnake roams by day in a variety of habitats, often with more intense activity at dawn and dusk. It switches to nocturnal ramblings in hot weather, spending the day underground. Like other kingsnakes, it includes venomous snakes such as rattlers, Copperheads, and coralsnakes in its diet without mishap, because it is resistant to the effects of their venom. It constricts powerfully to overcome its prey. Though secretive, it sometimes basks in the open and is a strong swimmer. The pattern on the Eastern Kingsnake varies from bold and distinctly chainlike to faint and irregular.

Milksnake

Lampropeltis triangulum L 24–35 in (60–90 cm)

Milksnakes get their common name from the far-fetched notion that they milk cows and deplete them of their milk. They are one of the most widespread North American snake species.

KEY FACTS

Body is gray or tan, with black bordered brown blotches or red/orange, black, and yellow/white rings.

+ **habitat:** Coastal areas, woods, fields, grasslands, farmland, and suburban areas

+ **range:** Southern Ontario and Quebec and most of U.S. east of the Rockies

+ **food:** Rodents, birds, lizards, and other snakes

The Milksnake is exceptionally skilled at getting around without being noticed, though it pursues the mice it favors around barnyards and abandoned buildings. This snake hides successfully, singly or in groups, under natural or man-made debris. Varieties of this species resemble Copperheads and coralsnakes. The Ringed Milksnake clearly shows the separation of the "caution colors" red and yellow (seen together in coralsnakes) by an intervening black ring. Like closely related kingsnakes, the Milksnake does eat venomous snakes. Some experts consider some of the currently recognized subspecies of Milksnake to be full species.

Smooth Greensnake

Opheodrys vernalis L 12–20 in (30–50 cm)

The Female Smooth Greensnake incubates her fertilized eggs internally for a varied amount of time. As a result, they may hatch in as few as four days after being laid.

KEY FACTS

Smooth body is grass green; very slender tail tapers; underside is white to yellowish.

+ **habitat:** Meadows, fields, fencerows, and grassy marshes

+ **range:** Extreme southern and eastern Canada and north-central and northeastern U.S.; scattered populations elsewhere

+ **food:** Insects and spiders

This attractive, leaf-green snake is nonvenomous and harmless. It prefers a grassy habitat with consistent moisture. If grabbed, it may put up a bit of a show by gaping, but it seldom bites. After death, this snake immediately turns a dull blue. This communal species may overwinter in large groups, and females may share an egg-laying site. Its southern counterpart, the Rough Greensnake (*O. aestivus*), is similar but longer, its scales are keeled, and it is more of a climber, earning the nickname "vine snake." It also periscopes to search for prey, lifting its head well above the vegetation to get a visual read on arthropod activity.

Eastern Ratsnake

Pantherophis alleghaniensis L 3.5–6 ft (1–1.8 m)

Formerly known as Black Ratsnakes, these snakes have a squarish body that has been compared to a loaf of bread, a feature that helps differentiate them from the round North American Racers.

KEY FACTS

Body is black above; throat is white; underside has checkerboard pattern.

+ **habitat:** Hardwood forests, canyons, swamps, farmland, fields, and barnyards

+ **range:** Southern New England and Ontario to Georgia, west to Louisiana, Oklahoma, and Wisconsin

+ **food:** Birds, eggs, salamanders, lizards, and rodents

One of the most common and largest snakes in the eastern United States, the Eastern Ratsnake is a powerful constrictor that often lies in wait along the beaten trails used by small rodents. It is also an adept climber, and it appropriates high tree hollows or drapes itself from a branch above a trail (often taking hikers by surprise). It frequently is found in abandoned buildings, where it cruises for mice and nesting birds. Females lay 5 to 30 oblong eggs in rotting logs, leaf litter, or under rocks; these hatch two to four months later. It often hibernates in groups containing individuals from other snake species, including Copperheads.

Red Cornsnake
Pantherophis guttatus L 30–47 in (75–120 cm)

Also known as the Red Ratsnake, the cornsnake ranks as one of the handsomest eastern species for its highly artistic markings, especially the elegant spearpoint pattern on top of its head.

KEY FACTS

Body is orange or brownish or grayish with black-edged reddish, brown, or gray blotches on back; underside has checkerboard pattern.

+ **habitat:** Woodlands, meadows, prairies, barnyards, and abandoned buildings

+ **range:** Southern New Jersey to Florida, west to Louisiana

+ **food:** Rodents, birds, and lizards

The Red Cornsnake keeps itself scarce, compared with its close relative the Eastern Ratsnake. It spends a lot of time underground in rodent burrows and other subterranean chambers, but it can also climb trees and building rafters. It prowls grain, hay, and similar storage buildings in search of rodents. The Red Cornsnake may adopt various defensive postures, including elevating a segment of its coils to appear significantly bigger to a predator. The Great Plains Ratsnake (*P. emoryi*), previously considered a subspecies of cornsnake, has a drabber pattern and mainly occurs in the southern plains and Texas.

Pinesnake
Pituophis melanoleucus L 4–5.6 ft (1.2–1.7 m)

A loud hiss, courtesy of modified vocal cords, a tail that vibrates loudly, and a flattened head indicate when a Pinesnake is displeased. Its bite is nonvenomous.

KEY FACTS

Body color is white, gray, yellowish, tan, brown, or rusty; pattern varies; scales are keeled: snout has enlarged scale.

+ **habitat:** Pine and pine-oak woods, sandy areas, and fields

+ **range:** Disjunct populations in mid-Atlantic U.S.; more continuous in Southeast

+ **food:** Mainly rodents

The appearance of this large constrictor varies considerably depending upon location. The mid-Atlantic Pinesnake is usually white with black blotches; Florida snakes are gray to brown with muddled blotches; and those in Louisiana have indistinct brown blotches near the head and sharp reddish ones near the tail. The Pinesnake spends a lot of time under the ground in burrows that it digs in loose soil or in abandoned rodent burrows. A female Pinesnake excavates a side chamber in the burrow as a nest for the 3 to 24 eggs that she will lay. This nesting chamber often is used year after year by a number of different females.

Southeastern Crowned Snake
Tantilla coronata L 8–10 in (20–25 cm)

The distinctive black crown serves as the main key to identification of the Southeastern Crowned Snake. So small that it is often overlooked, this snake is found even in urban backyards.

KEY FACTS

Body is tan to reddish tan; underside is whitish; has black head cap and chin; light collar and black neck blotch.

+ habitat: Pine woods, oak-hickory forest, scrubland, and grasslands

+ range: Southeastern U.S.; in Florida, only the panhandle

+ food: Earthworms, slugs, and insect larvae

The Southeastern Crowned Snake occupies a variety of habitats that differ significantly in moisture content. It leads a hidden life, feeding on subterranean prey and taking cover under rotting logs, leaf litter, and piles of debris. Its body color often blends with local soil coloration. It discharges a mild venom when dispatching prey, but the venom is not harmful to humans, and this shy, reticent snake hardly ever bites, even when provoked. A number of closely related species range across the southern United States. They can be distinguished from the Southeastern Crowned Snake largely by differences in their head patterns.

Common Wormsnake
Carhophis amoenus L 7.5–11 in (19–28 cm)

Unsurprisingly by its name, the slender, cylindrical Common Wormsnake looks much like a worm. It also counts the earthworm among its favorite food groups.

KEY FACTS

Body is brown, gray, or black with 13 rows of smooth scales; underside pink; head narrow with tiny eyes; sharp point on tail.

+ habitat: Fields, meadows, woodlands, farmland, and vacant lots

+ range: Southern New England to Georgia, west to the Mississippi River

+ food: Earthworms and soft-bodied insects

The Common Wormsnake prefers a moist habitat. Shy by nature, it will hide beneath rocks, rotting logs, boards, and debris. In cold or dry weather the species burrows, something it is well suited to do with its tapered head and smooth body. Like the earthworm, it may turn up when soil is turned over. When grasped the wormsnake resists, often releasing copious amounts of musk and feces to repel its captor. This species has both a spring and fall mating season. Females lay one to eight thin-shelled eggs that hatch in about seven weeks. The young, about 4 inches (10 cm) long, look even more like earthworms than the adults.

Ring-necked Snake

Diadophis punctatus L 10–30 in (25–75 cm)

The Ring-necked is abundant in much of the United States, but it is so secretive that it is not often spotted. The signature neck rings may be faint or missing in some subspecies.

KEY FACTS

Body is gray, olive, brown, or black above; red, orange, or yellow below with black spots; orange, yellow, or cream neck ring.

+ habitat: Forests, grasslands, hillsides, and upland deserts

+ range: Nova Scotia to Florida to west, except Northwest

+ food: Worms, slugs, salamanders, lizards, and young snakes

The red-bellied variety of the Ring-necked Snake is known to coil tightly when threatened, elevating its tail, exposing the bright color while carefully hiding its head. This may serve as a warning that it is unpalatable, although other snake species, such as coralsnakes, have no trouble polishing off a Ring-necked meal. When grasped, the Ring-necked defends itself vigorously, unloading urine, feces, and musk on an attacker. In many subspecies, the colorful undersides bear few or many black dots. Some experts consider some of the *D. punctatus* subspecies to be separate species, but other experts do not fully support this claim.

Red-bellied Mudsnake

Farancia abacura L 3.3–4.6 ft (1–1.4 m)

Shy and secretive mudsnakes often seclude themselves in aquatic vegetation or bury themselves in muddy bottoms, belying the abundance of this common snake.

KEY FACTS

Back is glossy black or gray; belly is red or pink that extends onto sides; black spots on pale chin and neck.

+ habitat: Swamps, lake edges, sluggish streams, and floodplains

+ range: Southern Virginia to Florida, west to Texas, north through Mississippi Valley

+ food: Salamanders and fish

At home in many lowland aquatic habitats, the Red-bellied Mudsnake captures slimy prey such as salamanders, which themselves commonly take young snakes. The mudsnake is otherwise fairly benign; even the sharp spine at the end of its tail, used to gain traction in slick mud, cannot cause much damage. The species hunts at night and often is seen crossing roads in the rain. Mating takes place in the spring. The female deposits a dozen to more than a hundred eggs in a hole in the soil. She may guard them for several months until they hatch. Young often stay close to the roots of aquatic vegetation.

Plains Hog-nosed Snake

Heterodon nasicus L 15–25 in (38–63 cm)

An upturned "hog snout" is characteristic of this group of snakes, which also is noted for over-the-top defensive behaviors. The Plains species is also known as the Western Hog-nosed Snake.

KEY FACTS

Body is tan, brown, or gray above with back blotches and side spots; large black blotches below.

+ habitat: Sandy and gravelly areas in scrub and grasslands

+ range: Plains from southern Canada to Arizona, New Mexico, and Texas; isolated populations elsewhere

+ food: Amphibians, rodents, lizards, and insects

The Plains Hog-nosed Snake is a bit less demonstrative in its defensive actions than the Eastern species and others in the group. It relies less on the puffing up and head-flattening routine and more on thanatosis, or playing dead. The Plains Hog-nosed will coil belly-up with head flung back, open-mouthed, and immobile. If righted, it will flop back into dead mode. This snake is active mainly during the morning and late afternoon, and burrows in loose soil to escape extremes in temperature. Like other hog-noses, it produces venom that immobilizes prey, but may produce only a mild reaction in humans, though it rarely bites.

Eastern Hog-nosed Snake

Heterodon platirhinos L 20–33 in (50–84 cm)

The Eastern Hog-nosed Snake resorts to an assortment of dramatic fake-outs—some taken from the playbooks of other snake species—to deter potential predators.

KEY FACTS

Color is variable, brown to gray with dark back blotches and side spots; turned-up snout; underside mottled.

+ habitat: Open woodlands, wooded hillsides, fields, and meadows with sandy soil

+ range: Southern New England to Florida to Texas, north to South Dakota

+ food: Toads and other frogs

A signature defense of the Eastern Hog-nosed Snake involves inflating the body with air, flattening the head and neck into a kind of cobra hood, hissing vociferously, and striking. This display has earned it the epithet "puff adder" for its resemblance to an African viper. If that routine doesn't work, it may segue into a play-dead posture—belly up, head back, and mouth open. This snake uses its mild venom to immobilize the toads that are its preferred prey. The Eastern species' snout is less sharply upturned than that of other hog-noses. A black form occurs when melanin obscures an individual's normal color and pattern.

Southern Watersnake
Nerodia fasciata L 22–41 in (56–105 cm)

Fear of the highly venomous Cottonmouths has made life precarious for the nonvenomous aquatic snakes, such as the Southern Watersnake, which often is killed as a matter of routine.

KEY FACTS

Body is gray, tan, olive, or black, often with crossbands, light lateral stripes, or no pattern; may have dark line from eye to jaw.

+ habitat: Freshwater and brackish habitats

+ range: Coastal plain from North Carolina to Florida to Texas and Mississippi Valley

+ food: Frogs and fish

Warm evening rains in the spring can bring the Southern Watersnake out of the water in numbers in search of potential mates. Like other *Nerodia* species (and aquatic snakes in general), the Southern Watersnake female bears live young, from a few to more than 50 at a time, in summer. This species can be distinguished from the Common Watersnake by the dark line angling down to the jaw from behind its eye, which the Common Watersnake lacks. In the Mississippi Valley, a broad-banded subspecies with fewer, wider bands separated by prominent yellow or red shapes carries the nicknames "yellow moccasin" and "pink flamingo snake."

Common Watersnake
Nerodia sipedon L 22–41 in (56–105 cm)

Also known as the Northern Watersnake, the Common Watersnake has long teeth that are adapted for holding fish. Its bite may introduce saliva with a mild anticoagulant.

KEY FACTS

Body is brown, red, gray, or blackish; dark bands at neck and blotches on back and sides farther back.

+ habitat: Freshwater and brackish habitats

+ range: Southern Ontario and Quebec to North Carolina, west to Mississippi and Colorado

+ food: Fish, amphibians, turtles, crustaceans, and small mammals

The Common Watersnake is often found sunning on banks and logs, or leisurely hanging from trees above water. It may quickly retreat to the water when disturbed. As this snake ages, its pattern darkens, becoming obscure. Frequently covered with mud, it is difficult to distinguish from the Southern Watersnake. It is often mistaken for the venomous Cottonmouth, or Water Moccasin, and killed as a result. But the Common Watersnake does not have the Cottonmouth's distinctive head shape, nor does it gape when disturbed. It will flatten its body, discharge lots of musk, and may defecate instead. Always use caution around watersnakes.

Graham's Crayfish Snake
Regina grahamii L 18–28 in (45–71 cm)

Crayfish snakes prefer their favorite prey to be freshly molted, with new and softer exoskeletons. They live where the small crustaceans, also known as crawdads, are abundant.

KEY FACTS

Body is olive or brown, with a yellow stripe along side and narrow black stripe below that; may have pale dorsal stripe.

+ **habitat:** Ponds, streams, bayous, sloughs, and ditches

+ **range:** Illinois and Iowa to Texas and Louisiana; northern Mississippi

+ **food:** Crayfish, other crustaceans, amphibians, and fish

The Graham's Crayfish Snake not only eats crayfish, it sometimes takes cover in their mud-dribbled burrows, known as chimneys. This largest crayfish snake also shelters under rocks and debris and may bask on banks and trees hanging low over the water. These secretive aquatic snakes usually forage at night during the summer heat. Females bear between 6 and 39 young in late summer. A form of this species in Iowa is brown above with obscured pattern details. The similar Queen Snake (*R. septemvittata*) has a narrower yellow stripe and bold stripes on its belly, compared with the mostly plain yellow underside of Graham's.

Dekay's Brownsnake
Storeria dekayi L 9–13 in (23–33 cm)

A small, elegant snake with a small-scale pattern, Dekay's Brownsnake used to be far more abundant, especially in urban settings, than it is now. It is closely related to the Red-bellied Snake.

KEY FACTS

Body is brown, often with reddish, yellowish, or olive tones; back has central light stripe with two rows of small dark spots.

+ **habitat:** Bogs, swamps, marshes, moist woods, and golf courses

+ **range:** Southern Minnesota, Quebec, and Maine south to Texas

+ **food:** Earthworms, slugs, snails, and insects

The Dekay's Brownsnake may seem to be far less abundant than it actually is, because it is very adept at hiding. If encountered and grabbed, it may musk, but it seldom bites. The paired lines of brown spots on the snake's back may nearly disappear toward the tail; the head may also be dark. The female bears 3 to 41 live young, which often have a thin, light ring around the neck. This species is named for James Ellsworth DeKay, a 19th-century New York doctor turned naturalist who collected the first specimen of the species. The genus is named after another doctor-naturalist, David Humphreys Storer of New England.

Red-bellied Snake

Storeria occipitomaculata L 8–10 in (20–25 cm)

The Red-bellied Snake has its own take on the threat-response gape seen in other snakes:
Instead of gaping, it curls its upper lip on one or both sides when disturbed.

KEY FACTS

Body is brown, gray, or black with a light dorsal stripe or 4 narrow brown stripes, or both.

+ **habitat:** Wooded hills and mountains and sphagnum bogs

+ **range:** Extreme southeastern Saskatchewan eastward and south through eastern U.S., except southern Florida

+ **food:** Earthworms, slugs, and insects

The Red-bellied Snake strongly resembles the closely related Dekay's Brownsnake—until it displays its colorful underside. This feature isn't universal within the species; some Red-bellieds have a gray or blue-black belly. The Red-belly also may have three light spots on the neck, which may be fused into a wide collar. This small snake prefers moist habitats and is more likely to be found at higher elevations than brownsnakes. The Red-bellied Snake usually confines itself to small areas where its moisture-loving prey are abundant and where it is difficult to spot, because of its secretive nature. Females give birth to 2 to 20 live young in summer.

Checkered Gartersnake

Thamnophis marcianus L 18–24 in (46–61 cm)

Gartersnakes get their name from the patterned garters that men once commonly wore to hold up their socks. As its name suggests, the Checkered Gartersnake has checkered sides.

KEY FACTS

Body is tan, olive, or brown, with 3 light stripes along back and bold checkerboard pattern on sides.

+ **habitat:** Arid and semiarid grasslands and sandy habitats, near a water source

+ **range:** Southeastern California, southern Arizona, and New Mexico to Texas, north to Kansas

+ **food:** Frogs, fish, and crayfish

The checkered pattern of the Checkered Gartersnake may distract from the lateral stripes it also wears as a gartersnake and may obscure them. Despite occupation of arid and semiarid lands, the species depends on water sources, such as streams, ponds, or irrigation canals, to provide the frogs and other aquatic life it eats. Many individuals forage on warm, moonless nights and travel along roadsides. Females bear smaller numbers of live young—usually between 6 and 18—than other gartersnake species. Gartersnakes and ribbonsnakes are closely related to watersnakes; in areas where the last are absent, the others are more aquatic.

Western Ribbonsnake

Thamnophis proximus L 20–30 in (50–75 cm)

Ribbonsnakes are basically slimmer-bodied gartersnakes with more defined patterns and, in some locations, more aquatic habits. Western Ribbonsnakes have a distinctive orange dorsal stripe.

KEY FACTS

Body is black, with central orange stripe on back, lighter side stripes; underside is unmarked.

+ habitat: Edges of lakes, marshes, ponds, ditches, cattle tanks, and streams

+ range: Southern Wisconsin and Indiana through Mississippi Valley to Texas and Louisiana

+ food: Frogs, tadpoles, and small fish

During rainy weather, the Western Ribbonsnake will sometimes leave its usual haunts and slither about in search of amphibian prey. In years of drought, though, this snake is much harder to find. When threatened, this species often takes to the water, where it swims on the surface to a safe haven of vegetation rather than submerging, as a watersnake might. The Western species is similar to the Eastern Ribbonsnake *(T. sauritis)*, but it often has two colorful spots on the top of the head that touch, whereas they don't touch in the Eastern Ribbonsnake. This species is also known as the Orange-striped Ribbonsnake.

Common Gartersnake

Thamnophis sirtalis L 18–26 in (45–66 cm)

The Common Gartersnake is literally here, there, and everywhere. It has the widest distribution of any snake species in the United States and Canada and can tolerate cold climates.

KEY FACTS

Body is olive, brown, or black; light dorsal stripe and 2 side stripes; often specks or checkering on sides.

+ habitat: Meadows, marshes, ditches, damp woods, and urban areas, among others

+ range: Southern Canada and all of U.S., except desert Southwest

+ food: Amphibians, worms, small fish, and mice

There are many subspecies and varieties of the Common Gartersnake in North America; appearance-wise, most are variations on the general theme of back and side stripes and side checkering, within a wide range of coloration. An all-black melanistic form also occurs. Behavior-wise, they are also similar. They occupy a large variety of habitats. Many hibernate communally in dens of hundreds of snakes. Females of this live-bearing species produce between 7 and 85 young a year. Gartersnakes can be feisty when threatened—flattening their body, discharging musk, and striking—but they usually lose their combativeness over time.

Lined Snake

Tropidoclonion lineatum L 8.5–15 in (22–38 in)

Though it belongs to a different genus, the Lined Snake shares characteristics with its gartersnake relatives, yet it is often smaller, and displays underneath two rows of half-moon shapes.

KEY FACTS

Body is grayish or olive brown; back has 3 yellowish stripes, with middle stripe most prominent.

+ **habitat:** Prairies, grasslands, woodland edges, and suburban and urban areas

+ **range:** Disjunct populations from South Dakota to Texas, west to New Mexico, east to Illinois

+ **food:** Mainly earthworms

The Lined Snake is so secretive that the known populations may not be as disconnected as they seem. Undiscovered populations may connect some of the disparate population into a more continuous range. At the same time, this is a very adaptable species, comfortable in suburbs and cities, where it can be found in cemeteries, backyards, and trash dumps, and under debris. These snakes come out in numbers in the evenings after warm rains to take advantage of readily accessible earthworm prey. Females give birth to between 2 and 12 young in summer. The Lined Snake may discharge musk and defecate in defense, but it seldom bites.

Smooth Earthsnake

Virginia valeriae L 7–10 in (18–25 cm)

Essentially a snake in a plain brown (or gray) wrapper, the Smooth Earthsnake lives a secretive life that goes along with its nondescript looks. The various earthsnake species look much alike.

KEY FACTS

Body is gray or reddish brown; often has faint dorsal stripe and line from eye to nostril.

+ **habitat:** Deciduous forests, timbered slopes, and grassy edges of streams and roads

+ **range:** New Jersey to northern Florida, west to eastern Kansas, Oklahoma, and Texas

+ **food:** Earthworms and soft insects

The small Smooth Earthsnake spends much time underground; it burrows with its pointed snout, going deeper in search of moisture in the face of drought. Subterranean life makes the earthworm its primary source of food. The earthsnake has no need to poison or constrict its pray; it swallows the worms alive. Rains may flush these snakes—and their prey—to the surface, where the snakes may take shelter under rocks or natural or human-made debris. When encountered, the Smooth Earthsnake is a benign presence, rarely attempting to bite, although it may discharge musk, defecate, or even play dead when stressed.

Harlequin Coralsnake
Micrurus fulvius L 20–30 in (50–75 cm)

Like all coralsnakes, the Harlequin announces its venomous status by adjacent rings with "caution" colors of red and yellow. In similar harmless snakes, black rings separate red from yellow.

KEY FACTS

Body has wide red and black rings separated by narrow yellow rings; snout is black.

+ **habitat:** Sandy areas, pine woods, pond and lake edges, and hardwood hammocks

+ **range:** North Carolina to southern Florida and along Gulf Coast to eastern Louisiana; also central Alabama

+ **food:** Small snakes and lizards

Despite its colorful "candy-stick" pattern, the Harlequin Coralsnake is extremely adept at keeping itself concealed. It doesn't require much in the way of cover—trash, leaves, bent grasses, boards, and fallen logs will do the trick—to stay out of sight. The species does not go looking for human encounters but will bite if disturbed; a bite requires immediate medical attention. Coralsnakes are the only venomous snakes in the United States and Canada that lay eggs; other venomous species, such as Copperheads and rattlers, give birth to live young. The Harlequin was formerly known as the Eastern Coralsnake.

Copperhead
Agkistrodon contortrix L 24–35 in (60–90 cm)

The Copperhead's triangular copper head is distinctive, and so are the hourglass shapes that span its back. It is a pit viper, a venomous snake with a heat-sensing pit between nostril and eye.

KEY FACTS

Body is tan or brown, with dark hourglass bands on the back; head distinct from body.

+ **habitat:** Rocky, wooded hillsides, swamp edges, and mountainous areas

+ **range:** Southern New England to northern Florida, west to Texas, north to southeastern Nebraska

+ **food:** Rodents, lizards, frogs, and large insects

This patient snake may wait motionless for prey in the same spot for days on end. Its heat-sensing pits provide pinpoint accuracy in detecting prey, even in the dark. In ambush mode, these snakes inadvertently may be disturbed by humans. A threatened Copperhead may vibrate its tail, making a whirring sound in dry vegetation. If it strikes, it may give a "dry," or venomless, bite or may inject a variable amount of venom from front fangs. Any bite requires medical attention; even dry ones can introduce bacteria. Young Copperheads have thin yellow or green tips on the tail that may serve as lures for lizards and amphibians.

Cottonmouth

Agkistrodon piscivorus L 30–47 in (75-120 cm)

A close relative of the Copperhead, the Cottonmouth (or Water Moccasin) is noted for the cotton-white interior of its mouth, which it shows off in a defensive display called "gaping."

KEY FACTS

Body is olive, brown, or black with no pattern or dark, serrated-edged crossbands; flat-topped head.

+ **habitat:** Swamps, lakes, rivers, canals, ditches, and rice fields

+ **range:** Southeastern Virginia to Florida Keys, west to central Texas, north to Illinois

+ **food:** Amphibians, fish, snakes, and birds

Avoid contact with the Cottonmouth; its venom is very dangerous, and the risk of secondary infection from a bite is also a major consequence. Nonvenomous watersnakes (*Nerodia* species) tend to scurry or swim quickly away when disturbed. The Cottonmouth, on the other hand, may stay put and mount a challenge by coiling and gaping, or it may slowly slither away. It may also vibrate its tail, which watersnakes do not. It often swims with its head elevated out of the water. Like all pit vipers, the Cottonmouth bears live young; the female produces 1 to 15, which are patterned and have yellow-tipped tails much like Copperheads.

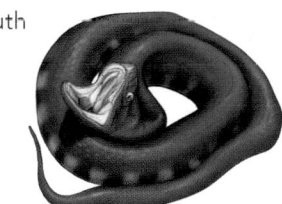

Eastern Diamond-backed Rattlesnake

Crotalus adamanteus L 33–71 in (84-180 cm)

North America's largest rattler, the Eastern Diamond-backed Rattlesnake is a snake to be avoided. Its venom is largely hemotoxic, destroying tissue and inhibiting blood coagulation.

KEY FACTS

Body is brown or black with large, dark diamonds bordered in cream or white scales; broad head.

+ **habitat:** Saw palmetto, pine and oak woods, and coastal areas

+ **range:** Southeastern North Carolina to Florida Keys and along Gulf to eastern Texas

+ **food:** Rodents, other small mammals, and birds

A rattlesnake's warning rattle is formed from attached hollow segments of keratin, which strike each other when the snake activates specialized muscles in its tail. The sound has been described as a buzzing, similar to the hum of cicadas. A young rattlesnake makes no noise because it has only a "training" rattle in the form of a terminal button on the tail; segments are added with each skin shedding until the snake reaches adult size. The Eastern Diamond-backed is often found in the vicinity of Gopher Tortoise burrows. Habitat loss and human persecution have reduced numbers of this species, especially in Florida.

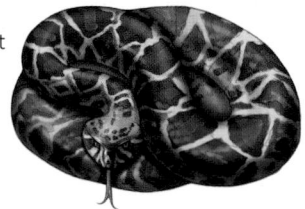

Western Diamond-backed Rattlesnake

Crotalus atrox L 30–91 in (75–230 cm)

The Western Diamond-backed is the typical Hollywood rattler, its head rising high above its coils in a face-off. Its tail buzzes an ominous warning, which alerts grazing animals to its presence.

KEY FACTS

Body is brown, gray, or pinkish above with brownish white-edged diamonds; large, distinct head.

+ **habitat:** Arid and semiarid areas with a variety of habitats, including roadsides and vacant lots

+ **range:** Southeastern California to central Arkansas southward

+ **food:** Mammals, lizards, birds, and amphibians

The Western Diamond-backed is active at dawn, dusk, and during the night in hot weather, but also can be found out and about during the day. It ranges from sea level to more than 8,000 feet (2,400 m). This large, aggressive western rattler has a high-volume venomous bite and is very dangerous, second only perhaps to the Mojave Rattlesnake, whose venom is largely neurotoxic. The feisty Western Diamond-backed is usually more than ready to stand its ground rather than retreat, and should never be approached. This species is sometimes called the coontail rattler due to the black and white or light gray bands on its tail.

Sidewinder

Crotalus cerastes L 17–33 in (43–84 cm)

The Sidewinder's signature S-shaped locomotion, which leaves a J-shaped pattern in the sand or dust, is a key to its identification. The species often is seen crossing roads at night.

KEY FACTS

Body is tan, pink, or gray with blotches down back; dark eye stripe; scales above eyes are hornlike.

+ **habitat:** Arid desert with scrub-topped sand hammocks, dunes, hardpan, and rocky slopes

+ **range:** Southern California to southwestern Utah southward

+ **food:** Small rodents, lizards, and birds

Sidewinding is a superb adaptation to life in the desert: It is a speedy way to get from one place to another, reducing slippage in fine sand and minimizing body contact with the scorching desert floor. Like many desert animals, the sidewinder is primarily nocturnal, and spends a large portion of the day in underground rodent burrows or perhaps coiled in a shallow depression at the base of a shrub. The Sidewinder's rattle and its erect supraorbital scales, or "horns," give it the nickname of "horned rattler." The horns seem to offer the snake's eyes a degree of protection when it crawls through underground burrows.

Timber Rattlesnake

Crotalus horridus L 36–60 in (90–152 cm)

Despite a species name of *horridus*, the Timber Rattlesnake is one of the calmer rattler species. It often makes no response to an encounter, but nevertheless should be given a wide berth.

KEY FACTS

Varied color and pattern: body is yellow, brown, gray, tan or pinkish; back with blotches or crossbands, may have dorsal stripe; tails dark.

+ habitat: Woodlands, rocky hillsides, swamps, and stream valleys

+ range: Southern Maine to northern Florida, west to Texas and Minnesota

+ food: Rodents and birds

The Timber Rattlesnake is a wait-and-ambush predator, staying quietly coiled in a location conducive to snagging a rodent passerby. It also gathers in groups to sun in rocky areas near hibernacula, or wintering dens, which it may share with other snakes including Copperheads. This is the only rattlesnake species in much of the Northeast, where it has undergone targeted elimination. It is more abundant in the South, where it is also known as the Canebrake Rattlesnake. Like many other rattlesnakes, Timbers often engage in "combat dances" in which two males sidle up to each other and wrestle repeatedly until one tires and leaves.

Prairie Rattlesnake

Crotalus viridis L 35–45 in (89–114 cm)

The Prairie Rattlesnake assumes the role of the Timber Rattlesnake in the Great Plains, but has a more aggressive demeanor and delivers a dangerously venomous bite.

KEY FACTS

Body is greenish or gray or brown; brown or black blotches, sometimes edged in white, extend down middle of back.

+ habitat: Grasslands, dunes, forests, and rocky areas

+ range: Great Plains (related species occur toward the west)

+ food: Small mammals and birds

Despite historically being persecuted throughout much of its range, the resilient Prairie Rattlesnake still occurs in abundance. Like other rattlesnake species, it may hibernate communally in winter in a den on a slope or ledge that faces south for warmth. Some Prairie Rattlers take to the ready-made convenience and shelter of mammal burrows for hibernation. The species is often found in the vicinity of prairie dog towns, but there is no evidence that the snake makes the communal rodent its prey. The species differs from the Western Diamond-backed by the absence of a black-and-white ringed tail.

Pygmy Rattlesnake
Sistrurus miliarius L 15–21 in (38–53 cm)

It may be small in size and make a faint rattle, but the Pygmy Rattlesnake nevertheless packs potent venom and delivers a dangerous bite.

KEY FACTS

Body is gray to brown to reddish; dark blotches on back; 2 to 3 rows of spots along sides.

+ habitat: Wet prairies, palmetto-pine woods, mixed forests, lakes, and marshes

+ range: South Carolina through Florida, west to eastern Texas and Oklahoma

+ food: Amphibians, lizards, and small rodents and snakes

Known as a ground rattler, the Pygmy Rattlesnake is noted for its slender tail and mini-rattle that creates a sound like a buzzing insect. Despite its size, it can show a temper, lashing out without cause, or it may not react at all. In the past, this species was found in the vicinity of sawmills, with attractions such as piles of sawdust and bark. It is still found in heavily littered lots and around dumps, as well as warming itself on country roadsides in the evening. Unlike rattlesnakes in the genus *Crotalus,* which have small scales on the top of the head, the Pygmy Rattlesnake and its relatives have nine large scales.

Hellbender
Cryptobranchus alleganiensis L 11.5–20 in (29–50 cm)

The Hellbender is a giant salamander, a large aquatic species that can measure almost two feet long. Its sliminess has earned it the nickname of "snot otter."

KEY FACTS

Body is gray to brown to black; may have spots; skin folds on sides; flat head.

+ habitat: Fast-flowing clear streams with rocky bottoms

+ range: New York to eastern Illinois to northern Mississippi, Alabama, and Georgia; disjunct populations in Missouri and Arkansas

+ food: Crayfish and other invertebrates

Aquatic, nocturnal, and secretive, the Hellbender doesn't get a lot of "face time" with humans. It has wrongly been labeled as aggressive and venomous, which it is not. The slightly smaller male takes on many parental duties before the first egg is laid. He makes a nest cavity under a rock or log. The female lays an egg mass, and the male fertilizes it. Then he protects the eggs until they hatch in about three months. Hellbender numbers have declined dramatically because of pollution, silting of streams from erosion, sand or gravel mining, and introduced predators, resulting in U.S. endangered species protection for the Ozark subspecies.

Mudpuppy

Necturus maculosus L 8–13 in (20–33 cm)

The Mudpuppy is an aquatic salamander that never loses its larval look or lifestyle. It does not metamorphose, and retains gills throughout its life.

KEY FACTS

Body is gray to brown, with blue-black spots; feathery maroon gills at neck; tail is compressed.

+ **habitat:** Lakes, ponds, and streams

+ **range:** Great Lakes region south to northern Louisiana, Mississippi, Alabama, and Georgia

+ **food:** Crayfish, other invertebrates, and small fish

The Mudpuppy, also known as the Water Dog, is one of the few salamander species to make noise. It's not much of a noise—basically a squeak—but enough to fuel the far-fetched comparison with a dog's bark. The length of its distinctive maroon gills varies according to the temperature and clarity of the water it lives in: The warmer and murkier the water, the longer the gills. The Mudpuppy lives on the bottom of its watery habitat and hunts at night for crayfish and other invertebrates. The female lays between 30 and 190 eggs, which she attaches one by one under rocks and guards until they hatch some five to nine weeks later.

Lesser Siren

Siren intermedia L 7–27 in (18–68 cm)

The eel-like Lesser Siren is a primitive aquatic salamander. It has small forelimbs and no rear limbs, and it retains gills throughout its life.

KEY FACTS

Body is gray to olive to black; may have black spotting; small front legs have 4 toes.

+ **habitat:** Swamps, ponds, and ditches

+ **range:** Coastal plain from southeastern Virginia to Florida, Gulf Coast to Texas, north along Mississippi Valley to Michigan

+ **food:** Aquatic invertebrates and small fish

With only two reduced limbs, the Lesser Siren has no real chance of relocating when its watery habitat dries up. It may take refuge in the muddy bottom, and when that dries out, it can secrete a protective covering around its body and wait for the rains. The Lesser Siren is one of the rare noise-making salamanders; its repertoire includes clicks and shrill yelps. The Lesser is usually shorter than the similar Great Siren (*S. lacertina*) and has fewer costal (rib) grooves along its sides. The Lesser usually has no more than 35 and the Greater Siren between 36 and 40, although the grooves are difficult to count in a live animal.

Three-toed Amphiuma
Amphiuma tridactylum L 18–30 in (45–75 cm)

The Three-toed Amphiuma is a denizen of mainly muddy waters of the Mississippi Delta region, preferring bodies of water with dense vegetation.

KEY FACTS

Body is gray, black or brown above and light underneath; tail is compressed; thin limbs.

+ **habitat:** Lakes, marshes, slow-moving streams, swamps, and ditches

+ **range:** Missouri and Kentucky south to Texas, Louisiana, Mississippi, and Alabama

+ **food:** Invertebrates, fish, snakes, and amphibians

The Three-toed Amphiuma sports gills as a larva, but loses them as an adult and retains only a gill slit on each side. This salamander has a slim and slimy eel-like body and swims in an undulating manner, like a snake. It emerges at night to forage, taking a wide range of invertebrate and vertebrate prey. The genus *Amphiuma* is represented by three species that have one, two, or three toes respectively; they all inhabit the southeastern United States. Unlike most other totally aquatic salamanders, amphiumas have internal fertilization. A few months after mating, the female lays a string of about 200 eggs in a cavity nest.

Red-spotted Newt
Notophthalmus viridescens L 2.25–4.75 in (5.7–12 cm)

The aquatic Red-spotted Newt usually goes through a land-based stage, when it is known as a Red Eft, between its larval and adult forms. This stage can last from one to seven years.

KEY FACTS

Smooth adult body is brown to green with black-circled red spots on back, and small black specks above and below.

+ **habitat:** Ponds, lakes, quiet streams, swamps, and moist woodlands

+ **range:** Southeastern Canada and eastern United States

+ **food:** Invertebrates, young amphibians, and amphibian eggs

During a Red-spotted Newt's larval stage, usually within a year of hatching from the egg, an individual transforms into a terrestrial Red Eft. It looks similar to the adult but can't reproduce and is often brilliantly colored, especially in moist upland habitats. This stage may allow the species to disperse to other locations, and it often is omitted in stressful environmental conditions. Eventually, the eft returns to the water and morphs into a full-fledged aquatic adult. The Red-spotted Newt gets a pass from a lot of predators because it secretes a toxin from its skin that produces irritation and makes it generally unpalatable.

Adult

Red Eft

Rough-skinned Newt

Taricha granulosa L 2.25–3.6 in (5.7–9.1 cm)

The Rough-skinned Newt typically has very grainy skin, but the male at breeding time has smooth skin when he goes courting at aquatic sites.

KEY FACTS

Warty body is tan to brown to black above; underside is yellow to orange.

+ habitat: Ponds, lakes, slow-moving streams, moist forests, and grasslands

+ range: Pacific coast from southeastern Alaska to northern California; possibly introduced populations in Idaho and Montana

+ food: Invertebrates

At any given time throughout their range, some Rough-skinned Newts are observing their breeding season. Mating involves a froglike mounting followed by the male's dismount and deposit of a sperm capsule that the female retrieves into her reproductive system. Then the female lays eggs and hides them singly on submerged vegetation. This highly toxic species seems to feel secure enough to roam about during the day. When threatened, it arches its head toward its curled tail and displays its colorful underside as a warning of the toxin embedded in its skin, a strong toxin related to that found in blowfish. These newts possess it in varying amounts.

Long-toed Salamander

Ambystoma macrodactylum L 1.6–3.5 in (4–9 cm)

The Long-toed Salamander, a species of mole salamander so called because of its burrowing habits, has long legs with four long, thin toes on each foot, one longer than the rest.

KEY FACTS

Body is dark brown to black; yellow dorsal stripe or varied blotch pattern; 12–13 costal (rib) grooves.

+ habitat: Varies from sagebrush to damp forests to mountain meadows, near ponds and lakes

+ range: Southeastern Alaska to northeastern California, east to Idaho, north to western Alberta

+ food: Insects and other invertebrates

The Long-toed Salamander occurs in a wide range of habitats ranging from sea level to 9,000 feet (2,700 m). It is often one of the first species to begin breeding in the early spring. A male approaches a female in the water and rubs his chin on her snout, releasing a pheromone to make her receptive. She follows him to a spot where he deposits a sperm packet. She stands over the packet and draws it up into her vent, fertilizing her eggs internally. She attaches her eggs in clusters underwater on stalks, sticks, or rocks. A Santa Cruz subspecies, with separate populations found in central California, is listed as endangered.

Spotted Salamander
Ambystoma maculatum L 4.3 –7.25 in (11–18 cm)

The Spotted Salamander adult is not often seen aboveground. As a mole salamander, it lives concealed under logs or underground in burrows of other species.

KEY FACTS

Stout body is gray to blue-black to black; 2 rows of orange-yellow spots on back from head to tail; usually 12 costal (rib) grooves.

+ **habitat:** Moist, mixed woodlands near ponds, streams, and vernal pools

+ **range:** Southeastern Canada south to Texas and Georgia

+ **food:** Insects and other invertebrates

The Spotted Salamander comes out at night to forage for insects and in the spring. With the first warming rains, salamanders journey en masse to vernal pools and ponds to breed. The female lays large, globular egg masses that are clear when they are laid but turn green when infiltrated with algae. The algae have a symbiotic relationship with the eggs. They infiltrate the cells of the embryos and photosynthesize, bringing energy and oxygen to the developing embryos. Spotted Salamander reproduction is affected by acid rain in some areas of the Northeast. Otherwise, the species is long-lived, with a potential life span of 20 years.

Marbled Salamander
Ambystoma opacum L 3.5–4.25 (9–11 cm)

The attractive, blunt-headed Marbled Salamander has either a gray or white pattern of bands on a shiny black background that gives the species a marbled effect above.

KEY FACTS

Body is black with bands across back that vary from white to gray; belly is black.

+ **habitat:** Woodlands and floodplains

+ **range:** Southern New England to northern Florida, west to Texas, north to Wisconsin

+ **food:** Insects and other invertebrates

The Marbled Salamander tends to be a bit more "out there" than other mole salamanders; there is a greater chance of spotting one, and the male, with his silvery white pattern, is quite distinctive. In general, though, these salamanders do spend a great deal of time in underground burrows. They also tend to do things a bit differently from other *Ambystoma* species. For one thing, they mate in the fall, on land. The female then deposits her eggs on dry ground in a shallow depression and guards them until the rains come and inundate them. The larvae often hatch in a few days or weeks, although eggs may overwinter if conditions are not right.

Eastern Tiger Salamander

Ambystoma tigrinum L 7–8.25 in (18–21 cm)

Large, stout, and sturdy, the Eastern Tiger Salamander is one of a group of salamanders with varied tigerlike markings on a dark background that inhabit the eastern and central United States.

KEY FACTS

Body is brown to black, with olive or yellowish spots on back that may reach sides.

+ **habitat:** Pine barrens, woodlands, and meadows

+ **range:** Mid-Atlantic and southeastern U.S. and central U.S., except southern Appalachians and lower Mississippi delta

+ **food:** Invertebrates, amphibians, and small mammals

The Eastern Tiger Salamander is one of the largest land salamanders in the world. It lives in areas with debris cover, and it appropriates burrows of other species or digs its own, up to 2 feet (0.6 m) below the surface. This enables related species of tiger salamanders in the West to stay moist while inhabiting arid regions. The Eastern Tiger Salamander breeds very early, often in deeper water than other salamander species. The female lays egg masses that attach to debris in permanent or temporary bodies of water. The species has suffered habitat loss and degradation and is listed as endangered in a number of states.

California Giant Salamander

Dicamptodon ensatus L 2.5–6.7 in (6.3–17 cm)

The California Giant Salamander is one of the rare noise-making salamanders. It can make low-pitched barking noises when grabbed or otherwise threatened.

KEY FACTS

Body is brown to purplish, with black mottling; skin is smooth.

+ **habitat:** Rivers and cool, moist forests

+ **range:** Southwestern British Columbia along the coast to central California, with separate populations in Idaho and Montana

+ **food:** Invertebrates, mice, snakes, and salamanders

The California Giant Salamander, which normally hides under rocks, logs, and in burrows along stream banks, is sometimes spotted walking the forest floor or otherwise out and about, crossing roads during rains. It also is known to climb low bushes and trees. This large predator eats a wide range of small animals, from insects to snakes. Larvae are just as predaceous, even cannibalizing smaller larvae and polishing off many tadpoles. The larvae have been known to dominate in numbers the total amphibian populations of stream headwaters in California. This species is also known as the Pacific Giant Salamander.

Green Salamander
Aneides aeneus L 3.5–5 in (9–13 cm)

The Green Salamander is eastern North America's only truly green salamander species.
The distinctive green pattern helps it blend with lichens in its customary habitats.

KEY FACTS

Body is brown to black above, with extensive greenish patches from head to tail; squared toes.

+ **habitat:** Mainly rock crevices in sandstone outcrops; also under bark

+ **range:** Southwestern Pennsylvania through mountains to Alabama

+ **food:** Ants, mosquitoes, and other invertebrates

The Green Salamander hibernates—often communally—deep within the rock crevices in sandstone outcrops. At the end of the winter, the salamanders disperse. Breeding occurs from spring to fall. It may be preceded by territorial disputes among males that involve biting and shoving. Females deposit clusters of 10 to 20 eggs by mucus strands from the upper rock surfaces of the crevices, and they remain to guard them against intruders. When the eggs hatch, the young look like miniature adults; there is no aquatic larval stage in this salamander. The young immediately seek the shelter of nearby mossy crevices and begin to hunt for insects on the cliff faces.

Northern Dusky Salamander
Desmognathus fuscus L 2.5–4.5 in (6–11 cm)

The Northern Dusky Salamander is one of a number of dusky salamander species that have similar characteristics and overlap in range to a large degree.

KEY FACTS

Body is gray to brown above, sometimes mottled; pale line from eye to jaw.

+ **habitat:** Rocky creeks, springs, seeps, floodplains, and sloughs

+ **range:** Southern New Brunswick and Quebec to Louisiana and Arkansas, except for southeastern coastal plain and Florida peninsula

+ **food:** Invertebrates and salamanders

The Northern Dusky Salamander is common and abundant in many parts of its range. It can be easy to spot but will often do a vanishing act by running or jumping and disappearing into debris. A compressed and keeled tail is another distinguishing characteristic of this species. The mottling or spotting on some salamanders is very variable and may become obscure with age. The female lays grape-like clusters of eggs on land, under logs or rocks, but near water. She tends to steer clear of streams with hungry fish. The Northern Dusky often is seen in the company of the Northern Red Salamander as well as other dusky salamanders.

Ensatina

Ensatina eschscholtzii L 1.5–3.5 in (4–9 cm)

Unusual among salamanders, the terrestrial Ensatina does not undergo an aquatic larval stage. The female lays eggs underground and broods them to keep them moist until they hatch.

KEY FACTS

Body color and pattern vary; tail constricts at base; yellow or orange upper legs.

+ **habitat:** Moist woodlands, shady canyons, grasslands, and chaparral

+ **range:** Southwestern British Columbia along coast to California, and in western Cascades and Sierra Nevada

+ **food:** Insects and other invertebrates

The Ensatina comes in a variety of colors and patterns that differ geographically. To base colors of brown or black are added different patterns of blotches or mottling in yellow, orange, or cream; other forms are plain brown above. Some experts argue that some of the subspecies should be regarded as separate species. When threatened, an Ensatina may stand upright and stiff-legged, arch its back, and swing its tail. If the tail is grabbed, it may break off at the narrow base. Ensatinas often are found among rocks, beneath rotting logs, and under bark. Like other lungless salamanders, Ensatinas breathe through their skin.

Long-tailed Salamander

Eurycea longicauda L 4–6.25 in (10–16 cm)

The Long-tailed Salamander sports an especially long tail that in extreme cases can measure up to two-thirds of an adult's total body length. Young have shorter tails.

KEY FACTS

Body is yellow to orange to red with dark spots on back; can have yellow dorsal stripe and dark side stripes; dark herringbone tail pattern.

+ **habitat:** Springs, seeps, stream banks, and caves

+ **range:** Southern New York to northern Alabama, west to eastern Oklahoma

+ **food:** Insects and other invertebrates

Like the many other brook salamanders, the Long-tailed Salamander hides by day under debris, stones, and boards, often in very large numbers. It ventures out at night to forage for insects and other invertebrates. Breeding occurs in fall and winter. The female deposits 60 to 100 eggs, attached to surfaces in water, around which she often coils to protect them until they hatch. Another distinctive feature of this species is its yellow underside. A dark-sided subspecies, *E. l. melanopleura*, is a denizen of caves, usually dwelling in the twilight zone. It occupies the western portion of this species' range, especially in the highlands of Missouri and Arkansas.

Texas Salamander
Eurycea neotenes L 2–4.5 in (5–11 cm)

As its scientific name suggests, the Texas Salamander is an example of a neotenic salamander, which remains at the larval stage throughout life and becomes reproductively mature in that form.

KEY FACTS

Body is brownish yellow with dark flecks on back; sides have double row of light flecks; retains red external gills on neck throughout life.

+ **habitat:** Streams in caves, springs, seeps, and streams

+ **range:** South-central Texas

+ **food:** Aquatic invertebrates

The Texas Salamander occupies a small area of south-central Texas: the edge of the Edwards Plateau and adjacent Balcones Escarpment, the latter a geologic fault that separates the plateau from the coastal plain and contains many springs and seeps. Permanent waters are important for a neotenic species that lives a wholly aquatic life; only rarely does it metamorphose to a terrestrial form. The Texas Salamander, one of several similar local brook salamanders, hides under and among the rocks on the bottom of its watery habitats. It has much to fear from fish and other predators; the young especially suffer from high mortality.

Spring Salamander
Gyrinophilus porphyriticus L 4.75–7.5 in (12–19 cm)

The large Spring Salamander is not named for the season but for its preferred habitat. This lungless salamander needs highly oxygenated water, as from springs, to breathe through its skin.

KEY FACTS

Body is pink to salmon to orange to red to brown; light ridge from nostril to eye, dark-bordered in some subspecies.

+ **habitat:** Springs, clear streams, and seeps

+ **range:** New England through Appalachians to Alabama and Georgia

+ **food:** Invertebrates and small salamanders

A long body and nasal ridge distinguish the Spring Salamander from the similar red and mud salamanders. The Spring Salamander is active year-round, coming out from cover to forage at night. In some areas, other salamanders are its favored prey. Breeding in the fall and spring, males and females may be seen writhing around together in the water in a courtship maneuver. Later, the female deposits 40 to 60 eggs under stones and logs and protects them. After the eggs hatch, the aquatic larval stage of this species may last up to four years, and the larvae are almost as large as adults when they metamorphose.

Northern Red Salamander

Pseudotriton ruber L 4–6 in (10–15 cm)

Northern Red Salamanders like their watery habitats to be clean and clear—no muddy water for this species, as a general rule.

KEY FACTS

Body is orange-red to red, with many rounded black spots; iris is yellow.

+ **habitat:** Woodlands, swamps, and meadows with streams, springs, and seeps

+ **range:** Southern New York to Indiana, south to Gulf Coast, except Atlantic coast south of Virginia and Florida peninsula

+ **food:** Earthworms and other invertebrates

A species in the large group of lungless salamanders that breathe through the skin and the lining of the mouth, the Northern Red Salamander seeks out moist places to hide, under rocks, rotting logs, leaf litter, and moss. Its red coloration is stronger in juveniles and fades as the animal matures, becoming a dull purplish brown, often with fused spots. The Red Salamander is similar in appearance to the Mud Salamander (*P. montanus*), which overlaps in range. The two are difficult to tell apart, but the Mud is browner, prefers a muddier habitat, and has brown eyes instead of the Red's yellow eyes.

Four-toed Salamander

Hemidactylium scutatum L 2–3.5 in (5–9 cm)

Look to the feet to see the features that identify this salamander: It has four toes on both front and hind feet, whereas most salamanders have five on the hind feet.

KEY FACTS

Body is reddish brown; sides are grayish; stark white belly has black spots.

+ **habitat:** Bogs, streams, and woodlands with sphagnum moss

+ **range:** Disjunct populations, mostly east of Mississippi River from Nova Scotia to the Florida Panhandle, except southern Atlantic coast

+ **food:** Insects and other invertebrates

The Four-toed Salamander breeds in late winter or early spring, especially on rainy evenings. The female attaches 20 to 50 eggs to sphagnum moss or other vegetation at the water's edge and protects them. When the eggs hatch, the larvae wiggle toward the water, which ideally has no fish. This species' tail constricts at the base, forming a natural point for breakage when the tail is grabbed. Separated from the body, the tail wiggles as a distraction, allowing the rest of the animal to make an escape. The tail will then regenerate. The need to be close to sphagnum moss may account in part for this species' disjointed distribution.

Eastern Red-backed Salamander

Plethodon cinereus L 2.5–4 in (6–10 cm)

The Eastern Red-backed Salamander is one of many small and slender red-backed woodland salamanders that often are the most abundant species in any given area.

KEY FACTS

Body is dark gray to black; center of back often has straight-edged red line.

+ habitat: Wooded ravines, hillsides, and woodlands

+ range: Southern Canada and Minnesota south to North Carolina

+ food: Invertebrates

Once thought to be a single wide-ranging species, the Eastern Red-backed is but one of many similar and closely related forms. Although the red-back feature usually is evident, it might be missing in some individuals. Related species have stripes that vary in width, color, and occurrence in individuals. This salamander hides under cover during the day and comes out at night to forage. Males find females by tracking pheromone trails. Mating occurs in spring and fall. The female remains with her one to five eggs laid under a log or in a burrow. Like many other woodland salamanders, this species lacks an aquatic larval stage.

Northern Slimy Salamander

Plethodon glutinosus L 4.75–6.75 in (12–17 cm)

Its scientific name indicates that the Northern Slimy Salamander is not gluten-free. This woodland salamander emits a sticky secretion through its skin that sticks to humans' hands like glue.

KEY FACTS

Body is shiny black, with small white spots above; large white, gray, or yellow spots on side.

+ habitat: Wooded ravines, hillsides, and woodlands

+ range: Southern New England to Illinois, south to Alabama and Georgia, with an isolated population in New Hampshire

+ food: Invertebrates

Many experts describe a complex of similar slimy salamanders formerly thought to be a subspecies of *P. glutinosus,* which have been accorded species status following DNA examination. These species have many traits in common and collectively inhabit much of the eastern United States. The Northern Slimy Salamander has the largest and northernmost range. This salamander hides out under stones, logs, debris, or in burrows during the day and comes out at night to forage. It mates in spring and fall. The female lays 6 to 36 eggs under a log or in a burrow. As in many woodland salamanders, there is no aquatic larval stage.

Eastern Spadefoot
Scaphiopus holbrookii L 1.7–2.2 in (4.4–5.7 cm)

Spadefoots have a small, horny "spade" on their hind feet that they use for digging their underground burrows in loose soil, sand, or gravel.

KEY FACTS

Body is olive to brown to black; may have 2 light lines running from eye down back.

+ habitat: Woods, brushlands, and farmland

+ range: Southern New England to Florida, west to Texas, north to Illinois, Indiana, and Ohio, with some populations separated from one another

+ food: Insects, other invertebrates

Spadefoots, sometimes called toads, are not true toads: Their skin is less bumpy, they have vertical pupils, and they lack visible parotoid glands. They exude a skin toxin that can cause an allergic reaction in people who handle them. The Eastern Spadefoot stays underground unless driven from its burrow by downpours or many days of rain. The breeding male's abrupt *whar* call has been compared to the squawk of a young crow. The male calls from water, but at times from within the burrow. Because this species mates often in very temporary pools, the eggs hatch in a few days and the tadpoles can morph in as few as two weeks.

Western Spadefoot
Spea hammondii L 1.5–2.5 in (3.8–6.3 cm)

The Western Spadefoot has very catlike eyes, with vertically elliptical pupils adorned with gold irises. Its signature digging spades are a very glossy black.

KEY FACTS

Body is olive to brown to gray; light stripes and blotches on back; wedge-shaped spades on hind feet.

+ habitat: Grassy areas, scrub, and mixed woodlands with sand and gravel

+ range: Northern California through Central Valley and foothills to Baja California

+ food: Insects, worms, and other invertebrates

Using the spades on its hind feet, the plump Western Spadefoot digs a burrow that provides constant temperature and humidity, in which it spends most of its life. This spadefoot comes to the surface mainly in search of a mate. The male's advertisement call is a short, loud trill that has been compared to a short snore and is made repeatedly while he floats on vegetation on the water's surface, often in a temporary pool. The skin exudes a mucus that smells like peanuts and can cause a severe allergic reaction in some people who handle these frogs. The reaction can include skin inflammation, itchy eyes, and runny nose.

American Bullfrog

Lithobates catesbeianus L 3.5–6 in (9–15 cm)

For many people, the American Bullfrog is the quintessential North American frog, with its bright gold bulging eyes, wide froggy smile, and familiar, resonant *jug-o-rum* call.

KEY FACTS

Body is green and may have mottling; legs may have dark bands; thin ridge around tympanum (external eardrum).

+ **habitat:** Ponds, lakes, and slow-moving streams with vegetation

+ **range:** Eastern and central U.S., parts of eastern Canada; introduced in western U.S.

+ **food:** Invertebrates and small vertebrates

North America's largest native frog, the American Bullfrog can make a meal of anything it can grab with its tongue and stuff into its gaping mouth using its front feet. Prey includes birds and other bullfrogs. When snagged by a snake or other predator, it emits a loud, eerily catlike wail. It is distinguished from the similar Green Frog by the lack of a ridge along its sides. Females usually have tympanums the size of their eyes; males' are larger than the eyes. This species has been introduced to areas outside its historical range, to the detriment of less aggressive local species—a prime example of an invasive species.

Pickerel Frog

Lithobates palustris L 1.7–3 in (4.4–7.6 cm)

The double row of rectangular blotches down the middle of its back distinguishes the Pickerel Frog from the closely related leopard frogs with their round or oval dorsal spots.

KEY FACTS

Body is tan, with 2 rows of rectangular blotches; yellow fold on sides; light jaw stripe; yellow on belly and under hind legs.

+ **habitat:** Woodland streams, ponds, lakes, meadows, and swamps

+ **range:** Southeastern Canada and eastern U.S., except extreme Southeast

+ **food:** Insects and other invertebrates

The Pickerel Frog emits a secretion through its skin that at best makes it unpalatable to some potential predators and at worst can kill other frogs that are confined with it, as in a terrarium. It is thought that the yellowish under-thigh wash may possibly confuse predators. In northern areas, this frog prefers cool, clear water, but in the South, it is often found in swamps and other turbid waters. When many frogs are dormant for the winter in mucky bottoms, the Pickerel Frog often may be out and about even if the pond has iced over. The male's call is a low, brief "snore" repeated at short intervals.

Northern Leopard Frog

Lithobates pipiens L 2–3.5 in (5–8.9 cm)

In the past, leopard frogs were often seen on a dissection tray in a high school or college biology class. Back then, the frogs also were more abundant in nature.

KEY FACTS

Body is brown or green, with 2 or 3 rows of rounded spots between ridges; light line on upper jaw.

+ **habitat:** Fresh and brackish bodies of water with dense vegetation

+ **range:** Central and eastern Canada, northern North America; introduced in West

+ **food:** Invertebrates, amphibians, snakes, and small mammals

The Northern Leopard Frog tolerates a range of water conditions and altitudes to 11,000 feet (3,400 m). It may leave the water in summer and travel to grassy areas, a habit that has given it the nickname "meadow frog." The male's call is a deep, guttural, rattling snore, often followed by clucking noises. This is repeated with a break of several seconds between sequences. The range of the Southern Leopard Frog (*L. sphenocephalus*) more or less ends where the northern species' range begins, although there is some overlap. A pointier snout and a light spot in the middle of the tympanum distinguishes the Southern from the Northern.

Wood Frog

Lithobates sylvaticus L 1.3–2.75 in (3.3–7 cm)

The hardy Wood Frog lives farther north than any other reptile or amphibian in North America, ranging past the Arctic Circle. This frog breeds explosively, laying eggs over only a few days.

KEY FACTS

Body is pink to brown to black; has dark eye mask; light stripe on upper jaw.

+ **habitat:** Woodlands, grasslands, and tundra with temporary or permanent water sources

+ **range:** Throughout much of northern North America and to southern Appalachians in the East

+ **food:** Insects and other invertebrates

The Wood Frog gets the breeding season going earlier than most other frog species. In much of its range, it emerges from hibernation in burrows or under woodland leaves in February or March. It starts calling and mating when ice may still be on the ponds. After the eggs are fertilized, the frogs don't linger; they retreat to their home bases. In summer, this frog may wander far from water. The male's call, delivered from double vocal sacs on its lower chest, is a raspy duck-like quack that doesn't carry very far. This frog can alter its coloration between light and dark, and it has a fairly transparent underside.

Northern Red-legged Frog
Rana aurora L 1.75–5.25 in (4.4–13 cm)

This western frog is native to coastal areas in southwestern Canada and northern California. The species often migrates considerable distances to breed.

KEY FACTS

Body is reddish brown to gray with black flecks; dorsolateral ridge: light jaw stripe; reddish on belly and under hind legs.

+ **habitat:** Moist woodlands, grasslands near ponds, streams, and other water sources

+ **range:** Southwestern British Columbia to northern California

+ **food:** Insects and other invertebrates

The male Northern Red-legged Frog's advertisement call is a series of stuttered guttural notes often performed under water. It also emits a low "chuckle." The female lays globular masses of eggs, which are attached to stems of aquatic vegetation. Outside of the short breeding season, the species may disperse up to several miles on land. Mark Twain immortalized the closely related and threatened *R. draytonii*, the California Red-legged Frog, in "The Celebrated Jumping Frog of Calaveras County." A competition is still held in Calaveras County each May. Until recently this frog was considered a subspecies of *R. aurora.*

Eastern Narrow-mouthed Toad
Gastrophryne carolinensis L 0.75–1.25 in (1.9–3.2 cm)

The Eastern Narrow-mouthed Toad represents a group of small, chubby, native amphibians with smooth skin, pointy little heads, and very untoadlike narrow mouths.

KEY FACTS

Body is gray to brown to red; light stripe on sides; center of back dark; fold of skin behind head; male has dark throat.

+ **habitat:** Edges and environs of bodies of water with good, moist cover

+ **range:** Southeastern United States

+ **food:** Ants and other small insects

The Eastern Narrow-mouthed Toad hides by day under natural and artificial cover, coming out at night to feed mainly on ants, termites, and beetles. It often stations itself at the opening to an anthill. The male calls to females with a sheep-like *baaaa.* The female has a sticky skin secretion that aids amplexus, the basic frog mating position. Eggs are laid as a film on the surface of the water and hatch in a few days. The similar Great Plains Narrow-mouthed Toad (*G. olivacea*) ranges from southern Nebraska through Texas, overlapping with the Eastern species. It lacks dark mottling on its belly, which is present in the eastern species.

American Toad

Anaxyrus americanus L 2–3.5 in (5–9 cm)

The common toad of the eastern United States and Canada, the American Toad lives in a multitude of different habitats, requiring access to a body of shallow water only at breeding time.

KEY FACTS

Body is olive to brown to red, with wart-dotted spots; may have pale stripe on back; parotoid gland usually separate from cranial crest.

+ **habitat:** Diverse habitats, from mountains to sea level, with moist shelter and insects

+ **range:** Eastern Canada and U.S.

+ **food:** Insects and other invertebrates

Short limbs, warty skin, and a lightning-quick, insect-zapping tongue characterize this iconic toad, which assists gardeners by keeping down unwelcome insect populations. The dark-throated male moves to shallow pools to breed, calling an invitation to females with a long, cricketlike trill. The larger female lays up to 20,000 eggs in long, double strands that have a partition between each egg. The American Toad and similar species were formerly placed in the genus *Bufo*, which now is reserved for toads in Eurasia and northern Africa. A dwarf subspecies of American Toad, *A. a. charlesmithi,* is about a third smaller.

Great Plains Toad

Anaxyrus cognatus L 2–3.5 in (5–9 cm)

A pattern of distinctive blotches helps identify the Great Plains Toad. It secretes a poison from the parotoid glands behind the eyes, which inflame the mouths of would-be predators.

KEY FACTS

Body is gray to olive to brown; has light-bordered dark blotches; cranial crests form hump on snout; crests touch parotoid glands.

+ **habitat:** Grasslands, brushlands, and farmland

+ **range:** Southeastern Alberta to Wisconsin, south through Great Plains to Southeast

+ **food:** Insects

With sharp tubercles, or projections, on its hind feet, the Great Plains Toad can burrow rapidly by backing into loose soil in the manner of spadefoots. After heavy rains in spring and summer, the males congregate in mostly temporary pools and create a din by calling to females with prolonged, loud, metallic trills that emanate from a sausage-shaped vocal sac. Farmers like this nocturnal toad because it relishes cutworms, moth caterpillars that feed voraciously on crops. Like many other toads, when threatened, this large species will inflate itself to even bigger proportions and close its eyes, sometimes also lowering its head.

Red-spotted Toad

Anaxyrus punctatus L 1.5–3 in (4–8 cm)

The Red-spotted Toad's mating chorus resonates in desert environments. Like other frogs and toads, a male may grasp another male in the chaos, causing the latter to vocalize a release call.

KEY FACTS

Body is greenish to grayish brown; usually covered with many small, reddish warts; head and back flat; parotoids are round.

+ **habitat:** Desert, rocky areas, grasslands, scrublands, oak woodlands, and arroyos

+ **range:** Southeastern California to central Texas

+ **food:** Insects

The Red-spotted Toad usually lacks the cranial crests found on the head of other toad species, and its round parotoid glands behind the eye are diagnostic. A flattened body allows this toad to hide among rocks and crevices. It is sometimes found in prairie dog burrows and can alter its basic coloration to blend with the environment. This species breeds in spring and summer, usually after rains. Males congregate at the edge of shallow pools and make a high-pitched, sustained trill that is more musical than the calls of many other toads. Unusually, the females lay their eggs singly, not in strings, at the bottom of mating pools.

Oak Toad

Anaxyrus quercicus L 0.75–1.3 in (1.9–3.3 cm)

The tiny Oak Toad is the smallest toad in North America. Still, in chorus at breeding pools, the males' high-pitched advertisement calls can create a torturous din.

KEY FACTS

Body is light gray to almost black; paired dark spots on back; light to orange stripe on back.

+ **habitat:** Sandy and loose soils in pine and oak scrub

+ **range:** Coastal plain from southeastern Virginia through Florida and Gulf Coast to eastern Texas, and into northeastern Alabama

+ **food:** Insects

Small size and a light middorsal stripe identify this attractive little toad, which has finely roughened skin instead of bulky warts, befitting its diminutive proportions. This species hunts insects by day, unlike the many nocturnal toad species, and hides the rest of the time, except during breeding season. Its pattern offers camouflage when it is out and about. The male's call is a high-pitched, repeated chirp that resembles the peeping of chicks. When the vocal sac is not being used, it takes the shape of a triangular apron that hangs low on the male's chest. The female Oak Toad lays her eggs in segments of two to five eggs.

Southern Toad
Anaxyrus terrestris L 1.5–3 in (3.8–7.6 cm)

Warts galore adorn the Southern Toad, along with a prominent cranial crest with knoblike extensions on the top of its head. It also has tubercles, or small projections, between its warts.

KEY FACTS

Body brown to red to black; can have mid-dorsal stripe; cranial crest has knobs.

+ **habitat:** Sandy areas, edges of ponds and lakes, wooded areas, and yards

+ **range:** Coastal plain from southern Virginia, throughout Florida, and along Gulf Coast to Louisiana

+ **food:** Insects and other small invertebrates

The nocturnal Southern Toad spends much of the day in a burrow it constructs, but comes out at night to forage. Like many other toads, it often is drawn to the insects that cluster around outdoor lights. This species mates from March to October. With a rounded vocal sac expanded to the max, the male produces a high-pitched, piercing trill. The female deposits long double strands of eggs in shallow fresh water. Sometimes the pool chosen is so ephemeral that it dries up before the eggs can hatch. In some parts of its range, the Southern Toad has suffered from introduction of the larger, more aggressive, and extremely opportunistic Cane Toad.

Cane Toad
Rhinella marina L 4–6 in (10–15 cm)

Introduced in the southern United States to control beetle larvae in the commercially important cane fields, the behemoth Cane Toad now competes with unfair advantage over native species.

KEY FACTS

Body is brown, often with dark mottling; large, pitted parotoid glands behind eyes.

+ **habitat:** Cane fields, open forests, grazing land, parks, gardens, and yards

+ **range:** Much of Florida and coastal Louisiana and Texas

+ **food:** A wide range of invertebrates and vertebrates, some plant matter

The Cane Toad, a native of South and Central America and Mexico, was brought to the United States with a specific job to do. But it didn't stay put in the cane fields, and now is an established presence in Texas, Louisiana, and Florida, a population derived from animal dealer escapees. The big-mouthed toad can swallow a wide range of prey, including native frogs and toads. Its huge parotoid glands contain a toxin that poisons other species, including curious pets. It can reproduce year-round, with the female laying thousands of eggs at a time. Its tadpoles compete with those of other species, such as the native Southern Toad.

Eastern Cricket Frog

Acris crepitans L 0.6–1.5 in (1.5–3.8 cm)

The Eastern Cricket Frog belongs to a genus of treefrogs that are less arboreal than other treefrogs. It lacks toe pads and is not a climber, relying on hopping to take it where it needs to go.

KEY FACTS

Body is greenish brown, yellow, red, or black; dark triangle between eyes; ragged stripe on thighs.

+ **habitat:** Shallow, sunny water with vegetation; quiet-stream banks

+ **range:** Southern New York to eastern Texas, except Appalachians, Atlantic coastal plain, and peninsular Florida

+ **food:** Insects

The genus name for this tiny frog derives from the Latin for "cricket," which is an apt image for its hopping pattern of locomotion and the nature of its call. There is a wide range of color and pattern within this species, and it is often necessary to closely examine the stripes on the rear of the thigh to distinguish it from other cricket frogs and from some treefrogs. The call of the Eastern Cricket Frog is a high-pitched, rapid clicking that sounds like a bunch of marbles being shaken in a bag. This small frog gets itself out of harm's way by engaging in a series of random hops, thwarting a predator's plan of attack.

Green Treefrog

Hyla cinerea L 1.25–2.25 in (3–6 cm)

Male Green Treefrogs often come together by the hundreds or even thousands at breeding time, creating a full-bodied chorus. Males often sing while clinging to vegetation in water.

KEY FACTS

Smooth body is yellow to green to almost gray; often with white or yellowish stripe on side; often tiny gold flecks on back.

+ **habitat:** Permanent bodies of water with vegetation

+ **range:** Delaware to Florida Keys and along Gulf Coast to Texas, north through Arkansas to Illinois

+ **food:** Insects and spiders

The Green Treefrog call has been likened to the sound of a cowbell, kind of a *quonk, quonk, quonk* quality. A chorus of male frogs produces sounds on different pitches. Folklore holds that this species' singing predicts rain, but it sings in advance of sunny weather as well. Large, sticky toe pads make it an agile climber, and the ability to alter its coloration helps it blend with the surroundings. It often hangs out on windows and doors to catch insects attracted to porch lights. The somewhat bigger female can lay up to a thousand eggs in small clusters on aquatic vegetation during a breeding period.

Gray Treefrog
Hyla versicolor L 1.25–2 in (3–5 cm)

The widespread Gray Treefrog is usually described with its look-alike relative, the Cope's Gray Treefrog *(Hyla chrysoscelis)*, as identification is difficult and the two have overlapping ranges.

KEY FACTS

Body is gray to green to almost white with rough skin on back; dark-edged light spot under eye; underside of thighs yellow-orange.

+ **habitat:** Trees and shrubs in or near water

+ **range:** Extreme southeastern Canada and eastern half of U.S. to northern Florida

+ **food:** Insects

The Gray Treefrog has a squat body and warty skin, quite a bit more toadlike than most treefrogs. It is a nocturnal species that usually remains in low trees and shrubs, coming to the ground only in the breeding season to call and mate. Cope's Gray Treefrog has a call that is usually a faster, higher pitched trill than the slower, more chirplike call of the Gray Treefrog. Even this distinction is tricky, because a lot depends on the ambient temperature at the time. Even if the two are calling in the same pond, it is difficult to distinguish them. The best way is to examine their significant chromosomal differences in a lab.

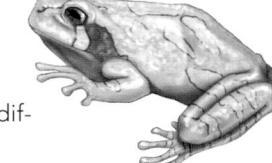

Cuban Treefrog
Osteopilus septentrionalis L 1.5–5.5 in (4–14 cm)

The Cuban Treefrog was introduced, probably inadvertently, from Cuba to southern Florida early in the 20th century. It is now the largest treefrog in the United States.

KEY FACTS

Body is bronze, green, or gray with warty skin; toe pad is as large as the tympanum (external eardrum).

+ **habitat:** Any source of consistent moisture, such as cellars, drains, fishponds, and well-watered potted plants

+ **range:** Southern Florida and Keys

+ **food:** Insects, spiders, and frogs

Enormous, almost comical, toe pads define the Cuban Treefrog, which wasted no time making itself comfortable in southern Florida. It likely came in through the Keys and is expanding successfully along Florida's coasts. It is a fierce and clever predator and takes advantage of all that the developed world offers, such as lighted billboards providing a smorgasbord of insects at night, when this species is active. Its call has been described as a snarly rasp; it has a sound suggesting that the frog is "ticked off." As with many introduced alien species the Cuban Treefrog's gain is a loss for native species—of habitat, food, and security.

Spring Peeper

Pseudacris crucifer L 0.25–1.25 in (0.6–3.1 cm)

The Spring Peeper is a well-known harbinger of spring. It congregates in large numbers around ponds and vernal pools to mate and lay eggs, singing a chorus of high-pitched whistles.

KEY FACTS

Body is tan to brown to gray; back often has a dark X shape; legs may have dark bars.

+ **habitat:** Wooded areas with permanent or temporary ponds and pools

+ **range:** Southeastern Canada and eastern half of U.S.

+ **food:** Insects, spiders, and other invertebrates

In northern winters, this frog hibernates under logs or loose bark. It emerges early in the spring, often earlier than other species, to begin the breeding season. The male calls from vegetation overhanging the water. The call sounds like an ascending whistle and sometimes contains a trill. Groups of less desirable males may position themselves near dominant males and intercept females heading toward the dominant ones. Females lay hundreds of eggs attached individually or in small clusters to underwater objects. In the South, the mating season runs through the winter. The Spring Peeper makes itself scarce at other seasons.

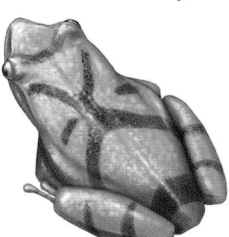

Pacific Treefrog

Pseudacris regilla L 0.75–2 in (2–5 cm)

Coming to a theater near you: The voice of the Pacific Treefrog is heard in many Hollywood films when nighttime nature sounds are required in a scene, regardless of the purported location.

KEY FACTS

Body is green to tan to red to black, often with dark blotches; dark line through eye; male has gray throat.

+ **habitat:** Forests, grasslands, shrublands, and farmland, near water sources

+ **range:** Southern British Columbia through California, east to Montana and Nevada

+ **food:** Insects and other invertebrates

A common treefrog found within a wide variety of habitats ranging from sea level to more than 11,000 feet (3,300 m), the Pacific Treefrog nevertheless is a ground dweller, sticking close to shrubs and grasses near water. It also occurs on some coastal islands. During the breeding season, which lasts from January through August, males congregate in temporary and permanent bodies of water to plead their case. The male's advertisement call is a two-part *ribbit* with a rising inflection, repeated continuously. Groups of this tiny frog can create quite a din over a prolonged period. The female lays eggs in clusters of 9 to 70 eggs at a time.

Western Chorus Frog

Pseudacris triseriata L 0.75–1.5 in (2–3.8 cm)

Tiny widespread chorus frogs get an early start on the breeding season, singing from the surface of ponds and vernal pools that may retain traces of ice.

KEY FACTS

Body is green-gray to brown to red; back has 3 variable dark stripes; dark stripe through eye.

+ **habitat:** Grassy areas, swamps, woodlands, farmland, and urban areas

+ **range:** Southern Ontario, western New York and Pennsylvania to Michigan, south to Ohio, Indiana, and southern Illinois

+ **food:** Invertebrates

Dorsal stripes on the Western Chorus Frog are usually unbroken, in sharp contrast to those of close relatives, several of which were recently considered subspecies of the Western. The Western Chorus Frog's upper lip also may have a thin white line. The male's advertisement call is a rising trill that is often described as the sound of raking a fingernail across the small-tooth end of a comb. He calls while perched on surface vegetation on the water, although as a general rule these frogs are difficult to spot. After mating, the female may lay up to a hundred eggs, attaching them in clusters to underwater vegetation.

African Clawed Frog

Xenopus laevis L males 2–2.5 in (5–6.3 cm) females 4–4.5 in (10–11 cm)

African Clawed Frogs can survive in just about any body of water, whether a natural stream, a fish farm pond, or a sewage system. They often thrive at the expense of native wildlife.

KEY FACTS

Mottled olive-brown body has smooth skin; no tongue or visible ears; webbed hind feet have 3 horny-tipped toes.

+ **habitat:** Natural and artificial bodies of standing and slow-moving water

+ **range:** Established in California; reported in many states

+ **food:** Aquatic invertebrates and vertebrates

The African Clawed Frog is misnamed: Its "claws" are actually toe tips that have formed a horny layer of dead skin cells. The males lacks a vocal sac, but it and the female still manage to call to each other. The eggs laid after mating can hatch within 36 hours into tadpoles with whisker-like barbels at the mouth. Released frogs account for the introduction of the species to the United States, and owning or dealing in this frog is outlawed, except by permit, in a number of states. Recently, the African Clawed Frog has been implicated in the spread of a chytrid fungus, to which clawed frogs are resistant, which can kill off entire amphibian populations.

A Humpback Whale and her calf, which has just surfaced, swim in the Pacific Ocean just off Hawaii.

4 | Aquatic Life

A Watery World of Wildlife

We are drawn to water and to the creatures that live in it. As evidence, the American and Canadian coasts are crowded with vacation homes, millions of people spend their holidays at the beach, and we are fascinated by aquatic wildlife. Maybe it's an attempt to return to our origins. Life on this planet evolved in the water before it started moving around on land. All living things require water to survive. In spite of all that, maybe the mystery of it grips us, the fact that creatures exist in the water in ways we are still trying to understand.

That Bullhead you first caught, or the Bluegill Sunfish—did you ever wonder where it fits in the world of fish? How about the Basking Shark and how it maintains its neutral buoyancy, or why Frilled Sea Anemones can lose their anchor but coral cannot? And what is the largest mammal ever to have lived on Earth? The pages in this chapter answer those questions and many more.

■ Life Underwater

Life underwater is something that we cannot truly experience with our submarine vehicles and personal diving gear. Fish have gills that derive oxygen from the water, and they can swim fast. Sperm Whales breathe air like we do, but they can dive thousands of feet into the ocean and hold their breath for an hour and a half. We can walk on land and breathe air like land mammals, and to some

extent, we can fly. But watery life occupies a niche that we cannot fully know.

Studies have been done on the way a dog regards its world, but few people know what an animal that depends on water for survival knows or responds to. These creatures have adaptations that we don't fully understand or even know about. Most of us think that fish cannot live out of the water for long, but snakeheads can wriggle from one body of water to another on dry land, without benefit of extracting oxygen from the water. Invertebrate sponges are unlike other animals in that they have no tissues, internal organs, or body symmetry.

Animals in water live in all kinds of temperatures, minerals, and depths. Brook Trout tend toward chilly, fast-moving freshwater streams. Manatees prefer slow-moving warm water. Herring live their entire lives in the sea. Salmon hatch in fresh water, move to salt water, and then when they are ready to spawn, return to the freshwater stream of their origin. Their final act after swimming upriver is to produce eggs and die. Males follow the females upriver and fertilize the eggs before dying themselves.

This chapter offers a sampling of the aquatic species of North America that many people encounter, and some that they don't. The next time you see catfish swimming, thank them for cleaning up the stream, which they do by eating green algae, carrion, and terrestrial insects. As the saying goes, "to understand is to appreciate."

■ Ever Present Danger

Life in the water thrives where death is ever present, where more creatures die than live to adulthood. Invertebrates lay thousands of eggs, many of which are eaten by other creatures, and some even by their parents. Small fish are eaten by larger fish. Some small fish gather in schools where the ever-shifting population gains its protection from being in a large group.

"Large fish eat smaller fish" is a truism, sometimes to a fault. A famous example is the fossil of a fish with a smaller fish protruding from its mouth, which it could not swallow. Both died. The larger the fish, the fewer young it seems to produce, because so many are eaten as eggs or young fish. A mature Menhaden can produce 362,000 eggs, while a Giant Oceanic Manta Ray of 3,000 pounds produces only one pup. The Manta Ray may grow to be huge and to prey on young fish. But it is, in turn, sometimes eaten by sharks or Killer Whales. Sand Tigers, which can reach a weight of 350 pounds, give live birth to few young because their pups eat each other in the womb.

■ People Threaten Life in the Water

Into the world of watery wildlife came humans, in ever-increasing numbers and ever-increasing

ANATOMY OF A FISH

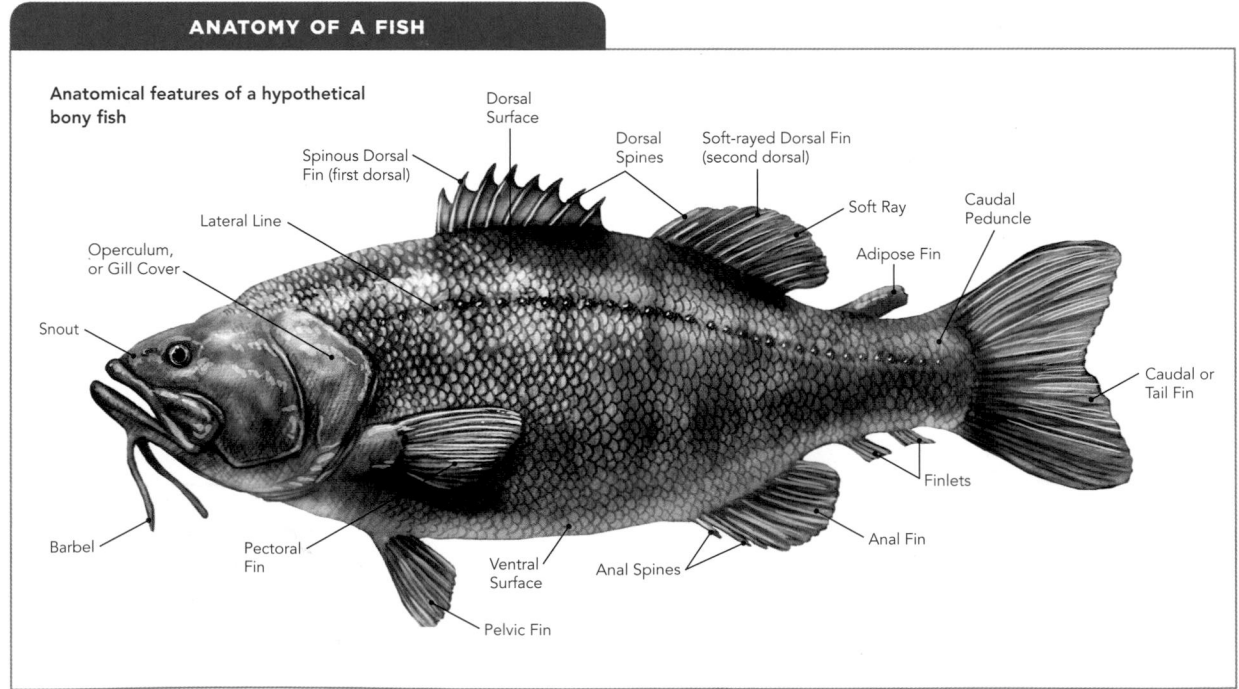

Anatomical features of a hypothetical bony fish

Dorsal Surface

Spinous Dorsal Fin (first dorsal)

Dorsal Spines

Soft-rayed Dorsal Fin (second dorsal)

Soft Ray

Caudal Peduncle

Lateral Line

Operculum, or Gill Cover

Adipose Fin

Snout

Caudal or Tail Fin

Barbel

Pectoral Fin

Ventral Surface

Anal Spines

Finlets

Anal Fin

Pelvic Fin

A pair of Purple Sea Stars move slowly over the Pacific floor near Canada.

appetites for things aquatic. An international flotilla hauls in nets full of fish for the dinner table, especially since cholesterol and chemical additives were discovered in the flesh of land animals. The annual catch of some species is listed in many tons. Cod have been traded on international markets since the Viking period around A.D. 800. The fish were so numerous that they contributed to the development of the North American East Coast. A wood carving of a "sacred cod" still hangs in the chambers of the Massachusetts House of Representatives. Today, many groups consider cod unsustainable due to overfishing.

The conservation group Greenpeace has also added Haddock to its "red list," because of overfishing. Atlantic Salmon was once commonly caught off the American East Coast but is now mostly farmed. Goliath Groupers were eaten widely in restaurants because they are fearless and easily caught, and they have become endangered.

The sea, once it casts its spell, holds one in its net of wonder forever.
—JACQUES YVES COUSTEAU

Our planet is made up of much more salt water than fresh (97.5 percent compared to 2.5 percent), so naturally there are more species in the sea than in streams or lakes. As overfishing seems to threaten species in the sea, human pollution, habitat destruction, and invasive species threaten freshwater wildlife. Drains still drip untreated garbage into rivers. Lakes and sloughs still disappear under housing developments, croplands, and parking lots. Lake Trout, which can reach 100 pounds (45 kg), were once an important commercial fish in the Great Lakes until foreign lampreys, pollution, and overharvest depleted their numbers. Invasive species have threatened many other species. Voracious Snakehead fish from the Far East have invaded eastern streams and, although tasty themselves, have gobbled up local fish.

Zebra Mussels have escaped from the ballast of ships and spread in streams all over the East, replacing local mollusks despite control efforts. Invasive plants have changed our habitat as well. Purple Loosestrife began in the Great Lakes and spread to ditches and watery sloughs over much of the East by releasing millions of seeds. The plant has an attractive blossom but replaces native plant populations that support wildlife and prevent erosion.

Sockeye Salmon swim upstream during a spawning run in the Adams River, British Columbia, Canada.

■ Local Species Need Our Help

The first colonists would have gone hungry for vegetables if the Native Americans had not taught them to drop a small fish into each plant, guaranteeing fertilizer for the plant and a larger harvest. Local mollusks fed Native Americans for centuries, judging from the mounds of shells the eaters left.

Native species of aquatic wildlife, both saltwater and freshwater, need our help. They have assisted in our survival for centuries. Now it's time for us to tend to their welfare, lest they be gone for good. This chapter helps us to recognize them.

GLOSSARY

+ brackish: water that is slightly less salty than seawater.

+ carapace: a bony shield, part of the exoskeleton, covering the head and thorax of an animal but not its abdomen.

+ cirrus (plural cirri): a slender animal appendage or projection that varies in shape; may be tentaclelike or fingerlike.

+ detritus: particles that result from disintegration, worn off of a larger body.

+ echinoderm: any of a phylum of radially symmetrical marine animals such as sea star and sea urchin.

+ ectoderm: the outermost of three primary layers in a developing embryo.

+ endoderm: the innermost of three primary layers of a developing embryo.

+ exoskeleton: an external supportive covering of an animal such as an arthropod.

+ genus: a category of biological classification that ranks between family and species.

+ gonad: sex organ that produces sperm or eggs.

+ hermaphrodite: an animal having both male and female reproductive organs.

+ mesoderm: the middle of three primary layers in a developing embryo.

+ nematocyst: a capsule in a cell that erupts when stimulated, extending a stinging thread used to catch prey.

+ parasite: an organism living in or on the body or another organism to the detriment of the latter.

+ phylum: a category of biological taxonomy of animals that ranks above class and below kingdom.

+ phytoplankton: plankton composed of plants.

+ plankton: term for all small organisms that live suspended in water.

+ ray: the appendage of an echinoderm or structural element in the fin of the fish.

+ species: a logical division of animals or plants within the genus category.

+ tentacle: an elongated flexible structure of an animal, especially invertebrates, usually on the head or around the mouth.

+ whorl: one of the turns of a snail's shell.

+ zooplankton: plankton composed of animals.

AQUATIC LIFE

||||||||||||||||||||||||||||||||||||||||||||||||

Lake Sturgeon

Acipenser fulvescens L up to 6 ft (1.8 m)

The Lake Sturgeon, like the Atlantic Sturgeon but smaller, is a primitive fish from the age of dinosaurs. It is covered with bony plates, not scales, and its skeleton is cartilage, not bone.

KEY FACTS

An extremely large and long-lived freshwater fish, the Lake Sturgeon is greenish gray and has a pointed snout with mouth underneath.

+ **range:** Great Lakes and Mississippi River

+ **diet:** Small clams, worms, leeches, and fish eggs

+ **name origin:** From French *estourjon,* meaning "the stirrer"

The Lake Sturgeon is one of the largest fish in North American fresh waters, with some individuals reaching a weight of more than 200 pounds (91 kg). It was once plentiful in the Great Lakes and the Mississippi River, and was killed as a nuisance because it damaged fishing gear. Overfishing, pollution, and slow rates of reproduction have greatly depleted its numbers, and today it is protected in several states and provinces. Its eggs, made into caviar, are considered a delicacy, and its meat is tasty. Some remain in their natal rivers for their first summer. The species is making a comeback due to protection efforts and improvements to its environment.

Alligator Gar

Atractosteus spatula L up to 10 ft (3 m)

The Alligator Gar is the largest gar species in North America. It is a predaceous carnivore. This primitive ray-finned fish has an elongated, torpedo-shaped body and a mouth full of teeth.

KEY FACTS

Once worldwide, it is now found only in the Americas; color is olive green or yellow.

+ **range:** Coastal waters west to Texas and Oklahoma and north into the Mississippi, lower Ohio, and Missouri Rivers

+ **diet:** Other fish and waterfowl, small turtles, and carrion

+ **name origin:** Resembles an alligator

Fierce in appearance, the Alligator Gar has never been known to attack people. In addition to a diminished range, their numbers have also decreased due to overharvesting. This warm-water gar prefers large, slow-moving streams but can exist in brackish water and even in salt water. This large freshwater fish is known to reach 10 feet (3 m) and 300 pounds (140 kg). It has few predators, but small gars are known to be eaten by other fish, and alligators sometimes attack adult fish. The Alligator Gar can live for hours outside the water because it has a buoyancy bladder that can extract oxygen from the above-water air.

Bluntnose Minnow
Pimephales notatus L 2.5 in (6 cm)

Perhaps the most abundant fish in North America, the Bluntnose Minnow is not very big and is not eaten by humans but is used as bait. It prefers shallow water with sandy or gravelly bottoms.

KEY FACTS

A dark line runs from head to tail, ending with a dark dot.

+ **range:** Streams, rivers, and lakes in eastern U.S.

+ **diet:** Aquatic insects, small crustaceans, Creek Chub eggs, algae, and decaying organisms

+ **name origin:** For its blunt, rounded head with a mouth that ends just below the snout

If you see a small fish in a small stream, it is probably a Bluntnose Minnow, which starts small like most fish but never grows beyond 2.5 or 3 inches (6.3–7.6 cm). It is one of the first fish to inhabit a stream and serves as prey for many species, including other fish and birds such as the Great Blue Heron and night herons. It is used as bait to catch larger fish and is sometimes farmed for this purpose. Its maximum life span is five years in captivity, but two years is more realistic in the wild. Males select the spawning site and it spawns repeatedly starting in May and continuing into August. It is sometimes introduced by bait bucket release.

Common Carp
Cyprinus carpio L up to 47 in (1.2 m)

Endemic to Asia, the Common Carp was introduced to North America and now exists coast to coast. It has large scales and a forward-protruding mouth with barbels, which it uses to find food.

KEY FACTS

The silvery-colored fish can grow to a large size.

+ **range:** Streams, ponds, lakes, and freshwater habitats in North America

+ **diet:** Almost anything, from minnows to aquatic insects, algae, and microorganisms

+ **name origin:** From Greek *cyprinus* and Latin *carpio*; both mean "carp"

The first immigrants could hardly believe that this country had no Common Carp when they settled here, for it was long a food fish and a garden decoration in Europe and Asia. Austria had 20,000 acres (8,000 ha) of carp ponds, and the ancient Romans practiced aquaculture with the fish. Today, China produces 70 percent of the domesticated carp in the world. The Common Carp has never reached such popularity in the United States where its tendency to muddy waters by rooting on the soft bottom for food has displaced sport fish. Hardy, it can gulp air at the surface and therefore survive in waters where other fish perish.

Creek Chub

Semotilus atromaculatus L up to 10 in (25 cm)

The Creek Chub, also called Horned Dace, is one of the most common minnows in eastern North America and is often used as a baitfish to catch larger specimens.

KEY FACTS

Appearance varies from greenish brown to dark on top; gradually turns cream-colored with a lateral dark line on the sides.

+ **range:** Slow-moving streams and coastal areas of large lakes

+ **diet:** Insect larvae, adult insects, and smaller fish

+ **name origin:** From its abundance in small streams

The Creek Chub is a medium-size minnow that can grow to 10 inches (25 cm). It is often used as baitfish for larger sport fish. Primarily a school fish, it rarely ventures farther than 165 feet (50 m) from its home. During spawning, the male develops large tubercules on its head. He digs a pit in a streambed, then piles small rocks at the upstream edge, hauling them in his mouth. He then curls around the front of the female in the pit, fertilizing the eggs and covering them with gravel. There is no further parental care of the young. The Creek Chub tolerates some pollution and clouded water.

Golden Shiner

Notemigonus crysoleucas L of adult up to 5 in (13 cm)

The Golden Shiner usually stays small, but in a few areas it can grow up to 12 inches (30 cm). It is native to eastern U.S. drainages but has been introduced in the West.

KEY FACTS

Gold and olive-silver in color, it has a large anal fin and dusky line extending its entire body.

+ **range:** Slow-moving water in eastern North America; found in reservoirs due to fishermen releasing bait

+ **diet:** Zooplankton, insects, plants, and algae

+ **name origin:** For its golden, shiny color

The Golden Shiner is primarily a small minnow used as food by larger fish and baitfish by sport anglers. It likes weedy areas to hide from larger enemies and is often found in the quietest part of the river. The Golden Shiner can reproduce in southern waters at one year of age, but first breeding is more commonly at three years in colder waters in Canada. It uses the nests of larger fish like Smallmouth Bass, laying its smaller eggs among bass eggs. Unlike birds such as cuckoos, which do the same, Golden Shiner young do not harm their host's young. It is tolerant of pollution and high temperatures.

Brown Bullhead

Ameiurus nebulosus L usually 10 in (25 cm)

Best known of the catfish family, the Brown Bullhead is smaller and bulkier, and less sleek than other catfish species. Also called the Horned Pout, it has barbels on its face for finding food.

KEY FACTS

Its coloring is olive brown with white underside.

+ **range:** Shallow, weedy, muddy water of streams, lakes, and ponds of eastern middle states, but has been introduced to western states

+ **diet:** Insect larvae, nymphs, crustaceans, mollusks, and snails

+ **name origin:** For its oversize head

The Bullhead is often caught by young Midwestern anglers, who quickly learn to detach the hook carefully lest they be horned by the sharp spikes on the fins. This species never gets as big as its cousin the Channel Catfish because it is genetically programmed to remain small. It is also distinguished from other catfish by its large head. Like other catfish, the Brown Bullhead is a cavity nester, hollowing out a nest under submerged logs or stumps or holes in the bottom; both sexes stay with the nest during incubation. The fish can best be caught in the morning or early evening on a wide variety of baits.

Channel Catfish

Ictalurus punctatus L up to 4.5 ft (1.4 m)

The Channel Catfish is common west to the Rockies. It has no scales, small eyes, and long feelers on its face called barbels with which it locates food. It is informally called the channel cat.

KEY FACTS

Its coloring is olive brown or slate blue and white on bottom.

+ **range:** Streams, lakes, and ponds all over U.S.

+ **diet:** Mostly insects, small crayfish, and other fish (young); the same, plus clams, mussels, and carrion (adults)

+ **name origin:** For feelers that resemble a cat's whiskers

The Channel Catfish is easily farmed and tastier to some people than trout. It is North America's most numerous catfish species. In the wild, it is the most fished catfish with millions of anglers concentrating on channel cats. The fish can live in murky water, thanks to the barbels on its face, which allow it to find food. In addition, it has taste buds all over its body, which give it exceptional abilities of smell and taste. It has a deeply forked tail. The world record size for the Channel Catfish is 58 pounds (26 kg), but the Blue Catfish, a close relative, can grow to more than 100 pounds (45 kg).

Brook Trout

Salvelinus fontinalis L 10–26 in (25–66 cm)

The only trout native to Eastern North America, this popular game fish, also called Brookie and Speckled Trout, is not the largest trout; the biggest on record was less than 15 pounds (6.8 kg).

KEY FACTS

Basic coloration is green or brown and a marbled lighter color on top and sides.

+ **range:** Clean, fast-moving cold water of small springs and creeks, plus lakes and ponds

+ **diet:** Insects

+ **name origin:** For habitat where it is usually found; *fontinalis* means "spring or fountain"

The presence of the Brook Trout shows that the water is clean, cold, and hasn't had its pH altered. Because of its tastiness and reputation as a game fish, it has been introduced around the world, and is also commercially farmed. Brook Trout spawn in gravelly areas, with one or more males fertilizing the eggs and the female then covering the nest. Eggs require oxygenation and hatch in two or three months, in nests usually in spring-fed streams or ponds and lakes with seepage. It has a sprinkling of red dots toward its flanks; the most distinctive coloration is the reddish color of the belly and lower fins.

Brown Trout

Salmo trutta L about 20 in (51 cm)

The Brown Trout is native to Europe and Asia but was introduced to the United States where it is widespread, and reduces native species by predation, displacement, and competition for food.

KEY FACTS

Freshwater version is brown with a creamy belly and spots over its upper body.

+ **range:** Clear, cold-water streams, lakes, and reservoirs

+ **diet:** Insects and larvae (young); smaller fish, frogs, and mice (adults)

+ **name origin:** Named for its color, with second Latin name meaning "trout"

Brown Trout eggs were introduced to North America in 1883, because the fish was considered a fine fighter. Self-sustaining populations are now in more than 30 states. Those transplanted to the Great Lakes are larger than those in streams, some growing to 20 pounds (9 kg). The lake dweller travels up tributary streams to spawn, but unlike oceangoing Pacific salmon, it does not always die after spawning. It is long-lived, some individuals to 20 years. Some Brown Trout migrate to salt water where they are known as Sea Trout. High water temperature or introduction of farm fertilizers causes losses to the species.

Cutthroat Trout

Oncorhynchus clarkii L 7–12 in (18–30 cm)

The Cutthroat Trout is found in the northwestern United States and Canada with reddish markings on its throat. Many subspecies occupy the Rocky Mountains and Cascade Range.

KEY FACTS

Gold to gray to green on top; always has red, pink, or orange on underside of mouth and gills.

+ range: Fast, cold streams in the Northwest from Alaska to northern California, and western Canada

+ diet: Mature insects

+ name origin: For red markings under the mouth and lower gills

This favorite of fly fishers is limited to cold streams of the West. It varies widely in coloration and size. It will interbreed with other species of trout, making identification difficult. Generally the Cutthroat Trout is known by the pink or red markings under its mouth and lower gills, which give it its name. Adults will rise to a dry fly in warm months, but a small spinner in chillier months, as adults will also eat small fish, leeches, and snails. Because Cutthroat Trout interbreed with other trout, they are known simply as "trout" in many places and are given local names such as Colorado trout.

Lake Trout

Salvelinus namaycush L 36 in (91 cm)

Though it is called a Lake Trout, this largest fish to be called a trout is not a true trout or a salmon but a freshwater char. It is also called Mackinaw, Lake Char, Touladi, Togue, and Gray Trout.

KEY FACTS

The fish has light spots on darker gray, with lower fins edged with white.

+ range: Deep, cold waters of large lakes.

+ diet: Other fish but also crustaceans, insects, and small mammals

+ name origin: Common name from deep lakes; *namaycush* is derived from the Cree language

The Lake Trout is a large fish in the Great Lakes, with some individuals weighing up to 50 pounds (23 kg), but most weigh about 10 pounds (4.5 kg). The Lake Trout has been fished commercially and for sport, although the advent of the parasitic Sea Lamprey drastically reduced its numbers between 1936 and 1965. Controls have now cut down on lamprey depredations, and Lake Trout populations are recovering. It is long-lived, with some individuals exceeding 25 years. During spawning, eggs are deposited after dark in shallow water and often take four to six months to hatch. Sexual maturity is reached in six or seven years.

Lake Whitefish
Coregonus clupeaformis L average 18 in (46 cm)

The Lake Whitefish likes cool water. It is sometimes called Humpback because of its small head compared with the rest of its body. Its mouth ends short of the tip of its snout.

KEY FACTS

The shape of its mouth allows it to feed on the bottom or at the surface.

+ **range:** Great Lakes and large lakes of northern U.S. and parts of Canada

+ **diet:** Small snails, clams, small fish, and insect larvae, and surface mayflies and midges

+ **name origin:** For its color and preference for lakes

The Lake Whitefish remains one of the most commercially valuable denizens of the cooler North American waters, despite its numbers being depleted in the early 1960s by overfishing and pollution. Environmental cleanup and careful fish management have largely remedied this and one can now buy Lake Whitefish smoked, pickled, or vacuum-packed in grocery stores across the U.S. The Lake Whitefish has a snout that overhangs its short lower jaw; as a result, the mouth opens in a slightly inferior position. This allows the fish to feed on lake bed bottoms or grab food particles off the top of the water.

Rainbow Trout
Oncorhynchus mykiss L 8–12 in (20–30 cm)

The torpedo-shaped Rainbow Trout, also known as Salmon Trout, is native to tributaries of the Pacific Ocean. It has been introduced for food or sport in every continent except Antarctica.

KEY FACTS

The fish has small black dots on its sides and fins.

+ **range:** West Coast rivers emptying into the Pacific

+ **diet:** Insects, small fish, leeches, snails, and algae

+ **name origin:** Pink streak that runs along its side; species name—*mykiss*—derives from the Kamchatkan name *mykizha*

The Rainbow Trout may be the most sought-after fish in North America. A close relative is the Golden Trout, which has a gold field instead of green. The Rainbow's fighting ability and tasty flesh have led it to be introduced in other parts of the U.S. and other countries, with effects largely negative. It eats small native fish and has introduced "whirling disease" into U.S. and foreign rivers. Most are freshwater fish, but some go into the ocean where they are known as Steelhead. They have silvery heads when oceangoing, but males sport red lines when they enter freshwater rivers to spawn.

Chain Pickerel

Esox niger average L 25 in (64 cm)

A species that preys on other fish, the Chain Pickerel has a mouthful of sharp teeth. This member of the pike family is also called the Southern Pike, Grass Pike, Jackfish, and Eastern Pickerel.

KEY FACTS

Color is dark green on top fading to lighter green, with a dark vertical line under eye.

+ range: Mississippi Valley north into lower Canada and south to the Florida Panhandle

+ diet: Other fish; also frogs, small mammals, and crayfish

+ name origin: For chainlike markings on its sides

The Chain Pickerel, a fish of the eastern U.S., puts up a tough fight when hooked. With its sharp teeth it prefers small fish, but sometimes seizes fish half its size, which it swallows by degrees. A warm-water, solitary fish, it can also tolerate cool water. Its body outline resembles that of the Northern Pike and may sometimes reach lengths of up to 30 inches (76 cm). Those who catch the fish with hook and line best take along a pair of pliers to extract the hook, for the fish does not give up easily. The Chain Pickerel is considered good eating, but great care must be taken to remove small bones.

Muskellunge

Esox masquinongy average L 48 in (122 cm)

Largest of the pike family, the Muskellunge, also known as Muskie, looks very like a Northern Pike or Chain Pickerel, but it is bigger. It moves little, preferring to wait for its prey to come by.

KEY FACTS

It has a single dorsal fin and green background with dark spots.

+ range: Clear cold water of Great Lakes and northern states; can withstand temperatures of 90°F (32°C)

+ diet: Mostly fish, sometimes frogs, crayfish, and small rodents

+ name origin: Ojibwa *mashkinoozhe* meaning "ugly pike"

The Muskellunge is considered a top game fish because of its fierceness and large size. Its typical length is from 28 to 48 inches (71 to 122 cm) and up to 36 pounds (16 kg), although some of 6 feet (1.8 m) and 70 pounds (32 kg) have been known. When caught, it often clears the water and shakes its head to throw off the hook. In summer it stays near the shore, hidden by shadows of overhanging foliage or a sunken log until prey comes by. An efficient killing machine, it lunges forward and grasps a meal with a mouth full of needle-sharp teeth. Its cheek and gill cover have scales. In winter it moves to deep water.

Northern Pike

Esox lucius L to 59 in (150 cm)

The Northern Pike resembles the Muskellunge and Chain Pickerel except for its markings and size difference. Weights of 55 pounds (25 kg) are larger than a pickerel and smaller than a Muskie.

KEY FACTS

Color is olive green lightening to yellow and white on belly; sides have beanlike markings.

+ range: Cool waters east of Rockies, especially in northern Midwest; occurs as far north as the Arctic

+ diet: Other fish; also water voles and ducklings

+ name origin: For the pikes, and it is usually found in the north

The Northern Pike, or just "Northern," puts up a hard fight when hooked. It can be caught through the ice because it is active in the winter. It can remain stationary by barely moving its dorsal and breast fins; it then bends its body to dash forward and catch prey. Its backward-slanting sharp teeth often immobilize smaller fish by catching them sideways, moving them headfirst to swallow them. Pike have been found dead by choking on fish as large as themselves. While it primarily lives east of the Rockies, some have been introduced to the western U.S. It is also found in cool waters of Russia and Europe.

Blackstripe Topminnow

Fundulus notatus L 2–3 in (5–8 cm)

A Midwestern species, the Blackstripe Topminnow occupies a narrower habitat than the Bluntnose Minnow. It is affected by human alteration of streams and usually occurs in lowlands.

KEY FACTS

A black stripe runs the length of this fish, and the mouth is slightly upturned.

+ range: Slow-moving water with vegetation

+ diet: Insects and insect larvae, tiny mollusks, spiders, and crustaceans

+ name origin: For the stripe along its body, and for its tendency to stay near the surface

Less ubiquitous than the Bluntnose Minnow, the Blackstripe Topminnow is an indication of a stream's health. Grazing livestock eat vegetation that the fish depends on for food. Oil seepage from machines may alter the surface of the water where these minnows eat. Irrigation may put them in danger during low water or drought. They can be identified by the gold spot on their flat heads. Blackstripes are found in wetlands and lakes where the vegetation seems critical to their survival. Males have dark vertical bars above and below their black stripes and yellow fins, and females have no extra bars and white fins.

Mosquitofish

Gambusia affinis L 1.6–2.8 in (4–7 cm)

A small fish that eats mosquito larvae and other harmful insect larvae, the Mosquitofish, also known as the Gambezi, has been introduced as part of an abatement program for pesky insects.

KEY FACTS

Both sexes are dull gray; female has a gravid belly and is slightly larger than male.

+ range: Thrives in still water including ditches, dredge impoundments, birdbaths

+ diet: Mosquito larvae, plankton, insects, and larval fish

+ name origin: For its voracious appetite for mosquito larvae

The male Mosquitofish forms his anal fin into a tube that deposits milt inside the female's genitalia. About 16 to 28 days later, the female gives birth to about 50 live young instead of producing eggs. Originally found in the Mississippi River drainage, this minnow has become the most widespread of freshwater fishes, exported all over the world as part of mosquito abatement programs. In some parts of California and Nevada, it is grown in aquariums and distributed free to residents who stock it in stagnant pools of water. The Mosquitofish is tolerant of high temperatures and low-oxygen environments.

White Bass

Morone chrysops L 10–12 in (25–30 cm)

The White Bass is a medium-size native, abundant in Midwestern waters. Written records place it there in the early 1800s. It is also known as Sand Bass, Barfish, Streaker, and Silver Bass.

KEY FACTS

A silvery fish, it has a golden eye and incomplete stripes that run horizontally on its sides.

+ range: Large reservoirs and rivers; goes up shallow streams to spawn

+ diet: Mostly freshwater copepods and sand fleas, with larger ones dining on worms and small fish

+ name origin: For its color

Medium in size—the world record is between 6 and 7 pounds (2.7–3.2 kg)—the carnivorous White Bass is abundant in its native Midwest and easily transplanted elsewhere. Kansas, for example, does not even have a daily creel limit, and limits elsewhere are up to 25. For anglers with hook and line, the fish will bite on worms and small fish if not frightened. In spawning, the White Bass creates no nest. The male bumps against a female's abdominal area after which she rises near to the surface and begins releasing eggs. Several males may fertilize them as they sink to the bottom and adhere to logs and vegetation.

Bluegill

Lepomis macrochirus L to 16 in (41 cm)

The Bluegill, a member of the sunfish family, has a small mouth and is easily identified by bright blue edging on its gill cover. It is also called Bluegill, Bream, Brim, and Copper Nose.

KEY FACTS

It has dark blue around its gills, head, and chin; abdomen is yellow and orange on breeding males.

+ **range:** Natural range east of Rocky Mountains

+ **diet:** Microscopic and near-microscopic plants and animals, such as worms, leeches, snails, and small fish

+ **name origin:** For blue around its gills

The Bluegill, which prefers warm water, is often teamed with bass in farm ponds, because the bass feed on young Bluegills and Bluegills eat bass eggs. Living in schools of 10 to 20, the saucer-shaped fish remains near the surface in the morning to stay warm. But it does not like direct sun, and often hides in the weeds. It retreats to deep water on the hot days of summer. The sunfish has beautiful coloring, with deep blue and purple on its face and gill flap. It is an important and abundant sport fish in the United States, and is easily caught while still fishing during the day as it stays near the surface.

Largemouth Bass

Micropterus salmoides L to 29 in (74 cm)

The Largemouth Bass is the top predator in many aquatic habitats; it grows to be the largest of the black bass family. It is also called Brown Bass, Widemouth Bass, Black Bass, and Bucketmouth.

KEY FACTS

Color is usually dark green on top, lightening farther down, with dark blotches in a line on its sides.

+ **range:** Lakes, ponds, sluggish streams, and swamps of the U.S.

+ **diet:** Other fish, insects, worms, mussels, and frogs

+ **name origin:** For the big mouth that extends beyond its eyes

This game fish is known to jump out of the water and shake its head when caught, trying to throw the hook from its mouth. An aggressive warm-water fish, the Largemouth Bass strikes at anything it thinks is alive, including rodents and birds. It responds well to being introduced to other locations. The fish hollows out a nest depression on the bottom, and after fertilizing the female's eggs, the male stays near the nest, attacking anything that approaches. The fry stay in tight schools until they reach an inch (2.5 cm) in length. As adults, they are mostly solitary, hiding in weeds and striking from ambush.

Rock Bass

Ambloplites rupestris L 6–8 in (15–20 cm)

Smaller than the Largemouth or Smallmouth, the Rock Bass rarely exceeds a pound (0.5 kg).
It can change color to match its surroundings, and it is a commercial species in the Great Lakes.

KEY FACTS

The Rock Bass has black dots all over and goggle eyes, and six spines in its anal fin instead of three like other sunfish.

+ range: Rocky, clear, cool lakes and stream pools

+ diet: Aquatic insects, crayfish, and small fish, including its young

+ name origin: For the rocky areas it inhabits

The Rock Bass is smaller than its relatives but strikes at the same bait—night crawlers and artificial lures. Once caught, it puts up a lively fight and is considered good eating. Its large eyes suggest it prefers hunting for food at night, and its chameleonlike ability to change colors to suit its surroundings probably accounts for its continued survival. Like other sunfish it hollows out a nest in the shallows, which the male guards, driving off other fish that approach, including the female once she has laid her eggs. It lives in groups and often associates with the Smallmouth Bass as an adult.

Smallmouth Bass

Micropterus dolomieu L to 20 in (51 cm)

The Smallmouth Bass is smaller than its cousin, the Largemouth Bass. It favors rocky outcrops, logs, and submerged trees, and is also called Bronzeback, Bronze Bass, and Bareback Bass.

KEY FACTS

It is green with brown vertical dark bands. Like the Largemouth, it has lots of fight and tasty flesh.

+ range: Cool streams and clear lakes in much of U.S. and Canada

+ diet: Smaller fish, crayfish, and insects

+ name origin: For its mouth, which is smaller than Largemouth Bass but still sizable

The Smallmouth Bass is a member of the sunfish family, and is more streamlined and about half the size of its relative, the Largemouth Bass *(M. salmoides)*. Introduced beyond its native range primarily in the central United States and Canada, it frequents cool-flowing streams and large clear lakes. It is a top game fish in the eastern United States. It is said to be, pound for pound, the most sporting of all freshwater fish. Although found in cooler waters than the Largemouth, it tolerates warmer waters than trout will, and it has replaced trout in many streams. Its mouth does not reach a vertical line with its eye.

White Crappie

Pomoxis annularis L to 8 in (20 cm)

With its close relative the Black Crappie *(P. nigromaculatus)* the White Crappie occupies the same habitat all over North America. The White Crappie is often considered our most tasty game fish.

KEY FACTS

It has a vertical line on its light body.

+ **range:** U.S. and Canada; prefers sluggish water, lakes with timber or brushy cover

+ **diet:** Very young Northern Pike, Muskellunge, and Walleye; also insects, zooplankton, and crustaceans

+ **name origin:** French Canadian *crapet,* applied to many kinds of sunfish

Both White and Black Crappies are fished in natural and man-made lakes, in every part of the United States and parts of Canada. They are not large fish; the state record for Minnesota is 5 pounds (2.3 kg) for Black Crappie and nearly 4 pounds (1.8 kg) for the White Crappie. They both spawn under sunken brush or against stumps. The male keeps the water around the eggs in constant motion with his pectoral fins, and savagely bites other fish that come close. The Black Crappie is darker than the White Crappie and has irregular black blotches. In spring, the two crappies occupy shallow coves but later move to deeper water.

Northern Snakehead

Channa argus L to 40 in (102 cm)

An invasive threat to native species, the Northern Snakehead is a carnivore that outcompetes and preys on local fish; it can live several days out of water and wriggle to other waterways.

KEY FACTS

It has a long dorsal and anal fins and large mouth extending behind the eye, and sharp teeth.

+ **range:** Widespread in North America, prefers stagnant water and slow, muddy streams with low oxygen content

+ **diet:** Other fish and aquatic vegetation

+ **name origin:** For its narrow snakelike head and mouth

The Northern Snakehead is the current bad guy of the fish world. It is native to China, Russia, and the Koreas, where it was an important food fish. It spread internationally and was first found in a pond in Maryland, in 2002. Since then, it has been found in the Potomac River and as far north as a park pond in British Columbia. Attempts have been made to eradicate the Snakehead by poisoning waterways, with instructions from the U.S. Fish and Wildlife Service on how to kill and preserve the fish if caught. Juveniles eat zooplankton and crustaceans, thus denying the food source to local species.

Atlantic Sturgeon

Acipenser oxyrinchus L to 14 ft (4.3 m)

The Atlantic Sturgeon lives up to 60 years and is one of the largest non-sharklike fish in salt water, reaching 800 pounds (360 kg). It attains some of the greatest ages of any fish in the world.

KEY FACTS

Its body is covered with bony plates, not scales like most other fish.

+ range: Original range, salt waters of New Brunswick, Canada, to Florida's east coast

+ diet: Worms, crustaceans, and mollusks

+ name origin: Old French *estourjon*, meaning "the stirrer"; Atlantic signifies its range

The Atlantic Sturgeon looks, and is, primitive, having five bony plates called "scutes" instead of scales. The round mouth on the bottom of its head is actually a tube that extends when it is eating food, which the fish locates by barbels in front of its mouth. It often stirs up mud with its long snout to find prey. It was once more plentiful, and individuals averaging 100 pounds (45 kg) were common, but overfishing and pollution have depleted the numbers. The Atlantic Sturgeon is not sexually mature until six or seven years old. It spends the first six years of its life in brackish waters where it is born.

American Eel

Anguilla rostrata L to 4 ft (1.2 m)

Unlike other North American fish, the highly adaptive American Eel lives most of its life in fresh water, traveling to salt water to spawn and die. It has a snakelike body and small pointed head.

KEY FACTS

In fresh water, it can be brown on top and yellowish tan below.

+ range: Coasts and in freshwater streams from western Greenland and eastern Newfoundland south to Gulf of Mexico, Panama, and Caribbean

+ diet: Dead fish, carrion, invertebrates, and insects

+ name origin: Middle English *ele*

The American Eel hunts at night and hides by day in the mud, sand, or gravel close to the shore. Despite its snakelike appearance and its covering of mucus, it was an important nutrient for aboriginal people. A European relative is very popular as a food fish across the Atlantic, but it is mostly used as bait by North American anglers seeking game fish. As a breeding adult, it swims to spawn in the Sargasso Sea in the west-central region of the North Atlantic Ocean. The young are transparent as they work their way back to the coast to enter freshwater streams, but they gain color when they become adults.

Atlantic Tarpon
Megalops atlanticus L to 8 ft (2.4 m)

The Atlantic Tarpon, or Silver King, is a large warm-water fish that grows to 350 pounds (160 kg) and can fill its swim bladder with air, like a primitive lung, to live in low-oxygen waters.

KEY FACTS

It has large, bony, shiny silver-colored sides, a darker greenish or bluish top, and a mouth that turns upward.

+ range: Chiefly in warm waters of the Caribbean and Gulf of Mexico; individuals are sometimes found as far north as Cape Cod

+ diet: Fish, shrimp, crabs, and fish

+ name origin: Unknown

The Atlantic Tarpon is targeted by sport fishers because of its size, fighting spirit when hooked, and the difficulty of catching it. It is very strong and makes spectacular leaps into the air. The fish can easily throw the hook; anglers are often asked, "How many tarpon did you jump?" instead of "How many tarpon did you catch?" Its meat is not very tasty, and there has never been a commercial fishery for it. The Tarpon hunts for food at night. It spawns far out at sea but often travels up freshwater rivers to feed; it prefers water of 72 to 82°F (22 to 28°C). Temperatures under 40°F (4°C) can be fatal.

American Shad
Alosa sapidissima L to 30 in (76 cm)

The American Shad, also called White Shad, is a saltwater fish that after four years in the ocean travels up freshwater streams to spawn, in January in the south and in July in the north.

KEY FACTS

Body has blue or green back, with large black dot on shoulders and other dots on body.

+ range: Atlantic from southern Labrador to northern Florida

+ diet: Zooplankton in deep ocean; large adults may eat worms and small fish

+ name origin: Old English *sceadd*, meaning "abundant"

The American Shad is usually between 3 and 5 pounds (1.4 to 2.3 kg) and spends most of its life in the sea. In spring it travels hundreds of miles up freshwater streams to spawn. A smaller relative—the Alewife—behaves similarly. During spawning it eats little or nothing. Females release eggs in the water column and males fertilize them, after which the eggs roll around the bottom. The young live in fresh water until the fall, when they go to the sea. After spawning northern adults return to salt water and live to spawn several times, but the southern Shad usually dies. Shad eggs, or roe, are considered a delicacy.

Atlantic Salmon

Salmo salar L 28–30 in (71–76 cm)

The Atlantic Salmon was plentiful off Atlantic shores 400 years ago, but overfishing and pollution have reduced its numbers. It is mostly farmed, but the Gulf of Maine has remnant populations.

KEY FACTS

The mature fish is silver-blue, with black spots mostly above lateral line.

+ **range:** North Atlantic

+ **diet:** Arctic squid and shrimp, amphipods, and small herring

+ **name origin:** Common name refers to its range; Latin *salmo* means "salmon" and *salar* means either "leaper" or "resident of salt water."

The Atlantic Salmon has been extirpated in the wild, although fish ladders and stocking attempt to bring it back. It is a medium-size fish with most weighing 8 to 12 pounds (3.6 to 5.4 kg), but a few individuals weigh 30 pounds (13.6 kg). Remnant wild populations go up freshwater rivers to spawn, and females sweep out a small trough in gravel with their tails and then cover the eggs after they are fertilized. When hatched, fry stay in the gravel for six weeks, living off their egg sac. In spawning, they turn green or red. Unlike Pacific Salmon, some adults may survive spawning and return to the ocean and spawn a second time.

Chinook Salmon/King Salmon

Oncorhynchus tshawytsha L to 58 in (147 cm)

Largest of the salmon family in North America, the Chinook Salmon or King Salmon spends most of its life at sea but returns to the freshwater stream of its birth to spawn, after which it dies.

KEY FACTS

Body has blue-green back with spots on sides and tail when at sea.

+ **range:** Pacific from Alaska to southern California

+ **diet:** Insects and crustaceans (young); jellyfish, sea stars, and other fish (adults)

+ **name origin:** For native people in Columbia River region where the salmon is found

The Chinook Salmon is the largest of the salmon family in North America. Individuals of 40 pounds (18 kg) are not unusual; some 5 feet (1.5 m) long and weighing 120 pounds (54 kg) have been reported. It matures at 36 inches (91 cm) and 30 pounds (13.6 kg). The salmon lives one to six years at sea before it returns to rapid waters of the Pacific Northwest to spawn. While spawning it turns red or partly red. Females prepare a nest, or redd, in gravel with the proper water flow. They then deposit eggs in several pockets of the redd, and males fertilize them. Adults guard the nest for a few days to a month before dying.

Coho Salmon/Silver Salmon
Oncorhynchus kisutch L 28 in (71 cm)

The Coho Salmon is similar to the Chinook and occupies the same area, but it is smaller. It is also called Silver Salmon. Like its larger relative, the Chinook, it spawns once and then dies.

KEY FACTS

The Coho in its oceanic phase has dark blue back and silver sides, with black spots on only the top half of its tail.

+ **range:** U.S. along the Pacific coast

+ **diet:** In freshwater stage, young eat plankton and insects; adults eat mostly smaller fish.

+ **name origin:** Russian *kizhugh*, for fish

The Coho Salmon is the backbone of the commercial salmon industry. Considered one of the tastiest of salmon, it has flesh that is deep red and low in fat. It is second to the Chinook Salmon in size, with an average of 7 to 11 pounds (3.2 to 5 kg), but ranging up to 36 pounds (16 kg). It spends some of its adult life at sea but returns to freshwater streams in the Northwest to spawn, preferring slower, more shallow streams and smaller gravel than its larger relatives. Spawning adults have a red color and hooked noses. Its native area also includes Russian waters, thus leading to its Russian-inspired name.

Pink Salmon/Humpback Salmon
Oncorhynchus gorbuscha L 18 in (46 cm)

The Pink Salmon is the smallest species in the genus of Pacific salmon. It prefers colder water than its cousins. Its young return quickly to the sea instead of growing for a while in fresh water.

KEY FACTS

It is bright silver in sea water.

+ **range:** From northern California to the Mackenzie River in Canada

+ **diet:** Shrimp and krill

+ **name origin:** "Pink" for the pink streak that extends along the sides of its body during spawning, and "humpback" for the hump that forms on the males

In spite of its small size, the Pink Salmon is a mainstay in the salmon market. It does not live as long as its cousins, adhering to a two-year life cycle. It prefers colder water, and its young do not pause in fresh water or brackish water estuaries but go directly to the sea. Some are even hatched in salt water. The female sweeps away a trough-shaped nest, or redd, with her tail, and while she deposits her eggs, several males fertilize them. Afterward, she covers the eggs with gravel and guards the nest for a few days until she dies. Pink Salmon turn darker with a pink streak along the belly when spawning.

Sockeye Salmon/Kokanee Salmon
Oncorhynchus nerka L 24–33 in (61–84 cm)

The Sockeye Salmon is one of the smaller Pacific salmon species and is known for its tasty meat. Also called Blueback Salmon, Silver Trout, and Redfish, it requires a freshwater lake to grow.

KEY FACTS

The fish has a blue back, silvery flanks, and black speckles while at sea.

+ **range:** Northern Pacific

+ **diet:** Mostly zooplankton; young may eat insects, and adults may eat small aquatic animals such as shrimp.

+ **name origin:** For *suk-kegh,* which means "redfish" in British Columbia's native Salish language

The Sockeye Salmon is a small salmon, usually weighing 5 to 15 pounds (2.3 to 6.8 kg). Its average life span in the wild is three to five years. Like all salmon, it is hatched in fresh water and spends more time there than other salmon—up to three years—before going out to sea as adults. It is the most delicious of the family, its orange-red meat coming from the orange krill it eats while in the ocean. Some landlocked salmon spend their entire lives in fresh water, where they are known primarily as Kokanee Salmon. Those landlocked fish are smaller than the Sockeye Salmon, rarely growing longer than 14 inches (36 cm).

Three-spined Stickleback/Tittlebat
Gasterosteus aculeatus L 2 in (5 cm)

The Three-spined Stickleback, a cool-water fish, is much studied because of its reproductive habits, genetic changes to adapt to a new environment, and ability to adapt to aquarium living.

KEY FACTS

The color is drab olive or silvery green, with breeding males turning red on throat, head, and belly.

+ **range:** Slow-moving waters such as rivers or lakes near coast

+ **diet:** Zooplankton, small crustaceans, insects, worms, and fish eggs

+ **name origin:** For sharp spines on its dorsal fin

The Three-spined Stickleback is the subject of scientific study. It lives in sea water but goes upstream to spawn. Some live permanently in fresh water because they have been cut off from the sea, an interesting adaptability. During reproduction, the male builds a half dome, then creates a tunnel through it. Females swim through the tunnel depositing eggs, and the male fertilizes them. He then chases the females away and guards the eggs, caring for the young until they leave the nest. The stickleback's sharp spines on its dorsal fin—usually three—can be locked erect, making it hard to swallow by predators.

Longhorn Sculpin
Myoxocephalus octodecemspinosus L 10–14 in (25–36 cm)

The Longhorn Sculpin, also known as Hornpout and Horndog, has a large head, slim body, and sharp spines on its head, and fins containing venom that causes a painful puncture wound.

KEY FACTS

It has three or four dark crossbars on its side.

+ range: Coastal waters of eastern North America from eastern Newfoundland, northern Gulf of St. Lawrence, south to Virginia

+ diet: Shrimp, crabs, mollusks, and fish

+ name origin: For the long spine along the gill cover that spreads when the fish is alarmed

The omnivorous Longhorn Sculpin can vary its color according to its surroundings, from dark olive to pale green-yellow to green-brown. The sculpin has no known commercial value and is considered a nuisance fish by anglers because it takes any kind of bait and spreads its fins and spines when landing, making it difficult to remove the hook. It grunts when pulled out of the water. The Longhorn Sculpin is comfortable in temperatures up to 65°F (18°C), but goes to cold deep sea water in winter. It is abundant in shoal harbors and bays, and it goes up onto the flats at high tide and then leaves the flats at low tide.

Striped Bass
Morone saxatilis L nearly 48 in (122 cm)

The Striped Bass, also called Striper and Rockfish, is an important game fish from the St. Lawrence to Florida. It was once so numerous that early colonists used it as fertilizer.

KEY FACTS

It is identified by dark horizontal stripes that run its length.

+ range: Native to East Coast ocean waters; goes up freshwater streams to spawn

+ diet: Smaller fish and eels, crustaceans, and invertebrates such as crabs, lobster, and squid

+ name origin: For the stripes on its otherwise silvery body

By the early 1980s, Striped Bass numbers had been reduced by overfishing and pollution. Through good management, its population has been restored, and sport and commercial fishermen alike now catch it. Like salmon, it goes up freshwater streams to spawn, but unlike most salmon it returns to the ocean alive. The Striped Bass lives some 30 years and reaches weights of 80 pounds (36 kg). The heaviest on record is 125 pounds (57 kg). Although naturally seagoing, some populations have been landlocked by introduction or trapped by dam building. It has also been introduced in the San Francisco Bay region.

Whale Shark

Rhincodon typus L 40 ft (12 m)

About the size of a bus, the Whale Shark is the biggest fish. The slow-moving creature is not dangerous to humans. Its large mouth has tiny teeth, and it filter-feeds only on small prey.

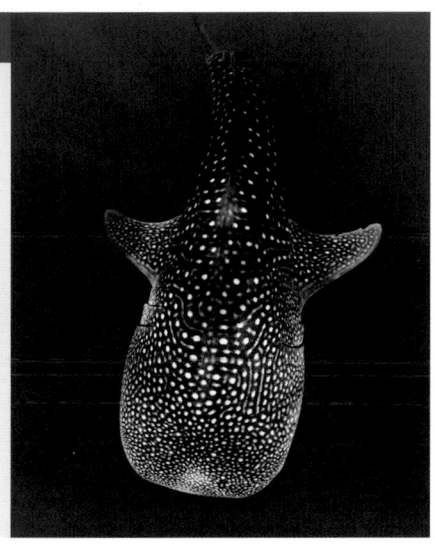

KEY FACTS

The whale shark has a flat head with gray or brown on the back, white spots on the sides, and white on the belly.

+ **range:** Warm water over the world, near coasts, and occasionally near the mouths of estuaries and rivers

+ **diet:** Plankton and small fish

+ **name origin:** For its size, like a whale's

The Whale Shark is the largest nonmammalian animal on Earth, but fortunately for humans, it prefers very small food. A docile shark, it will tolerate divers hitching a ride by hanging on to it. This shark species eats near the surface, entering lagoons and coral atolls to find food. Migratory, it sometimes appears in schools, the largest recorded being about 400 off the Yucatán coast of Mexico. The Whale Shark gives birth to live young, which nevertheless develop in an egg case within the mother's uterus. It lives an estimated 60 to 100 years, and it is considered vulnerable because it is still hunted in parts of Asia.

Sand Tiger Shark

Carcharias taurus L 9.8–11 ft (3–3.4 m)

The Sand Tiger Shark inhabits coastal waters worldwide. Despite its fearsome appearance, it is considered harmless to humans. It has a pointed head and two dorsal fins of almost equal size.

KEY FACTS

Body is bronze on top and paler below.

+ **range:** Coastal waters along both our East and West Coasts, as well as worldwide

+ **diet:** Large, bony fish, other sharks, rays, squid, and crustaceans

+ **name origin:** For the upper-body spots and the fact that it is found close to sandy shores

The Sand Tiger Shark has two features that are not found in other sharks: It has no eyelids, and it can gulp in air, suspending itself quietly in the water column to attack potential food. A night-feeder, it is known to hunt in groups while snatching large schools of fish or large prey. Although it prefers coastal shallows, it sometimes swims in water 600 feet (183 m) deep. It gives live birth, but the young that mature first eat the other babies in their mother's womb. The Sand Tiger Shark swims with its mouth open, displaying sharp teeth pointed in all directions, but it is generally considered harmless to humans.

Common Thresher Shark

Alopias vulpinus L to 20 ft (6 m)

The Thresher Shark, despite its size, poses little danger to humans because of its small teeth and timid disposition. It is identified by the long upper tail, which can be half as long as its body.

KEY FACTS

The white of its belly extends in a band over the base of the pectoral fins.

+ range: Worldwide in warm and temperate waters

+ diet: Mostly small fish such as herring and anchovies

+ name origin: "Thresher" probably from whipping and incapacitating its prey with the long tail

The Thresher Shark has a bad reputation that it probably doesn't deserve. It doesn't usually attack humans but is hunted by anglers because of the fierce fight it puts up when hooked. It is slow to reproduce, which leads it to be classified as vulnerable to extinction. It spends time both close to shore and in waters 1,800 feet (550 m) deep. Females give live birth to four pups at a time; before birth, the young at first feed on undeveloped eggs inside the mother that she continues to produce. Commercial fishermen highly prize the meat. The fins are made into shark fin soup and the skin is made into leather.

Basking Shark

Cetorhinus maximus L 20–26 ft (6–8 m)

Second biggest fish, the Basking Shark has a huge mouth with tiny teeth. Its most noticeable feature is gill slits that run from the top of its body to below its mouth, nearly encircling the head.

KEY FACTS

Body is dark gray to nearly black; undersides may be the same color or nearly white. The tip of its snout is conical.

+ range: Temperate to Arctic sea waters, near the shore

+ diet: Plankton, fish eggs, and larvae

+ name origin: For swimming near the surface as though basking in the sun

Like the Whale Shark, the Basking Shark is docile and poses little threat to humans. It takes in food by swimming with its mouth wide open, but unlike the Whale Shark, the Basking Shark has no pump to suck in food. Its gill rakers can filter up to 2,000 tons of water in an hour. Its liver accounts for up to one-fourth of its body weight and is rich in a low-density hydrocarbon that helps give the Basking Shark near-neutral buoyancy. The Basking Shark spends the winter in the deep water and migrates in spring to the surface in the company of others of its kind, but of the same sex and the same size.

Great White Shark

Carcharodon carcharias L to 20 ft (6 m)

The Great White Shark is mostly known for its size, with individuals reaching 5,000 pounds (2,200 kg). It is responsible for most attacks on humans, although humans are not preferred prey.

KEY FACTS

It is gray above and white below, and mottled brown and blue on sides, making it hard to see in profile.

+ **range:** Coastal waters of the world's oceans

+ **diet:** Most anything, including things it cannot digest

+ **name origin:** For its size and because it seems to be white when it clears the water

The Great White Shark is the nightmare of swimmers, thanks to the novel *Jaws* by Peter Benchley, and the shark's reputation as being a predator on humans. Actually, humans are not the preferred prey of Great Whites, which often mistake people for other mammals. This carnivore prefers high-fat mammals such as Fur Seals, and it generally strikes from below at speeds that make the shark clear the water. It has rows of serrated teeth, and when it grasps prey it shakes its head to break off chunks of meat. The Great White is the largest predatory shark in the ocean and has been known to attack the huge Blue Whale.

Leopard Shark

Triakis semifasciata L 4–5 ft (1.2–1.5 m)

The Leopard Shark, despite its fearsome name, is harmless to humans. This slender and adaptable species inhabits shallow water in summer. It is also known as Zebra Shark and Tiger Shark.

KEY FACTS

This fish has silvery-bronze skin with dark ovals in a neat row across its back.

+ **range:** Seawaters from Coos Bay, Oregon, to Mazatlán, Mexico; commonly seen off the California coast

+ **diet:** Clams, spoon worms, crabs, shrimp, and small fish

+ **name origin:** For leopardlike markings on its back

The Leopard Shark sometimes alarms clueless divers, but it rarely attacks people. It comes in with the tide over mudflats, eating small crustaceans, invertebrates, and small fish, then moves out as the tide recedes. It is rarely found deeper than 65 feet (20 m), and it has a small home range, seldom straying far from its preferred area. The Leopard Shark's meat tastes good, and its numbers could be vulnerable to fishing, but a minimum size limit in California—36 inches (91 cm)—seems to keep its population stable. Because of its good looks and adaptability to captivity, it is a popular fish in large public aquariums.

Bull Shark

Carcharhinus leucas L to 8 ft (2.4 m)

The Bull Shark is aggressive and probably responsible for attacks on humans blamed on other species. It spends most of its life in salt water, but has traveled up rivers such as the Mississippi.

KEY FACTS

Body is gray on top and white below; females are larger than males.

+ **range:** Coastal waters worldwide

+ **diet:** Carnivorous, eating almost anything: bony fish, other sharks, crustaceans, sting rays, and turtles

+ **name origin:** Origin of "shark" unknown, but used in English since the 16th century

The Bull Shark is a large, vicious fish that lives in warm coastal waters and sometimes attacks humans. It has large, sharp teeth and bites in the same way as the White Shark. The Bull Shark is wider and heavier than most sharks. It is typically a solitary hunter but often cruises the shallows in pairs, accelerating when it sees prey. The Bull Shark prefers turbid waters because its victims cannot see it attack. It gives birth to live young during late summer and early fall, often in brackish, near-shore waters. The young are more than 27 inches (69 cm) long at birth and reach sexual maturity in ten years.

Spiny Dogfish

Squalus acanthias L 2.6–3.3 ft (0.8–1 m)

The Spiny Dogfish is a small shark of temperate waters along shores. It is probably the world's most abundant shark, though overfishing has reduced its numbers. It can live in brackish water.

KEY FACTS

It has grayish brown skin on its upper side, with white spots along its back, and is lighter below.

+ **range:** Shallow waters along coasts around the world

+ **diet:** Mostly fish; also squid and octopus

+ **name origin:** For the spines on its dorsal fins, and because it hunts in packs

The Spiny Dogfish is not generally dangerous to humans, but it has sharp teeth and sharp dorsal spines attached to a mild poisonous gland, which can be harmful when it is handled. The dogfish is eaten as human food in Europe and in less expensive versions of shark fin soup in China. Some spiny dogfish, such as those in the North Atlantic, have declined because of overfishing. It mates in winter, the male inserting a "clasper" into the female and injecting sperm. Two to 11 pups are born about two years after gestation. Males achieve maturity about 11 years of age and females in about 20 years. It lives for 25 to 100 years.

Little Skate

Raja erinacea L 16–20 in (41–51 cm)

The disk-shaped Little Skate, grayish or brown on top and light gray or white on bottom, looks like a small stingray and is often used to bait lobster traps and is studied in medical research.

KEY FACTS

Its width measures about 1.2 times its length.

+ **range:** Gravelly or sandy bottoms from Nova Scotia to North Carolina

+ **diet:** Small crustaceans, amphipods, worms, and small fish

+ **name origin:** From the Old Norse word *skata*, which probably refers to the way it moves in short bursts

The Little Skate is a kind of ray that lives and feeds on the sea bottom in deeper water than the stingray, sucking in food through its mouth. It does not fly with its wings but "punts," thrusting itself forward on leglike fins and gliding for a while before settling again on the sand or gravel bottom. It is not considered overfished, although there is some commercial use of it as food: when circular cuts are made in its flat fins, and these are sold as scallops. Its egg cases often wash ashore and are known as mermaids' purses. The young are hatched outside the female and are miniatures of the adults.

Smalltooth Sawfish

Pristis pectinata L to 25 ft (7.6 m)

The Smalltooth Sawfish looks like a shark with its long body and two tall dorsal fins, but it is a bottom-dwelling ray whose tooth-lined bill captures and kills prey. It is also called Wide Sawfish.

KEY FACTS

The fish has a bill lined with long, sharp teeth on both sides and a flat belly.

+ **range:** Shallow tropical and subtropical waters in coastal Florida; originally found in great numbers from Long Island to Brazil

+ **diet:** Smaller fish

+ **name origin:** For the bill, which looks like a saw

The Smalltooth Sawfish was the first fish put on the United States endangered species list. Anglers killed and mounted it because of its unusual tooth-lined bill, which extends a fourth of the fish's length. Commercial fishers also killed it because its sharp bill tore nets. Consequently, the fish is now mostly found off Florida. The Smalltooth swims into schools of small fish, thrashing its bill back and forth and eating the fish that are killed and wounded. It also uses the bill to dig for clams and mussels in the sea bottom. The young are born live and with skin covering their bills so they do not injure the mother.

Southern Stingray

Dasyatis americana L to 6 ft (1.8 m) for females and 2.5 ft (0.8 m) for males

The Southern Stingray is diamond-shaped and flat. It sits on the bottom of warm, shallow coastal waters. Its tail has a stinger, which is not fatal to humans but causes a painful wound.

KEY FACTS

Body is dark gray, green, or black on top and nearly white on the bottom.

+ range: Tropical or subtropical shallow waters near shore

+ diet: Bivalves, worms, crustaceans, and small fish on the bottom of the ocean

+ name origin: For the stinger near the end of its tail, which it uses only for defense

The Southern Stingray is the enemy of waders, who sometimes step on it near the shore and receive a painful sting from the barbed end of the tail. It will not use the stinger unless provoked. Its eyes are on the top of its head, and behind the head are two holes, or spiracles, which allow it to take in water and expel it from the gills on its underside. The suction of its mouth has been compared to a vacuum cleaner. This stingray species copulates competitively, with the dominant male biting the female and injecting sperm into the cloaca. The eggs are hatched inside the female's body, and two to ten pups are born live.

Reef Manta Ray

Manta alfredi L 10–11.5 ft (3–3.5 m)

The Reef Manta Ray, also called Devilfish, is believed to be second in size only to the Pelagic or Giant Manta Ray. Small in the *Manta* genus, it can nevertheless grow to 18 feet (5.5 m) across.

KEY FACTS

The large, disklike fish is dark on top and pale on its belly.

+ range: Tropical and subtropical waters of world's major oceans; also in temperate seas

+ diet: Plankton and fish larvae

+ name origin: *Manta* from Spanish for "blanket"; "Devilfish" probably for its fearsome appearance

The Reef Manta Ray gets a bad rap. Though it is huge, it has a gentle disposition and eats mostly plankton and fish larvae. It feeds by opening its large mouth and moving water through its gills, which strain out food with gill rakers. It has no stinger on the end of its tail like a stingray and poses no threat to swimmers. Though sharks and Killer Whales eat it, its greatest threat is from humans, who catch it because of its enormous size and because it puts up a fight when hooked or speared. Its gill plates are used in soup. The Reef Manta Ray bears only one baby every two years, which is born live.

Bonefish
Albula vulpes L to 35 in (89 cm)

The Bonefish, also known as Gray Ghost, is a tropical saltwater fish prized by anglers for its speed and fighting ability. It is not commonly eaten because of its many small bones.

KEY FACTS

It has a streamlined shape, and it is mostly silver with some gray on the back, and dusky fins.

+ **range:** Worldwide in warm seas

+ **diet:** Worms, smaller fish, mollusks, and crustaceans

+ **name origin:** For the small bones in its flesh; the scientific name means "whitish fox."

The Bonefish is considered extremely fast and wily; the slightest ripple by a fly fisherman will make it flee. Bonefish come into shallow water to feed and then retreat to deep water. The streamlined shape of the bottom-feeding fish assists it in retreat from capture. Anglers pursue it in water as shallow as 8 inches (20 cm) and at most several feet deep. The fishers can sometimes locate it by the mud the fish stirs up looking for food. The Bonefish is a catch-and-release fish because it is so valuable to sport fishers. This fish travels in schools until it reaches large size, when it may move about alone.

Spotted Moray Eel
Gymnothorax moringa L to 7 ft (2 m)

The Spotted Moray Eel is medium to large size, with a snakelike body; it has a mouth full of sharp teeth, and its bite can dangerous; fortunately, it is not aggressive toward people.

KEY FACTS

Body is white or pale yellow, with overlapping dark brown spots.

+ **range:** Atlantic coast from North Carolina and Bermuda to Brazil; some in the Gulf of Mexico and Caribbean

+ **diet:** Feeds at night on the bottom on crustaceans and smaller fish

+ **name origin:** For the spots that cover it

The Moray Eel looks dangerous, but it need not be. It hides during daytime in holes with only its head protruding, and when divers see it, it appears to be aggressive, opening and shutting its mouth, which has sharp teeth. This eel has small gills, so it must open and shut its mouth to move water over the gills. Nevertheless, it has a vicious bite when threatened, so one had best bypass it. The Moray hunts during the night and rests during the day, but if food comes near, it may make lightning strikes from its hole and secure the prey with its teeth. It is not captured by humans, because it is not a food fish.

Atlantic Herring
Clupea harengus L to 18 in (46 cm)

The Atlantic Herring, an important part of New England and Canadian commercial fisheries, is strictly a saltwater fish and congregates in huge schools in cold waters of the East Coast.

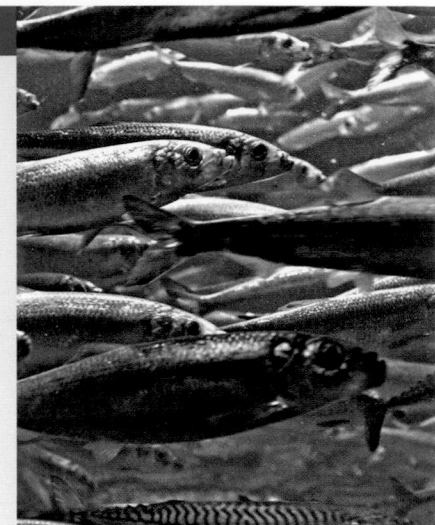

KEY FACTS

The Atlantic Herring is silvery in color with a blue or greenish blue back.

+ range: In summer, near shore in the Gulf of Maine and the Gulf of St. Lawrence; in winter, deep water in southern New England and mid-Atlantic states

+ diet: Zooplankton

+ name origin: Old German *harinc*

The Herring in general is the most abundant fish on Earth, and the Atlantic Herring travels in schools of billions. It is food for many predators such as seals and whales, but it has a great sensitivity to their presence; holes have been seen in large schools with a predator in the middle. The cold-water Atlantic Herring population crashed in 1970. A close relative, the Pacific Herring, was greatly depleted in the early 1990s and is still recovering. In the case of the Atlantic Herring, foreign fisheries have been phased out, limits have been put on the catch, and today the fish is harvested sustainably.

Atlantic Menhaden
Brevoortia tyrannus L 10–12 in (25–30 cm)

The Atlantic Menhaden, also called Pogy, Hard-head, and Fatback, is a small fish that travels in schools of thousands. Early Americans used its greasy flesh as fertilizer, oil, and food.

KEY FACTS

The Atlantic Menhaden has a scaleless head with a large mouth and flat body that is wider than it is long.

+ range: Historically numerous from Nova Scotia to central Florida

+ diet: Plankton

+ name origin: Native American for "that which manures"; the Pilgrims probably planted menhaden with corn.

Once the mainstay of the U.S. fishing industry, the Atlantic Menhaden has subsequently been overfished and now must be protected. Thousands of ships once sought this small fish; factories that processed them into fertilizer and oil lined the Atlantic coast. Today, limits are put on their catch, and the major companies in the U.S. have dwindled to one. Reduced numbers may be responsible for the sometimes toxic "red tides" that block sunlight, because the Menhadens' efficient filtering system kept the ocean water clear. An adult fish may filter four to seven gallons of water when swimming with its mouth open.

Northern Anchovy

Engraulis mordax L 8 in (20 cm)

The Northern Anchovy, also called Pinhead, is used as bait for game fish and for human food. With the collapse of the Pacific Sardine industry, anchovies are now a multimillion-dollar business.

KEY FACTS

It is blue-green on the back and silvery below; adults have a silver stripe on side.

+ **range:** From Baja California north to the Queen Charlotte Islands

+ **diet:** Planktonic crustaceans and fish larvae

+ **name derivation:** "Anchovy" from Spanish *anchova* for a type of anchovy found in the Mediterranean

The Northern Anchovy swims in schools for protection because it is prey to other fish. An anchovy's life is not long, but some four-year-olds have been found, and the maximum age known is seven. It is a meal for many other fish and for seabirds; in California, Northern Anchovy populations have been linked to the abundance of Brown Pelicans. Northern Anchovies can be distinguished from Pacific Sardines because these Sardines have dark spots on the body. Schools of sardines move in darting stops and starts, but anchovies form a funnel or swirl; when feeding, they swim with their mouths open to gather microscopic morsels.

Pacific Sardine

Sardinops sagax L 9.5 in (24 cm)

The Pacific Sardine is a small marine fish that had an important use as a nutritious and easily transportable food for soldiers in World War I. It has been a canned food fish in peacetime.

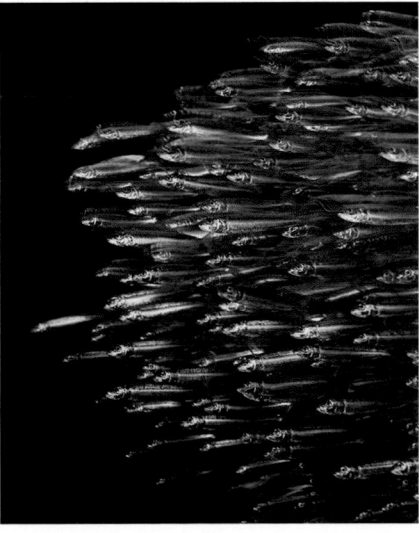

KEY FACTS

It is silvery on sides and belly with dark blue or green on back.

+ **range:** Gulf of Alaska to Baja California in Mexico in large schools; Pacific Sardines are food for many birds, marine mammals, and other fish.

+ **diet:** Zooplankton through a filtering system

+ **name origin:** Old French *sardine*, from Latin *sardina*

Pacific Sardines became a U.S. fishery as a result of a search for food for World War I troops. By the 1940s, Pacific Sardines made up 25 percent of the U.S. catch. Because of overfishing and the cooling of ocean waters, the population crashed for 40 years. By the late 1980s, sardine numbers had increased, and the fish is now considered out of danger. Pacific Sardines gather as adults in schools of thousands, the high numbers aiding their reproductive success and providing safety from predators. The Pacific Sardine migrates north in the spring and returns to southern California waters for the fall and winter.

Atlantic Cod

Gadus morhua L to 6.5 ft (2 m)

A cold-water fish that was once a mainstay of U.S. commercial fisheries, Atlantic Cod numbers have been seriously depleted by overfishing. Despite regulations, the population has not recovered.

KEY FACTS

This species is colored brown to green, with spots on the side turning silver toward the rear; lateral line clearly visible.

+ **range:** Along shore from Cape Hatteras, North Carolina, north to Greenland and Iceland

+ **diet:** Formerly, small fish; increasingly a bottom-feeder due to depletion of food

+ **name origin:** Uncertain

So successful as a food, the Atlantic Cod has been overfished in North America. In the 1990s, the species underwent a 95 percent reduction in its population. Despite fishing closures, gear restrictions, and minimum size requirements, it has not recovered its numbers. Today some of this species have adapted to become bottom-feeders with a lower mouth and a visible hump on the back. Atlantic Cod is currently listed as "vulnerable" on the International Union for Conservation of Nature and Natural Resources "red list" of threatened species. In spawning, females release their eggs in open water and males compete to fertilize them.

Haddock

Melanogrammus aeglefinus L 20–24 in (51–61 cm)

The Haddock is closely related to the Cod, and is a popular food fish. It prefers cool waters and rarely enters estuaries and river mouths, moving to the deep ocean in winter.

KEY FACTS

The fish has a dark line that runs laterally.

+ **range:** Deep cool waters off Georges Bank in New England in summer; deep water off North Carolina in winter

+ **diet:** Invertebrates such as small crabs, sea worms, clams, sea cucumbers, and occasionally squid

+ **name origin:** Middle English *haddok*

The Haddock is similar to Cod except for the dark lateral line and the dark blotch above its dorsal fin. Historically, Haddock was one of the most abundant fish in North American waters, and the Georges Bank was one of the most productive Haddock grounds in the world. It is one of the world's most important food fish and tastes similar to Cod. Spurred by new markets for food fish, however, commercial fisheries depleted the population by 82 percent, a major impetus in forming multinational groups in management of fisheries. Catch limits have stabilized the population but Haddock numbers are still in trouble.

Pollock

Pollachius virens L of nine-year-old 30 in (76 cm)

The Pollock is a gluttonous fish that feeds on the young of others, especially Cod. A cool-water fish, it has become important as food for people since the decline of Cod and Haddock.

KEY FACTS

The fish has a defined, lateral line running down its sides.

+ **range:** No farther south than Chesapeake Bay; moves off Massachusetts and stays until late January when it moves south again

+ **diet:** Fish and crustaceans such as shrimp, at the surface and intermediate depths

+ **name origin:** Uncertain

Pollock was known primarily as cat food until the decline of Cod and Haddock when Pollock replaced them in the human diet. It is stronger in taste than either and when minced, it is used as fish fingers and imitation crabmeat. It is vulnerable to parasites such as the female cod worm, which metamorphoses into a wormlike body clinging to Pollock gills. The fish may reach nearly 4 feet (1.2 m) long and weigh more than 40 pounds (18 kg). Pollock represent more than 10 percent of all fish in global fisheries. The Pollock looks like a Cod, with its three dorsal fins, but it is smaller when an adult, lacks spots, and has gray flesh.

American Goosefish/Monkfish

Lophius americanus L to 4 ft (1.2 m)

The American Goosefish is unlike any other fish in American waters. Its most distinctive feature is a mouth that is so big that, closed, it still shows most of the long, spiky teeth in its lower jaw.

KEY FACTS

The fish is chocolate brown on top, with white on the bottom.

+ **range:** Deep water off the continental shelf on East Coast of North America

+ **diet:** Mostly other fish, but will eat anything edible

+ **name origin:** Probably named for its appearance

The American Goosefish, not known until the 19th century, is so strange looking that it is called Monkfish to make it more palatable to buyers. It is also called the Angler because of two dangling appendages on the top of its head, which probably lure its smaller fish prey near. This fish's large head, enormous mouth, and pectoral fins set well back on its body give it the appearance of a tadpole, and it is a slow swimmer. The Goosefish mostly stays on the sea bottom, awaiting prey to swim by. Anglers have been surprised to sometimes catch it when it has latched on to a fish that is already caught and is being reeled in.

Striped Mullet

Mugil cephalus L 10–14 in (25–36 cm)

The most common mullet, the warm-water Flathead Mullet is also known as Jumping Jack and Black Mullet. It migrates in sizable schools and is an escape artist, often jumping out of nets.

KEY FACTS

Body is dark above and silvery on sides, with conspicuous stripes formed by spots at the base of the scales.

+ **range:** Very shallow water in southeastern U.S.

+ **diet:** Algae, detritus, and other tiny marine forms in mud or sand on the bottom

+ **name origin:** For its head, which appears flat from above

The Flathead Mullet is the most numerous of all mullets. It jumps out of the water frequently, for which it is called Jumping Jack. It adjusts easily to living in fresh water and moves far up freshwater creeks. The Flathead Mullet has been caught hundreds of miles from the sea but is most abundant in extremely shallow water. It is important to the environment because it cleans up estuaries and is an important food for larger finfish. The Flathead Mullet is not listed as endangered or threatened in any state in the southeastern U.S. Because of its fecundity and wide habitat tolerance, it is an abundant species.

Rainbow Smelt

Osmerus mordax L of adult 7–9 in (18–23 cm)

The Rainbow Smelt is a cold-water species that can winter under coastal ice because it has an antifreeze protein in its body. It is food for larger fish, mammals, and birds—and people.

KEY FACTS

It has a pale green back, and purple, blue, and pink on the sides, with a light underside.

+ **range:** Atlantic drainages between New Jersey and Labrador to the Arctic, and Pacific drainages south to Vancouver Island

+ **diet:** Crustaceans, sea worms, insects, and small fish

+ **name origin:** Norwegian *smelte*

Most Rainbow Smelt do not live more than a year; they are either eaten by predators or caught by people. Mortality for this species amounts to 72 percent of the total population annually. Rainbow Smelt may reach 13 or 14 inches (33 to 36 cm) in length, but it does not live beyond five years. It has a colorful body that is slender and cylindrical, and travels in schools less than a mile from shore and in water less than 19 feet (6 m) deep. In summer, it moves to deeper, cooler waters, and in autumn it travels to bays and estuaries where it actively feeds until winter. It spawns in early or late spring in freshwater streams.

Lined Seahorse

Hippocamus erectus L 6 in (15 cm)

One of the most unusual-looking sea creatures, the Lined Seahorse swims erect with an arched neck that resembles a small horse. It is also called Northern Seahorse and Spotted Seahorse.

KEY FACTS

The male is larger than the female and has a longer tail.

+ range: Weed beds in shallow Atlantic waters from Nova Scotia to Venezuela in summer; at depths to 240 feet (73 m) in winter

+ diet: Crustaceans, mollusks, and zooplankton

+ name origin: For lines on throat and resemblance to a tiny horse

The Lined Seahorse has bony plates, a long snout, and a prehensile tail. It is a weak swimmer, requiring it to hide in the vegetation to escape its predators, which include rays, crabs, and sea turtles. It is monogamous, choosing a mate with whom it will spend its short life and performing a dance to affirm its faith each day. If it loses a mate, it will take time choosing a new one. Males have an incubation pocket on their abdomen into which the females deposit eggs. Males then care for the young, attaching themselves with their prehensile tails to vegetation and releasing the young into the water column.

Red Lionfish

Pterois volitans L to 17 in (43 cm)

The Red Lionfish, also called Scorpionfish, Turkeyfish, and Zebrafish, is venomous. Originally found in the Indo-Pacific seas, it is now invading the Caribbean and our East Coast.

KEY FACTS

This species' beautiful but venomous spines, striped red, white, maroon, and brown, stick out from its fins.

+ range: Coral and rocky reefs, caves, drop-offs, and sandy areas

+ diet: Small fish and crustaceans

+ name origin: For feathery-looking spined fins that resemble a lion's mane

Beware of the Lionfish, which was once a creature of the Indo-Pacific but was accidentally introduced to North American waters. Its venom is not necessarily fatal to humans, but it causes a painful sting and induces headaches, vomiting, and breathing difficulties. This species hunts at night and swims with stateliness, as though it knows its venom makes it unpalatable to all but the largest predators. Lionfish spread rapidly; they are believed to reproduce daily and their eggs hatch within 36 hours after fertilization. A close relative, the Common Lionfish (*P. miles*), is slightly smaller and less reddish.

Atlantic Goliath Grouper

Epinephelus itajara L to 8.2 ft (2.5 m)

The Atlantic Goliath Grouper can grow to 800 pounds (363 kg) but is usually 400 pounds (180 kg) when mature. Also called Jewfish, it prefers rocky or coral waters and mud bottoms.

KEY FACTS

Its body is gray but mottled.

+ **range:** Shallow tropical waters around coral in the Florida Keys, the Bahamas, and the Caribbean

+ **diet:** Crustaceans, other fish, octopuses, and young sea turtles

+ **name origin:** Portuguese word for "fish," taken from an indigenous South American language

The Atlantic Goliath Grouper is endangered because its flesh is tasty, and it is easily caught. It is curious and virtually fearless, making it an easy target for spearfishers. Once considered in trouble, its population is believed to be recovering. Some scientists remain concerned because it is a slow-growing animal, so possession and marketing of the Goliath Grouper is prohibited no matter what its size. Most groupers are hermaphroditic, changing from female to male at an early age, but this has not been confirmed as happening in the Goliath Grouper. It has a large mouth and swallows fish prey whole.

Black Sea Bass

Centropristis striata L 13 in (33 cm)

A type of grouper, the Black Sea Bass is small, usually 1.5 pounds (0.7 kg). It hangs around rubble, either natural or man-made, at the bottom of the ocean.

KEY FACTS

Body is dark brown or black, taller than long, with a large mouth and a fatty hump in front of its dorsal fin.

+ **range:** Near coast from Maine to Florida and occasionally in the eastern Gulf of Mexico

+ **diet:** Small fish, crustaceans, and shellfish

+ **name origin:** Middle English *base*

An omnivorous bottom-feeder, the Black Sea Bass first reaches sexual maturity as a female; as the young continue to grow, some become males. The fish summers at depths about 120 feet (37 m) and winters at 180 to 426 feet (55–130 m), where the temperature is 46°F (8°C). It is usually 1.5 pounds (0.7 kg), with a maximum of 9.5 pounds (4.3 kg). It has one sharp spine on the dorsal fin, which is usually folded against the body unless the fish is being aggressive to other sea bass. It can be found around shipwrecks or a natural rocky bottom, where it usually rests stationary or cruises slowly.

Bluefish
Pomatomus saltatrix L 7–24 in (18–61 cm)

The Bluefish is partly misnamed, as it is green and silvery gray on its belly and blue only on its back; it is a hard fighter and a vicious predator, and is even known to attack bathers.

KEY FACTS

The fish has conical canine teeth that line the upper and lower mouth.

+ **range:** Inshore and offshore areas of Atlantic coast

+ **diet:** Smaller fish such as Menhaden, Herring, Mackerel, and Alewives, as well as squid and sand eels

+ **name origin:** For blue color most people see when it is in the water

Bluefish are popular with anglers because they grow large and battle well. They are abundant from Argentina to Cape Cod but occur as far north as Nova Scotia. They rarely exceed 20 pounds (9 kg); the record weight caught in North America is 31 pounds 12 ounces (14.4 kg). They reach sexual maturity at two years of age. The amount of eggs in a female depends on their size, with a 21-incher producing 900,000 eggs and a 23-incher bearing 1.1 million. Spawning occurs over the continental shelf when water temperatures reach 64° to 74°F (18°–23°C). Survival of the young depends on whether currents carry them out to sea or toward shore.

Common Dolphinfish/Mahimahi/Dorado
Coryphaena hippurus L 24–36 in (61–91 cm)

The Dolphinfish is often confused with the dolphin because of its blunt, large head and fast swimming; restaurants call it Mahimahi to avoid confusion and make it seem more palatable.

KEY FACTS

It has bright blues and greens on the back and gold on sides.

+ **range:** Waters of Puerto Rico, Gulf of Mexico, and Atlantic coast of Florida

+ **diet:** Flying fish and other smaller fish, crabs, and squid

+ **name origin:** For its resemblance to the mammal dolphin; Mahimahi means "very strong" in Hawaiian

The Dolphinfish is one of the most beautiful fish in the ocean, with its bright blues and greens on the back and gold on its sides. Upon death, it assumes a common, gray color. The Dolphinfish is one of the fastest swimmers because of its streamlined shape—a large head tapering to a slim, small tail. The warm-water creature is fast-growing, and males and females become sexually mature in about four or five months. It stays near the surface and lives four to five years. The Dolphinfish is highly popular in restaurants, which rename it Mahimahi so that diners can be sure to distinguish it from the mammals named dolphins.

California Yellowtail

Seriola lalandi L to 36 in (91 cm)

The California Yellowtail is a type of amberjack that likes warm water, is large and fast, and is known more as a game fish than a food fish. It is also called Forktail, Mossback, and White Salmon.

KEY FACTS

Color is blue on back, light blue and white on belly; bronze strip on the side turns yellow at the tail.

+ **range:** Off California and Hawaii, around offshore islands, on rocky bottoms

+ **diet:** Small fish, crustaceans, and squid

+ **name origin:** For the yellow on its tail, and the name for fish ("jack")

The California Yellowtail is known as the "murderer of the sea" because of its voracious appetite and its behavior of charging into a school of smaller fish. It is known more as a game fish because of its tendency to fight well, but is also called "Hamachi" in sushi bars and sold to the Japanese for sashimi. The Yellowtail is not threatened, but its numbers are a concern as Japanese commercial fisheries move into United States waters to catch it. It spawns slowly, not being sexually mature until three years, but is believed to live up to 30 years. Unlike other fish, it feeds more frequently during spawning.

Florida Pompano

Trachinotus carolinus L 17 in (43 cm)

The Florida Pompano is a member of the jack family; a warm-water marine species, it is mostly found in Florida but also occurs elsewhere along the East Coast, depending on the season.

KEY FACTS

It is greenish gray on back, silver on the sides, with gold on the throat, belly, and anal fins.

+ **range:** When water is warm, from Massachusetts to Brazil

+ **diet:** Mollusks and clams offshore of sandy beaches

+ **name origin:** Spanish meaning "vine leaf," referring to their flat shape

The Florida Pompano has a compressed body and blunt snout. This bottom-feeder is a fast swimmer, and it is fished by both sport anglers and fisheries. It is so tasty that Mark Twain called it "as delicious as the less criminal forms of sin." So great is the demand for it at restaurants that it is farmed in aquaculture. Commercial fisheries were well established in the early 20th century but have declined since then. Lower commercial catches are not due to overfishing but to aquaculture and the rising numbers of sport anglers. It makes its home in surf flats, and likes to stay away from clear water areas.

Red Snapper

Lutjanus campechanus L 24 in (61 cm)

The Red Snapper, also called a Pargo, occurs worldwide and is a prized food fish. It usually inhabits waters 30 to 200 feet (9–60 m) deep, but some have been reported at 300 feet (91 m).

KEY FACTS

It is a deep rosy red color with a dark fringe around dorsal and tail fins.

+ **range:** Near structures such as oil rigs, artificial reefs, and shipwrecks off the southeast Atlantic states and in the Gulf of Mexico

+ **diet:** Crab, shrimp, and small fish

+ **name origin:** For its color; origin of "snapper" is uncertain

The Red Snapper is shaped much like other snappers, with a sloped profile, medium to large scales, a spiny dorsal fin, and a laterally compressed body. The fish may grow to 50 pounds (23 kg), but most are 2 to 5 pounds (1 to 2.3 kg) when caught. The Snapper normally congregates in schools near ledges and man-made structures. The species is long lived, the oldest reported age being 57 years. It is one of the most sought-after fish, both commercially and by anglers. Its flesh is tasty, and the species is carefully regulated to ensure a continuing population. Despite the "snapper" name, it is a nibbler at bait on a line.

Sheepshead

Archosargus probatocephalus L 10–20 in (25–51 cm)

The Sheepshead, a southern inshore species, is cylindrically shaped, with sharp dorsal spines and a hard mouth with several rows of stubby teeth. Only a portion of the eggs is laid at one time.

KEY FACTS

It is colored silver, with dark bars that run vertically on its body from front to back.

+ **range:** Along the Atlantic coast of North America

+ **diet:** Oysters, clams, and other bivalves; barnacles, fiddler crabs, and other crustaceans

+ **name origin:** For its blunt face, which looks like a sheep's

The Sheepshead may be short and fat, but it is among the gamest of saltwater fish. Commonly found around oyster bars, seawalls, and in tidal creeks in summer, it winters in deep water but spawns close to shore in late winter and early spring. The young look so like the adults that they can be readily identified at three-quarters of an inch. They grow to 15 pounds (6.8 kg), but average 3 or 4 pounds (1.4 to 1.8 kg). The Sheepshead has a hard mouth and strong, blunt teeth. Anglers use a small hook to catch the famed nibblers. As the saying about them indicates: "Anglers must strike just before they bite."

Red Drum

Sciaenops ocellatus L at maturity 28–33 in (71–84 cm)

The Red Drum, also known as Channel Bass and Redfish, is found in the Atlantic Ocean. It reproduces with a cousin, the Black Drum, and the young of this hybridization are robust.

KEY FACTS

It is red on top, white on its belly.

+ **range:** Along the coast of the U.S. from Florida to the Gulf of Mexico

+ **diet:** Shrimp, crabs, and sand dollars in summer and fall; smaller fish in other seasons

+ **name origin:** Color on its back, and its drumming sound when alarmed and when breeding

The Red Drum is a popular game fish that is tasty when under 15 pounds (6.8 kg). Above that, it becomes tough. In the 1960s, Cajun-style blackened redfish (Red Drum) became popular, and by the 1980s, the catch had declined. Red Drum had not been overfished commercially, but had diminished possibly because of recreational fishing. The eyelike spot near the Red Drum's tail is thought to be a ruse to fool predator fish into striking at its backside, allowing the fish to escape. It prefers rocky areas including man-made structures such as jetties and oil rigs. The largest on record was 94 pounds (43 kg), caught in 1984.

Southern Kingfish

Menticirrhus americanus L 12 in (30 cm)

The Southern Kingfish is a bottom-feeder. Also known as the Southern Kingcroaker, Whiting, Carolina Whiting, and Sea Mullet, it prefers muddy or sandy bottoms of shallow waters.

KEY FACTS

Its top and sides are silvery gray; it has seven or eight faint dark bars running vertically.

+ **range:** Western Atlantic to Florida in shallow coastal water and estuaries

+ **diet:** Small crustaceans such as shrimp and crabs

+ **name origin:** May derive from a California species known as "Chenfish"

The abundant Southern Kingfish is a tastier food fish than many other species that are better known. Part of its high numbers may be the result of where it is usually found—in shallow, subtropical waters where fishery ships do not usually go. The kingfish moves offshore to spawn, from April to September, and the young move to estuaries and near beaches to mature. Heavy mortality may be caused by shrimp trawlers, which nab adults as a bycatch. Its mouth is small, and the fish has a single barbel on its chin. Aside from crustaceans, it eats worms, mollusks, and small fish, and it scavenges for detritus and carrion.

Spotted Seatrout
Cynoscion nebulosus L 19–25 in (48–64 cm)

The Spotted Sea Trout is not a trout but a drum. Also known as Speckled Trout, Spotted Weakfish, Salmon, or Black Trout, it lives along the coast of the southeastern states.

KEY FACTS

The fish has spots scattered on it, including on its dorsal and caudal fins.

+ range: Near coasts in the Southeast and in estuaries and lagoons; occasionally up freshwater rivers

+ diet: Shrimp and other crustaceans, and other fish

+ name origin: For its similarity to a Brown Trout

The Spotted Sea Trout is important as a sport fish because it is found in fairly shallow waters along the East Coast and far up coastal rivers. It is fast-growing, and about half of the young are sexually mature at the end of the first year. Spotted Seatrout reproduce in grassy areas in estuaries and have a long spawning season, from spring through summer. The males make a grunting noise to attract females, which then lay eggs. The eggs sink after they are fertilized when the salinity is low, which may keep them from drifting where the salinity is too low for the young to survive. The Sea Trout has canine teeth.

Banded Butterflyfish
Chaetodon striatus L to 6 in (15 cm)

The Banded Butterflyfish is disk shaped, with a pointed snout and prominent spines on its back. It is also known as Butterbun, Banded Mariposa, Portuguese Butterfly, and School Mistress.

KEY FACTS

It is white with dark vertical bands that begin at the dorsal fin.

+ range: Coral reefs from Massachusetts to Rio de Janeiro, including Gulf of Mexico and Caribbean Sea

+ diet: Marine worms, coral polyps, crustaceans, and mollusk eggs

+ name origin: For its color pattern, which looks like a butterfly

The Banded Butterflyfish reproduces quickly, doubling its numbers in less than 15 months. It is not a food fish and is not endangered, although there is some depredation for the aquarium trade. It is mostly seen moving in pairs and may be monogamous. The larvae are born in less than a day and are nearly transparent; they develop a large black spot that looks like an eye, confusing predators as to which end is their front. The Banded Butterflyfish is active during the day but hides from predators at night in reef crevices or sea grass. It occasionally cleans grunts, parrotfish, and surgeonfish.

Great Barracuda

Sphyraena barracuda L 24–40 in (61–102 cm)

The Great Barracuda is a large warm-water fish that ambushes smaller fish, with two rows of sharp teeth. It sometimes strikes people wearing bright, sparkling objects, mistaking them for prey.

KEY FACTS

Head is large and pike-like in appearance; elongated body is dark green or blue with chalky white below.

+ range: Open ocean from Florida to Brazil; north to Massachusetts in summer when the water is warmer

+ diet: Mostly other fish, occasionally squid

+ name origin: Uncertain

Beware of the Great Barracuda. Individuals 6 feet (1.8 m) long and weighing more than 100 pounds (45 kg) have been caught. Most are half that size, but they have few enemies. Their usual food consists of smaller fish, which they find in schools, striking suddenly. They have been observed herding a school of fish into shallow water where they guard them until hungry. Spearfishers are discouraged from hunting when Great Barracudas are around because the voracious fish may mistake a fish impaled on a spear as wounded prey. The Great Barracuda may be distinguished from other barracudas by dark spots on its sides.

Black Surfperch/Butterlips

Embiotoca jacksoni L 15 in (38 cm)

The Black Surfperch, despite its name, is rarely black but has a variety of colors including orange, gray, red, green, and brown. The colors change according to the environment.

KEY FACTS

This species is shaped like a football with thick lips.

+ range: Northern California to Baja California

+ diet: Amphipods; also crabs, brittle stars, and worms

+ name origin: Habit of being in the surf; *jacksoni* for A. C. Jackson who discovered that it gives live birth

The squat shape of the Black Surfperch allows it to swim in the surf at depths of around 3 feet (0.9 m) without being tossed around. Unlike most fish, it gives live birth, and the female nourishes the new young from her body. The Surfperch eats by gathering huge amounts of mud from the bottom, separates the food and swallows it, and ejects the nonnutritional remainder. It lives six to nine years, normally reaches 4 to 5 pounds (1.8 to 2.3 kg), and is prey to harbor seals and cormorants. It is also a "cleaner fish," removing parasites from the skin of other fish, which tolerate it because they want the parasites gone.

Hogfish

Lachnolaimus maximus L to 3 ft (0.9 m)

The Hogfish, whose color changes with size, is a wrasse with sharp teeth and an elongated snout, which it uses to root in the sediment on the bottom for food.

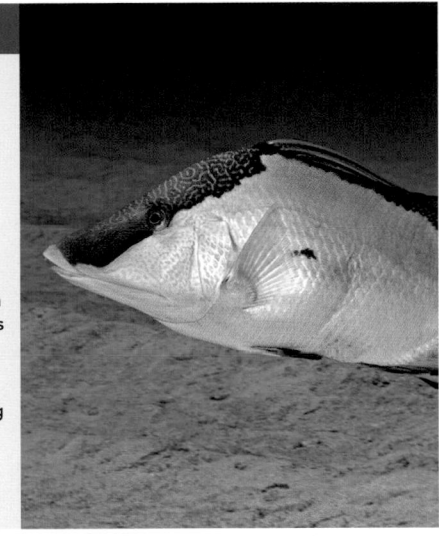

KEY FACTS

The male has a dark streak that runs down the top of the head and the ends of the upper lip.

+ range: Florida Keys, Caribbean Sea, and Gulf of Mexico

+ diet: Anything from clams, snails, or urchins

+ name origin: For its elongated snout and its habit of rooting in silt on the bottom for food

The warm-water Hogfish roots around on the bottom for food with its long snout and protruding lips, and crushes food with its strong, canine teeth. It changes sex, beginning as a female but changing to a male as it grows to a maximum size of 3 feet (0.9 m) and approximately 22 pounds (10 kg). Females can reproduce at 8 inches (20 cm) but do not usually reach maturity until they are larger. They spawn off the coast of Florida from September until April. Bag limits and bag size have both been placed on the fish to protect it from overfishing because it is listed as "vulnerable" by international organizations.

Atlantic Sailfish

Istiophorus albicans L 10 ft (3 m)

The Atlantic Sailfish is related to the Marlin, but with a larger dorsal fin extending high above its back and along most of its body. One of the smaller billfish, it can weigh 200 pounds (91 kg).

KEY FACTS

Body is blue on the back and white with brown spots below.

+ range: North Atlantic in summer; returns to Gulf of Mexico, Caribbean Sea, and off Florida coast in cold weather

+ diet: Smaller fish such as mackerel, jacks; also squid and octopus

+ name origin: For its range and its dorsal fin

The Atlantic Sailfish is highly migratory, changing habitats according to temperature and wind conditions. Its meat is tough, and it has no commercial value—but it is highly prized by recreational fishers. Popular fishing locations include Bermuda, Puerto Rico, and the Gulf of Mexico. The sailfish's pelvic fins are long, extending along more than a third of its body. If the fish wants to go fast, the pelvic fins can streamline it by receding into a pair of grooves on its underside. The vertical lines on the lighter parts of its body are made up of many light blue dots. A lateral line on either side is readily visible.

Black Marlin

Makaira indica L to 15 ft (4.6 m)

The Black Marlin may be the fastest fish, up to 80 mph (129 kph). It has an elongated body and a stout bill, and reaches 15 feet (4.6 m) long and 1,700 pounds (771 kg), making it one of the largest marlins.

KEY FACTS

Body is dark blue on the upper side, has a silver-white belly, and faint vertical stripes running down its side.

+ **range:** Pacific Ocean, usually near surface

+ **diet:** Dolphinfish, squid, cuttlefish, octopus; prefer small tuna

+ **name origin:** "Marlinspike," because of its sharp bill

The Black Marlin is a large fish that is found in tropical and subtropical waters. It has a blue-black dorsal fin that is the smallest proportionately to the size of any marlin; when erect, the fin is less than 50 percent of its entire body height. The Black Marlin is prized mostly by sport fishers, who value its large size and fighting ability. Black Marlins have been caught that weigh 1,500 pounds (680 kg). They have sharp bills in their upper jaw, and examinations of stomach contents indicate that they use the bill to slash and kill the fish they eat. They also use their spears to defend themselves from predators.

Atlantic Bluefin Tuna

Thunnus thynnus L 6.5 ft (2 m)

The Atlantic Bluefin Tuna is one of the largest, fastest, and most migratory fish in the sea; its meat is highly prized and its numbers are greatly depleted. Humans have eaten it for centuries.

KEY FACTS

Color is metallic blue on top and silver-white on bottom.

+ **range:** Off Newfoundland and Iceland to Gulf of Mexico and Mediterranean

+ **diet:** Mostly fish, crustaceans, squid eels, and kelp; also filter-feeds on zooplankton and eats almost constantly

+ **name origin:** Likely from Spanish *atún*

The Atlantic Bluefin Tuna is one of the largest and fastest fish in the ocean. It can shoot through the water at 43 mph (69 kph) and reach sizes well over 1,000 pounds (450 kg). The dorsal and pectoral fins can retract into slots to reduce drag. The Bluefin Tuna is warm-blooded and can adjust to cold waters or tropical equatorial waters. Its flesh is tasty, which accounts for its serious population decline, increased recently because of its value in sushi. Its coloring helps camouflage the animal from above and below. A close relative, the Pacific Bluefin Tuna, does not appear to be threatened despite being overfished.

Atlantic Mackerel

Scomber scombrus L to 19 in (48 cm)

The Atlantic Mackerel, also called Boston Mackerel, has been caught in U.S. waters since colonial times. It rarely reaches 8 pounds (3.6 kg), and it travels in large schools.

KEY FACTS

It has a greenish blue upper body with numerous wavy vertical bars.

+ range: Overwinters in open ocean; moves close to shore in spring to spawn

+ diet: Shrimp and small fish

+ name origin: French *maquereau*, meaning "pimp" in a belief that mackerel schools pimp for herring schools it often joins

Early colonists in North America fished for Mackerel, which they considered a valuable commodity. The fish was usually traded to other parts of the world where it was an important food fish, though it has rarely been relished in the United States because of its strong, rich taste. It has a long, thin body that allows it to swim fast. Mackerels are very low in mercury, and they are considered by many to be an important "brain food" with positive effects for patients with Alzheimer's. A close relative, the Pacific Chub Mackerel, has markings similar to those of the Atlantic Mackerel but has dusky blotches on its lower side.

Yellowfin Tuna

Thunnus albacares L to 110 in (279 cm)

The Yellowfin Tuna, called Ahi in Hawaiian, is a cousin of the Atlantic and Pacific Bluefin Tuna, but less than half their size. Distinctive features are the yellow second dorsal fin and anal fin.

KEY FACTS

It is dark metallic blue on top and white underneath. Most are caught commercially in purse seines.

+ range: Tropical and subtropical waters from 40°N to 35°S

+ diet: Fish such as flying fish, Dolphinfish, and mackerel, plus octopus, lobster, and crab

+ name origin: For its bright yellow fins

The Yellowfin is much sought after by commercial fisheries, which have greatly depleted its numbers. The fish will approach coastlines in warm weather and seems to have an affinity for floating objects, under which it may find shade. Others swim at greater depths because fewer of its predators are there; those swimming in the deep tend to scatter more easily. Like Bluefins, the Yellowfin is a fast swimmer, but it does not migrate as much as its larger relatives. It is a strong schooler, and it is often not in schools its own kind but in gatherings of similar-size denizens of the sea, including other tuna.

Pacific Halibut

Hippoglossus stenolepis L 4–8 ft (1.2–2.4 m)

The Pacific Halibut is one of the largest oceanic bony fish, up to 8 feet (2.4 m) long. It swims on its side, with both eyes on the same side, but it is a strong swimmer and highly migratory.

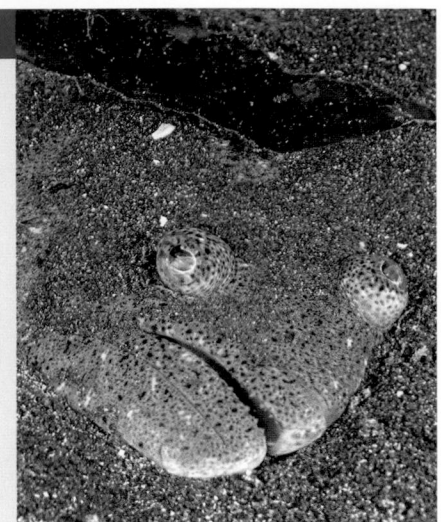

KEY FACTS

Its color depends on the color of the sea floor.

+ **range:** Primarily the continental shelf of Alaska and Canada

+ **diet:** Smaller fish such as Pacific cod, and invertebrates such as octopus, crab, and shrimp

+ **name origin:** *Haly* (holy) and *butte* (flat fish) because it was eaten on Catholic holy days

The Pacific Halibut grows to enormous size, sometimes weighing more than 500 pounds (227 kg). It is flat and flounderlike, and generally strikes from ambush. It lies on the bottom of the ocean, unmoving and waiting for other fish in the cold waters, its color resembling the sea floor. Its eyes begin on each side of its head like most fish, but at an early age one eye migrates to the other, top side. The Pacific Halibut is slow to mature. The males are able to reproduce after 6 or 7 years, and the females after 8 to 12 years. Eggs hatch in about 15 days, and the free-floating larvae drift for up to 6 months.

Summer Flounder

Paralichthys dentatus L 15–20 in (38–51 cm)

The Summer Flounder, also known as Fluke and Sole, is a flat fish with eyes on the left side of its head; it burrows into the bottom mud with only its head exposed, and strikes at passing fish.

KEY FACTS

It has eyelike spots on the top part of its body.

+ **range:** Winters off our Northeast coast; moves to shallow waters to spawn from Maine to Florida in early spring

+ **diet:** Smaller invertebrates and very small fish

+ **name origin:** French *floundre*, which probably came from Scandinavian *flythra*

The Summer Flounder changes color to blend into its environment on the sea floor, and it has important differences from the Winter Flounder. The Summer Flounder has both eyes on the left side of its head, and the Winter Flounder has them on the right. The fish prefer different temperatures—Summer Flounder staying in the deep ocean water in winter and moving to the shallows in summer, and Winter Flounder doing the opposite. The Summer Flounder has sharp teeth on the upper and lower part of its jaw, and the Winter Flounder has none. Both burrow into the mud with only their heads exposed to catch and devour prey.

Northern Puffer/Blowfish

Spheroides maculatus L 8–10 in (20–25 cm)

The Northern Puffer or Blowfish is strangely shaped, with a defense mechanism of taking water or air into its modified stomach and puffing into a ball. It is also called Sugar Toad and Sea Squab.

KEY FACTS

Coloring can be yellow, brown, or olive, with black spots on its back, sides, and cheeks.

+ range: Protected water from Florida to Newfoundland; winters in deep water

+ diet: Primarily shellfish, crushing them with its beak

+ name origin: For its defense of turning itself into a round ball

The Northern Puffer, or Blowfish, is a club-shaped fish with small prickles all over its body. It has the unusual defense of inflating a chamber derived from its stomach , which it puffs into a round shape. Many larger fish are reluctant to swallow the prickly ball. Nevertheless, people in the Chesapeake Bay region eat this fish as a delicacy. If thrown back in the water while inflated, it floats on its back until it assumes danger is past, then it quickly deflates and swims away. It spawns from May through August in shallow water, the eggs adhering to the bottom. The male guards the eggs until they hatch.

Blue Whale

Balaenoptera musculus L 82–105 ft (25–32 m)

The Blue Whale is an extremely large mammal that belongs to the suborder of baleen whales. It is the largest animal ever to have lived on Earth and can grow to more than 100 feet (30 m) long.

KEY FACTS

Its body is bluish gray on top and lighter on the bottom. Despite its large size, it is preyed upon by Orcas.

+ range: All the world's oceans

+ diet: Tiny shrimp-like animals called krill, which it filters from the water through baleen

+ name origin: For its blue appearance in the water

The Blue Whale is not only the largest animal currently on Earth, but it is also the largest animal ever to have lived on Earth. Its total weight can be 200 tons (181,000 kg), with its tongue weighing as much as an elephant, and its heart weighing as much as a car. The Blue Whale cruises the ocean at an average of 5 mph (8 kph), but it can accelerate to 20 mph (32 kph). It is one of the loudest animals on Earth, emitting a series of pulses, groans, and moans that another Blue Whale can hear up to 1,000 miles (1,600 km) away. This whale travels alone or in pairs. An adult can eat 4 tons (3,600 kg) of krill a day.

Bowhead Whale

Balaena mysticetus L 66 ft (20 m)

The Bowhead Whale is a stocky baleen whale of the right whale family, second in weight only to the Blue Whale, though shorter than several other species. It has the largest mouth of any animal.

KEY FACTS

It has a dark body and no dorsal fin, like most whales, dolphins, and porpoises.

+ **range:** Cold water of high latitudes; American whalers called it the "polar whale"

+ **diet:** Zooplankton, other invertebrates, and fish

+ **name origin:** For a lower jaw that is shaped like an archer's bow

The Arctic Ocean and other cold seas are home to the Bowhead Whale, which spends its summers in ice-free waters. The head for which it is named takes up 40 percent of its body length; the mouth is 10 feet (3 m) wide and 20 feet (6 m) deep. It uses its huge head to break through thick ice, breathing through two blowholes like all baleen whales. The Bowhead Whale has extremely long baleen plates, up to 14 feet (4.3 m). This whale species lives in a pod and has thick blubber because of its Arctic range. It weighs about 2,000 pounds (900 kg) at birth. The Bowhead Whale is endangered through-out its range.

Gray Whale

Eschrichtius robustus L 50 ft (15 m)

The Gray Whale, sole member of its family, Eschrichtiidae, is known for fighting back when it is hunted. Its gray and white mottling are scars from parasites that drop off in the cold water.

KEY FACTS

This species is dark-skinned and gray.

+ **range:** Mainly in shallow waters of North Pacific in summer, migrating to warmer waters in winter

+ **diet:** Mostly crus-taceans, turning on its side to grub them out of the mud

+ **name origin:** For its slate gray color and gray-white blotches

The Gray Whale is known for leaping out of the water, called breaching, but the reasons remain a mystery. Suppositions are that it is an expression of well-being, frustration at having parasites on its back, or a warning to other whales. The Gray Whale is popular with whale-watchers for breaching and because it always swims through the same pass in the Aleutians on its migration route north. In autumn, it passes along the west coast of Canada, the United States, and Mexico on its way to the waters of Baja California. Calves drink 50 to 80 gallons (190 to 300 L) of milk each day, which is 53 percent fat.

Humpback Whale

Megaptera novaeangliae L to 60 ft (18 m)

A favorite of whale-watchers, the Humpback Whale is a baleen whale with unusually long dorsal fins and a knobby head. Its lengthy annual migration is up to 16,000 miles (26,000 km).

KEY FACTS

It is primarily dark gray with areas of white.

+ **range:** All the major oceans, wintering at the Equator in warm water and summering at subpolar latitudes

+ **diet:** Tiny crustaceans (mostly krill), plankton, and small fish, eating only in summer

+ **name origin:** For the hump on its back

The Humpback Whale can be acrobatic, leaping and slapping the water with its dorsal fins and tail. It can consume about 3,000 pounds (1,400 kg) of food every day, ingesting large amounts of water and filtering nutrients with its baleen. Humpback Whales are singers, producing a song that lasts 10 to 20 minutes, which it repeats for hours at a time. In the winter it gives birth to a one-ton calf. Hunted to the brink of extinction, it is recovering due to a moratorium on hunting announced in 1966. Its numbers are still endangered by entanglements in fishing nets, collisions with ships, and noise pollution.

North Atlantic Right Whale

Eubalaena glacialis L to 50 ft (15 m)

The North Atlantic Right Whale lives in the North Atlantic Ocean and is one of three right whale species. The others are in the North Pacific and in southern waters. The three rarely mix.

KEY FACTS

It has dark gray or black skin, a V-shaped blowhole, and white on its head from numerous whale lice.

+ **range:** Northern waters of the East Coast

+ **diet:** Mainly small organisms called zooplankton

+ **name origin:** The "right whale" to whalers because it was easy to harvest

Whalers loved the Right Whale because it is easy to kill and floats after it is dead. The Right Whale tends to stay where the food is, near peninsulas, bays, and continental shelves. Scientists say that when its food is not concentrated, the Right Whale loses interest in feeding. Calves are born in warm waters and spend the first few months of their life traveling and playing with each other and older siblings. They stay with their mother for about a year, longer than Blue or Gray Whales. When presented with danger from orcas or sharks, Right Whales form a circle with their tails pointed outward and thrash them.

Fin Whale
Balaenoptera physalus L 75–85 ft (23-26 m)

The Fin Whale is a sleek baleen whale, the second longest whale in the world. It has two subspecies, one in the North Atlantic and one in the Southern Hemisphere.

KEY FACTS

It is black or gray on the back and sides with a white underside.

+ **range:** All the world's seas, but away from ice packs and most dense in temperate and cool waters

+ **diet:** Krill, small schooling fish, and squid

+ **name origin:** Perhaps because its fin is larger than a Blue Whale's fin

The Fin Whale lives for 80 to 90 years. It sometimes crossbreeds with the Blue Whale, and hybrids have been reported that show characteristics of each species. It is one of the fastest whales, because of its long, thin shape; it is able to sustain speeds of 23 mph (37 kph) while migrating, and bursts up to 29 mph (47 kph) have been recorded. Orcas are its only nonhuman predator. The Fin Whale is gregarious, traveling in social groups, and is sometimes seen feeding alongside other types of whales including Humpbacks and Minkes. A single Fin Whale can eat up to 2 tons (1,800 kg) of food a day.

Common Minke Whale
Balaenoptera acutorostrata L 22.5 ft (6.9 m)

The Minke Whale is the smallest baleen whale except for the Pygmy Right Whale, and is the most abundant. It is also called the Sharp-headed Finner and the Little Piked Whale.

KEY FACTS

The Minke is very dark above and lighter below. Calves are usually darker than adults. It has a tall dorsal fin.

+ **range:** All major seas except the polar seas

+ **diet:** Plankton, krill, and small fish

+ **name origin:** Possibly refers to a Norwegian whaler named Meincke, who mistook it for a Blue Whale

Although it is the second smallest baleen whale, the Minke still averages 25 feet (7.6 m) long and weighs about 6 tons (5,400 kg). The newborn calf weighs a half ton and swims to the surface within 10 seconds for its first breath. It can swim within a half hour of being born and lives about 20 years. The Minke Whale can dive for about 25 minutes but normally makes dives of 10 to 12 minutes. The 50 to 70 grooves on its throat start just below its chin and run almost to its midsection. It exhales just before surfacing to take a breath, resulting in a blow that is shorter than those of other whales.

Sperm Whale

Physeter macrocephalus L 67 ft (20 m)

The Sperm Whale, also called Cachalot (French for "big head"), is the largest toothed whale. It is easily recognized by its huge head, blunt forehead, and small jaw well back from the forehead.

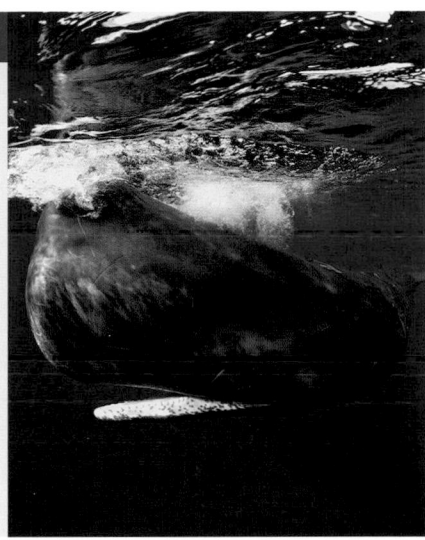

KEY FACTS

Blue-black in color, it has thick flukes, which it raises before a deep dive.

+ **range:** All oceans, with females staying in temperate and tropical waters and males traveling to high latitudes

+ **diet:** Mostly squid deep; also octopus and fish

+ **name origin:** For spermaceti, a waxy substance found in its head

The classic whale form is that of the Sperm Whale, perhaps because it is so abundant and because it was the mainstay of the early whaling industry. The Sperm Whale is not a threatened species, despite intensive hunting. It has the biggest brain of any animal. It makes a loud sound, emitting "clangs" that may be used for communication or, bounced off objects, may tell the whale how far away an object is, whether it is moving, and what shape it is. This species prefers ice-free waters, can dive to 3,280 feet (1,000 m) for squid, and can hold its breath for up to 90 minutes by reducing its metabolism to conserve oxygen.

Beluga Whale

Delphinapterus leucas L to 20 ft (6 m)

The Beluga Whale is also called White Whale, White Porpoise, and Sea Canary. It is a small toothed whale with a beaked mouth, rounded head, and flexible neck vertebrae.

KEY FACTS

This species has paddlelike flippers and notched tails. Calves are born gray or brown but change to white when they mature at about five years.

+ **range:** Arctic waters, migrating south when the sea freezes over

+ **diet:** Squid, fish, crustaceans, and marine worms

+ **name origin:** *Belukha,* Russian for "white"

Few people are likely to see the Beluga Whale because it lives in the far north. It is a smallish whale, ranging from 18 to 20 feet (5.5 to 6 m) in length, and has no dorsal fin, which helps it when it swims under the ice. Social animals, Beluga Whales live together in pods, communicating with a series of clicks, whistles, and clangs. They can also mimic other sounds. Belugas are prey to Polar Bears, Orcas, and northern people who have long hunted them. Commercial fisheries brought some populations to near collapse, and although the continuing harvest has been small, the animals have not recovered their numbers.

Narwhal

Monodon monoceros L to 12 ft (3.6 m)

The Narwhal has a single ivory tusk that protrudes from the front of its head that measures up to 8 feet (2.4 m) long. The Narwhal does not migrate, but swims and hunts under the ice in winter.

KEY FACTS

It is mottled blackish brown over a white background.

+ **range:** Arctic Ocean; sighted off Greenland and in Hudson Bay

+ **diet:** Fish, squid, shrimp, and other aquatic fare

+ **name origin:** Old Norse *nar,* meaning "corpse," because its mottled color reminded some people of a drowned sailor

The Narwhal is the unicorn of the sea because of the single horn extending from its forehead. The horn is actually a tooth that grows through the upper lip of the animal, and its use is not known. Because the horn is longer in males, it may be used in battling rival suitors during mating. The creature lives in pods of hundreds and even several thousand. It often is trapped by shifting pack ice and falls prey to Polar Bears, Walruses, and Inuit hunters, who kill it for the tusks and skin, which contains vitamin C. In winter, the Narwhal moves closer to the coast, and to deeper water in summer.

Killer Whale/Orca

Orcinus orca L 25 ft (7.6 m)

The Killer Whale, or Orca, is a large predatory porpoise found in all the major seas. It hunts other sea creatures and has no predators itself. Orcas often work in teams to overcome larger prey.

KEY FACTS

This dolphin has distinctive coloring, with a black back and white chest and sides.

+ **range:** Cold coastal waters at high latitudes, rather than deep oceans

+ **diet:** Fish, seals, penguins, seabirds, and large whales

+ **name origin:** For its vicious nature and attacks on large mammals

The Killer Whale is the hunter of the sea. It is the largest of the dolphins and one of the world's most powerful predators. It kills with its sharp teeth, by slapping the victim on the ice, ramming it, striking the prey with its tail, or breaching and landing on it. Sometimes the Killer Whale beaches itself to grab prey before wriggling back into the water, unusual behavior for a water dweller. It has been known to snatch seals right off the ice. Highly intelligent, Killer Whales travel in pods, communicating with a series of clicks and other vocalizations. Humans have never extensively hunted them.

Short-finned Pilot Whale

Globicephala macrorhynchus L to 20 ft (6.1 m)

The Pilot Whale is one of the largest oceanic dolphins and is highly intelligent. The very similar Long-finned Pilot Whale *(G. melas)* occurs only on our Atlantic coast.

KEY FACTS

The animal is dark gray or black with pale markings on its underside. Highly adaptive, it does well in captivity.

+ **range:** All oceans worldwide; prefers warmer waters

+ **diet:** Mostly squid and some fish

+ **name origin:** From observers who thought its pods were directed by a leader

A carnivore like the Killer Whale, the Pilot Whale does not eat seals and large whales, although it is only slightly smaller than an Orca. It is easily trained and is often found in captivity. Pilot Whales have been taught by the U.S. Navy to retrieve items from ocean depths. Highly gregarious, it lives and travels in groups, and it is the most common dolphin to become stranded on a beach. Pods include some females that nurse other young Pilot Whales, even though they are not their mothers. Although they have been hunted for their meat, bone, oil, and for fertilizer, they are not considered endangered.

Common Bottlenose Dolphin

Tursiops truncatus L to 13 ft (4 m)

The Common Bottlenose Dolphin is highly trainable and is often seen in aquatic parks. Largest of the beaked dolphins, it breathes through a blowhole, although the beak looks like a nose.

KEY FACTS

This dolphin species is gray and has a short, well-defined rostrum, or snout. It is sleek and swims rapidly.

+ **range:** Oceans worldwide, usually in tropical, subtropical, and temperate waters

+ **diet:** Eels, squid, shrimp, and fish

+ **name origin:** For its snout, which resembles a bottle

If you think of a dolphin, you probably picture the Common Bottlenose Dolphin. Very intelligent, it can be trained to do tricks in aquatic parks. Its brain is larger than a human brain. In the wild, it locates prey with a type of sonar using clicks, waiting for the return click to indicate the location of the prey and determine its size. The dolphin does not chew food but swallows it whole. It uses body language and sound as communication, emitting a series of squeaks and whistles that come from its blowhole. The dolphin's head contains an oily substance that acts as an acoustic lens and protects the brain case.

Common Dolphin

Delphinus delphis L 6.2–8.2 ft (1.9–2.5 m)

The Common Dolphin is not the dolphin everyone in North America imagines. It is a tricolored animal that is highly sociable but prone to illness if kept in captivity.

KEY FACTS

This species is black on its back, white on the underside, and yellow buff on the sides to just behind the dorsal fin.

+ **range:** Atlantic and Pacific Oceans, short of the Arctic Ocean to northern South America

+ **diet:** Schooling fishes and squid

+ **name origin:** Greek *delphis*, meaning "womb"

Three species are called the Common Dolphin, but the Short-beaked version is considered most in this account. It prefers deep water and rarely comes close to shore, whereas the Long-beaked Dolphin (*D. capensis*) prefers warmer waters near the shore. A third (*D. tropicalis*) has an extremely long and sharp beak and is found in the Red Sea and Indian Ocean. Except for their beaks, all have similar markings and similar characteristics. The Common Dolphin is a fast swimmer and enjoys riding the bow waves of boats. It was the dolphin most often represented in the art and literature of ancient Greece and Rome.

Atlantic Spotted Dolphin

Stenella frontalis L 8.2 ft (2.5 m)

A warm-water species, the Atlantic Spotted Dolphin is relatively small, with a sharp beak. It travels in groups of up to a thousand. The similar Pantropical species (*S. attenuata*) occurs worldwide.

KEY FACTS

This dolphin begins life with dark spots around weaning; as it grows older, the spots become lighter and more numerous.

+ **range:** Atlantic and Gulf coasts

+ **diet:** Eels, invertebrates, and herring

+ **name origin:** For the spots on mature adults

When you hear about a dolphin that has drowned in nets meant to catch fish, it was probably a Spotted Dolphin. This species' numbers were threatened by seines until safe fishing nets were developed that allowed it to escape. Now it is considered at lower risk. The Spotted Dolphin is a fast swimmer, acrobatic like most dolphins, and gregarious to the point of schooling with other species of dolphins, especially spinners. Calves nurse from one to two years and begin taking solid food at three to six months. Like other dolphins, the Spotted Dolphin will stay with and support wounded or sick dolphins until they die.

Harbor Porpoise

Phocoena phocoena L 5 ft (1.5 m)

The Harbor Porpoise, also called the "Puffer" and "Puffing Pig," is shy and elusive but stays close to shore. It is related to the dolphin, and among the smallest cetaceans.

KEY FACTS

Its back is dark gray, with lighter gray sides; stomach is nearly white; it is often speckled.

+ **range:** North Atlantic and West Africa

+ **diet:** Schooling fish, squid, and crustaceans

+ **name origin:** "Porpoise" from a compound of Latin *porcus* ("pig") and *piscus* ("fish"); "harbor" from where it is often found

The Harbor Porpoise inhabits fjords, bays, estuaries, and harbors. The cool-water animal is among the smallest of its relatives, the dolphins, and very stout in front of its dorsal fin. It tends to hunt alone but will seek food with others of its kind, herding their fish prey in tight groups. These porpoises are sometimes attacked and killed by larger Bottlenose Dolphins in competition for food. It mates promiscuously, and the male produces large amounts of sperm, perhaps to overwhelm the sperm of other males. Unlike other dolphins, it is not gregarious, acting alone or gathering in pods of not more than five.

Manatee

Trichechus manatus L 8–13 ft (2.4–4 m)

The Manatee is slow-moving and stout, spending its life in tropical waters. It is not related to dolphins. Its large prehensile lip is used to gather food. It is also called a Sea Cow.

KEY FACTS

This mammal has thick, gray wrinkled skin.

+ **range:** Warm waters in the tropics, from southern Florida to South America

+ **diet:** More than 60 types of grasses, leaves, and algae

+ **name origin:** *Manati*, meaning "breast," from the language of the Taino, a pre-Columbian people of the Caribbean

The Manatee is a docile, blimplike animal, which has no natural predator. It is endangered due to collisions with the propellers of ships and boats. Fifty scars have been found on one individual, and the creature is sometimes cut in half by large ships. It holds its flippers close to its body as the broad, strong tail waves up and down to push it at 5 mph (8 kph), but it can cover short distances at 15 mph (24 kph). It is known to tolerate human swimmers in the wild, but most avoid people, who previously hunted them. Sailors once mistook Manatees for mermaids because of their humanlike breasts.

Sponges

Phylum Porifera L 0.04 in–3.3 ft (0.1 cm–1 m)

Sponges are multicelled animals whose bodies' pores and channels allow water to circulate, taking in food and oxygen and eliminating waste. People have used them for thousands of years.

KEY FACTS

Sponges have maximum efficiency moving water through the central cavity.

+ **range:** Tropical salt water, ranging from tidal zones to great depths

+ **diet:** Mostly nutrients such as bacteria and leaves; some crustaceans

+ **name origin:** Greek *spongos*, meaning "to soak up"

A sponge has no nervous, digestive, or circulatory systems. Its body cavity allows water to circulate through its pores and channels, from which food items are extracted. There are approximately 5,000 to 10,000 known species of sponge, all inhabiting water. Most sponges release sperm cells that fertilize ova released by the same species. Sponges come in a wide variety of colors, depending on the food they ingest. They have been used for bathing, cleaning agents, contraceptives, and toilet paper. Sponges almost became extinct from overuse, but some species are increasing after having been replaced by synthetic materials.

True Corals

Class Anthozoa L varies widely

True Corals are small organisms called polyps, which extrude rocklike calcareous skeletons. Colonies of these creatures form large reefs and islands in warm seas.

KEY FACTS

True Corals have a wide variety of shapes and colors. They are generally a strong pink, but can be moderate red or reddish orange.

+ **range:** Relatively warm, nutritious seas worldwide

+ **diet:** Plankton and small fish

+ **name origin:** Greek *korallion*, derived from *kyphos*, which means "bent, curved, humped"

Divers love corals because structures created by the polyps present a dazzling array of shapes and colors. There are hundreds of species of True Corals, named for their resemblance to familiar shapes such as staghorn, orange-cup, and clubbed finger. The polyps are tiny but not microscopic animals that pull food into their digestive systems with tentacles ringing their central mouths. Most attach themselves to others of their species, forming reefs or entire islands. Each individual has a cup into which it can retreat when in danger. Coral reefs are dying worldwide, and pollution is suspected as one of the causes.

Fire Corals

Family Millepora L varies widely

Fire Corals are not true corals, though they mimic coral and attach to coral reefs. They are related to jellyfish and anemones, and their polyps have tentacles with stinging cells that capture prey.

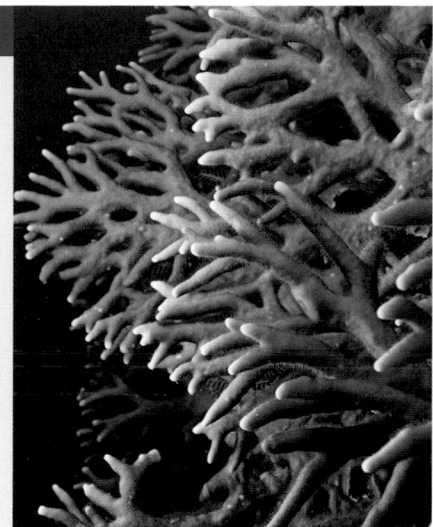

KEY FACTS

Fire Corals have a bright yellow-green and brown skeletal covering.

+ range: Form extensive outcrops on projecting reefs in warm-water seas

+ diet: Nutrients such as bacteria and plankton

+ name origin: Latin *Millepora* means "thousand pores," for the skeletal holes that contain polyps

Contact by humans is mostly accidental, because the Fire Coral's waving fans resemble seaweed and are bumped by divers. Once contact has been made, the stinging cells can produce an intense pain that lasts up to two weeks. The Fire Corals also have a sharp exoskeleton that can scrape the skin. They face the same threats that coral reefs are facing, which include poor management practices that release sediment, harmful nutrients, and pollutants into the ocean, stressing the fragile reef systems. Many fish hide in corals of all kinds, and corals are damaged when collectors deliberately smash them to reach the fish.

Frilled Sea Anemone

Metridium senile L 3–4 in (7.6–10 cm)

One of the largest and commonest anemones, the Frilled Sea Anemone has a smooth, cylindrical stem ending in some 1,000 tentacles that contain stinging cells it uses to capture prey.

KEY FACTS

It may be white, brick red-orange, or brown.

+ range: From Alaska to California and from New Jersey to Labrador; some off the coast of Europe and the Mediterranean

+ diet: Mostly plankton; also fish, squid, and dead animals

+ name origin: For the tentacles at the end of its long stem

The Frilled Sea Anemone can produce eggs or sperm when it is young and change sex when it is older. It can also deliberately break off parts of itself, which can reproduce. It often forms groups, but these are distinct individuals, not a colony like corals. When alarmed, the anemone fires stinging cells and when further threatened may break off its anchor and drift to another location. It tends to adhere to structures like pilings, wharves, or floats, but also attaches to boulders and the shells of oysters and crabs. It is the most common of these cylindrical animals that attach themselves to human or natural structures.

Portuguese Man-of-War

Physalia physalis Float 12 in by 5 in (30 cm by 13 cm); tentacles hang 165 ft (50 m) below float

The Portuguese Man-of-War may be confused with a jellyfish, but it includes several organisms working together. The sting of the tentacles, one of the organisms, is painful but rarely fatal.

KEY FACTS

The "sail" or "float" above the water is a rainbow of purple, blue, and pink.

+ **range:** Warm seas worldwide

+ **diet:** Small fish and other marine prey

+ **name origin:** For the float, which resembles an old warship

The Portuguese Man-of-War is not an "it," but a "they." It consists of four organisms, or polyps, working together. One is a gas-filled bladder, which floats on the water; the second is a set of long tentacles hanging below the float; the third is a digestive polyp that processes food; the fourth contains reproductive organs. One colony produces sperm that fertilizes the eggs of another colony. The tentacles, filled with venom-filled nematocysts, paralyze and kill prey, and draw it to the digestive polyp. Portuguese Men-of-War sometimes occur in groups of a thousand. When they wash onto a beach, don't touch the tentacles.

Moon Jellyfish

Aurelia aurita 10–16 in (25–41 cm) wide

The Moon Jellyfish, also called Common Jellyfish, is translucent with no gills, lungs, or trachea; it captures food with dangling tentacles with stingers, as it drifts with the current.

KEY FACTS

A Moon Jellyfish has its inner organs exposed to the world.

+ **range:** Most oceans worldwide, including our Atlantic coast north to the edge of Arctic waters and south to Argentina

+ **diet:** Mollusks, crustaceans, and small organisms

+ **name origin:** For its moon shape and jellylike appearance

The Moon Jellyfish is the bane of swimmers. A human can receive a painful sting by swimming into the tentacles, which it uses to capture food and to ward off predators. Its sting is not as painful as that of the Portuguese Man-of-War, but is irritating. The Moon Jellyfish lives six months and is eaten by Ocean Sunfish, Leatherback Sea Turtles, and seabirds. It can be identified by four horseshoe-shaped gonads, which can be seen from the top through its translucent body. It stings its prey with nematocysts and digests it in a gastrovascular cavity. This jellyfish's distribution depends on the flow of oceanic currents.

Lion's Mane Jellyfish

Cyanea capillata body 4 in to 8 ft (10 cm to 2.4 m) wide

The largest jellyfish, the Lion's Mane differs from Moon Jellyfish in that it has more tentacles, inhabits mostly cold waters, and achieves great size. Its tentacles extend to 120 feet (37 m).

KEY FACTS

Larger ones are crimson or purple; smaller ones are light orange or tan.

+ **range:** Cold, boreal waters of Arctic, Atlantic, and Pacific oceans; bays of the eastern U.S.

+ **diet:** Zooplankton, small fish, and Moon Jellyfish

+ **name origin:** For thick, hanging tentacles that look like a lion's mane

Zoologists do not agree on the taxonomy of the Lion's Mane Jellyfish, some suggesting that all species should be treated as one. Despite its large size, the Lion's Mane has a sting less painful than the sting of the Portuguese Man-of-War, causing only temporary pain and localized redness. This jellyfish swims beneath the surface, sometimes as deep as 60 feet (18 m), and is capable of a weak, pulsating motion. It snares its prey in its extremely sticky tentacles, moving food toward its digestive system. Predators of the Lion's Mane Jellyfish are seabirds, Ocean Sunfish, and sea turtles. This jellyfish lives about one year.

Atlantic Sea Nettle

Chrysaora quinquecirrha up to 10 in (25 cm) wide

The Atlantic Sea Nettle is a jellyfish that can give a painful but nonfatal sting to humans. It is usually pale, pinkish, or yellowish, often with stripes radiating from the center of the bell.

KEY FACTS

This sea nettle has two stages: free swimming when it moves and a polyp stage when it is immobile.

+ **range:** Coastal waters from Canada to Florida; sometimes observed in bays and lagoons near the Atlantic shore

+ **diet:** Marine worms, small jellies, and brine shrimp

+ **name origin:** For its sting

The Sea Nettle is bothersome to swimmers, but its sting is rarely fatal unless an allergic reaction sets in. Its tentacles contain venomous harpoons that it can shoot at prey or predators, resulting in a painful wound. The tentacles are nearly twice the size of the bell, or main body, hanging nearly 20 inches (51 cm). Spider Crabs eat the Sea Nettle, as do certain fishes. The crabs have been observed living among the tentacles, using them for protection, transportation, and food. The Atlantic Sea Nettle reproduces asexually. It is smaller than the Pacific Sea Nettle, but it has more variable coloration.

Freshwater Leech

Macrobdella decora L at least 2 in (5 cm)

The Freshwater Leech, flat with suckers at each end, is a segmented worm occupying fresh waters. An individual has both sex organs, and its main diet is blood.

KEY FACTS

This leech is brown-ish green with black and red spots and five pairs of eyes. An excellent swimmer, it comes to the host and attaches itself.

+ range: Streams, lakes, and marshes of the U.S.

+ diet: Blood of fish, birds, reptiles, and mammals

+ name origin: Greek *lahhi*, meaning "physician"

The Freshwater Leech, known to suck the blood of its prey, was used in medicine for 2,500 years, up to and including the 19th century in North America. Physicians applied it to patients to decrease swelling and to remove "bad blood." The leech senses the vibration of prey in the water, attaching to it with suckers at each end. It releases saliva that acts as an anesthetic; the victim can neither sense the cutting of flesh or the withdrawal of blood. Once sated, the leech releases itself. It needs to eat perhaps only twice a year. Remove a leech with a fingernail, dragging it under the suckers to break the connection.

Marine Worms

Class Polychaeta L to 4 in (10 cm)

Marine worms are the dominant animals on ocean bottoms. Many species are strikingly beautiful, colored red, pink, green, or a combination, and they come in many shapes and sizes.

KEY FACTS

About 10,000 species have been described. Some live in the ocean trench where they were discovered off the Galápagos Islands of Chile.

+ range: At varying depths in oceans all over the world

+ diet: Small nutrients

+ name origin: For its elongated shape and seawater habitat

The forgotten animals of the sea, marine worms include many species of segmented creatures that are overlooked but are very important to the marine environment. The absence of these ubiquitous creatures may mean that pollution has invaded a location, because many cannot live in areas that are flooded with toxic chemicals. As both adults and larvae, they are important food items for a number of commercial fish, and recreational fishermen often use them as bait. These worms have specialized tentacles they use to exchange oxygen and release carbon dioxide. The tentacles are also used for reproduction.

Pacific Littleneck Clam

Leukoma staminea 2.3 in (6 cm)

The Pacific Littleneck Clam occurs only in the Pacific Ocean, from the Aleutians to Baja California. Also called Rock Cockle, Rock Clam, and Ribbed Carpet Shell, it is edible for humans.

KEY FACTS

The shell is chalky, yellowish, or brown with little ridges.

+ **range:** Pacific coast, to depths of 34 feet (10 m); stays near shore in gravelly sediments and in bays and inlets

+ **diet:** Phytoplankton and other bits of food

+ **name origin:** For its range and its small valves, which extend to find food

The Pacific Littleneck Clam is abundant along the West Coast, and it is prized as food. It often has tapeworms inside its shell, but these cannot infect humans. The clam is distinguished by teeth near the hinge at the back of the shell. It filters food from the water. Black valves protrude from the front of the shell, and they have tiny teeth. The clam usually lives about ten years. Generally inhabiting shallow water, the Pacific Littleneck is harvested with rakes, shovels, garden forks, and trowels. It occurs in such profusion in the Hood Canal in Puget Sound that the individual clams almost touch one another.

Northern Quahog

Mercenaria mercenaria up to 6 in (15 cm)

The Northern Quahog is edible, and is the clam referred to in "clam digging." It buries itself in sediment and extrudes two valves, one to take in food and the other to eliminate waste.

KEY FACTS

The Quahog appears dun-colored or tan when alive and white when steamed.

+ **range:** Along eastern shore of Canada and U.S. from Nova Scotia to Florida; introduced in Gulf of Mexico and U.S. West Coast

+ **diet:** Algae and other food particles from the water

+ **name origin:** Algonquin word *poquauhock*

Anyone who has dug for clams on the East Coast is familiar with the Northern Quahog, also called the hard-shell clam and chowder clam. A big industry has sprung up around clams, which are served in many restaurants on the half shell or in stews. The age of a Quahog can be determined by counting the rings on its shell. A filter feeder, it draws its food in by one valve and passes it over gills, where the food is filtered out. Quahogs lie out of sight in shallow water sediment. Its Latin name means "wages," which comes from the practice of the Algonquin people of stringing shells together to produce money called wampum.

Pacific Razor Clam

Siliqua patula 3–6 in (7.6–15 cm)

The Pacific Razor Clam is a meaty, edible bivalve with a shape that is generally longer than either the Northern Quahog or the Pacific Littleneck.

KEY FACTS

It is generally brown and is sought by clam diggers.

+ **range:** Surf-pounded beaches and protected inlets on the West Coast from the Aleutians to central California

+ **diet:** Food particles filtered from salt water

+ **name origin:** For its shape like an old-fashioned straight razor

The Razor Clam occurs in depths to 30 feet (9 m), but it can be dug by amateurs in zones where tides are 3 to 4 feet (0.9–1.2 m), using a shovel or special clam pump that draws it out by suction. It can be located by a hole or a slight depression in the sand. Some leave necks exposed when the tide recedes. Those along the Washington coast and southward live to five years; in colder waters northward, they grow larger and live as long as 15 years. The Pacific Razor Clam is similar to the smaller Atlantic Razor Clam (*S. costata*) found on the East Coast. Predators include Dungeness Crabs, shorebirds, and numerous fish.

Atlantic Jackknife Clam

Ensis directus up to 7 in (18 cm)

The Atlantic Jackknife Clam is edible but is rarely caught because it rapidly digs deep into the mud. Also called Bamboo Clam and American Jackknife Clam, it lives in intertidal zones.

KEY FACTS

When alive, it is white or brown; when dead and washed up onto the shore, it is tan.

+ **range:** Atlantic coast from the Canadian Maritimes to South Carolina

+ **diet:** Particles suspended in the water

+ **name origin:** For its sharp edge and shape, which resembles an old jackknife

There is a demand for Atlantic Jackknife Clams, because the meat is sweet and tender, but they are hard to catch. This filter feeder is a fast digger, and few commercial collecting operations exist. The oval burrow hole can be seen at low tide, but when someone inserts a shovel into the sand or mud, a small spurt of water indicates that the clam is digging, too. With mucus, it traps food particles that pass over its gills, moving them to a digestive tract. The sexes are separate, and spawning usually takes place in June. Predators include several kinds of shorebirds, moon snails, the Green Crab, and the Milky Ribbon Worm.

Pacific Geoduck

Panopea generosa Shell to 8 in (20 cm), siphon to 4 ft (1.2 m)

The Pacific Geoduck is the largest burrowing clam and one of the longest-lived animals; one was documented at 168 years old. It is also called Mud Duck, King Clam, and Elephant-trunk Clam.

KEY FACTS

This geoduck species is white with brown near the edge of the shell; the siphon is so extensive that it cannot be withdrawn completely into the shell.

+ range: West Coast from Alaska to Baja California

+ diet: Plankton

+ name origin: Corruption of a Native American word meaning "dig deep"

The Pacific Geoduck was not harvested until 1970, because its texture was not appetizing, but a market has turned up recently in Asia. Divers must harvest the Geoduck, because it lives in deep water, and its long siphon enables it to remain deeper within the sea bottom than most clams. Farms of Geoducks have started in Washington and British Columbia. A filter feeder, this clam species draws in water through its siphon and extrudes its waste from the same siphon. It has few predators except humans, although Sea Otters and Dogfish are able to dislodge them, and seals and sea stars feed on the exposed siphons.

Eastern Oyster

Crassotrea virginica up to 8 in (20 cm)

The Eastern Oyster is a highly commercial bivalve, prized for its meat. Only one percent of the wild population remains from the numbers that the European colonists found when they arrived.

KEY FACTS

Rough shell is white, grayish, or tan.

+ range: Native to East Coast in water from 28° to 90°F (-2°–32°C)

+ diet: Extracted algae and food particles from attaching to a surface underwater, sucking in water, and expelling it afterward

+ name origin: Greek *ostrakon*, for its resemblance to a pottery shard

Native Americans ate oysters, leaving great piles of their shells. Visibility in the Chesapeake Bay in those times would reach 20 feet (6 m). Female oysters produce from 75 to 150 million eggs, but only one in a thousand survive. The eggs are released into the water column to be fertilized and develop into larvae. The larvae can swim, but eventually they develop a harder shell and sink to the bottom. They cannot attach to mud or sand, but they cement themselves to a hard surface—including other oysters. Oyster beds were once about 50 feet (15 m) thick. This oyster is farmed in Washington's Puget Sound.

Marine Mussels

Family Mytilidae 2–4 in (5–10 cm)

The family Mytilidae includes a number of saltwater mussel species, which are smooth-shelled, with one end pointed and the other round. Some species, such as the Blue Mussel, are edible.

KEY FACTS

Mussel colors vary from blue, black, brown, green, or red.

+ **range:** Worldwide, usually in colder waters, in clusters attached to rocks

+ **diet:** Current passes sea water over enlarged gills that sieve out food particles.

+ **name origin:** Latin *musculus*—"muscle"— with which mussels adhere to rocks

The family Mytilidae is the worldwide family of mussels, some edible and some not. They have a common tear shape, with one end pointed and the other rounded. Some are glossy, as if they are varnished. Mussels attach to rocks and piers, where the surf brings ample oxygen and food, which passes over the gills and is filtered out for nourishment. This bivalve is subject to the dangerous "red tide," algae that bloom in warm temperatures. People eating infected mussels can be poisoned. The most common in the family Mytilidae is the Blue Mussel. The Zebra Mussel that has invaded North America is not related to these mussels.

Bay Scallop

Argopecten irradians 3 in (8 cm)

The Bay Scallop is an edible bivalve with a scalloped shell and 18 pairs of eyes that sense shadows and movement. It swims on the bottom of bays and estuaries of our East Coast.

KEY FACTS

Shell color ranges from drab gray to yellow or reddish-brown. The inside of the shell is white.

+ **range:** Tidal zones in bays and estuaries, on eelgrass, sand, and mud, and offshore in moderately deep water

+ **diet:** Food particles

+ **name origin:** Old French *escalope*, meaning "this shell"

The Bay Scallop does not adhere to a surface or bury itself in the sea bottom, extending a valve to filter food from the water. It moves about freely, its 18 pairs of blue eyes sensing predators, which it escapes by clapping its two shells together to push out a jet of water that propels it. It does this using a large abductor muscle, the meat extracted for human food. Its population has been decimated by killing of sharks, which eat Manta Rays, which in turn eat scallops. More rays means fewer of their scallop prey. Another factor is an increase of "brown tides," an excess of phytoplankton, which kills scallops.

Red Abalone
Haliotis rufescens 12 in (30 cm)

The Red Abalone is a large snail and is the largest abalone in the world. It is found in depths of up to 100 feet (30 m) in rocky areas that have kelp and is known for its beautiful iridescent shell.

KEY FACTS

The external shell is usually brick red.

+ **range:** West Coast from Alaska to Baja California

+ **diet:** Primarily sea urchins, abalones, clams, and mussels; also other mollusks and crustaceans

+ **name origin:** The English version of *abulón*, the American Spanish name given to the abalone

The interior of the Red Abalone shell displays many colors, and the shells have been found in archaeological digs dating to 12,000 years ago. The Chumash and other California natives have made a number of beads, ornaments, and other artifacts from the shell. Red Abalone is the only abalone that is legally (though with restrictions) harvested in California, but abalones taken by sport fishermen cannot be sold. Abalone numbers swelled when Sea Otters, the main predators, were greatly reduced, then crashed because of hunting and destruction of kelp beds by industrial pollution and sewage. Today, its numbers are believed to be recovering.

Common Periwinkle
Littorina littorea L 0.5–1.5 in (13–38 mm)

The Common Periwinkle is an edible snail with gills and a dark shell, sometimes banded, which has six or seven whorls pointed at the top. It is also called Winkle, Buckie, Willick, and Crickle.

KEY FACTS

The shell is dark brown or black to grayish.

+ **range:** Native to coasts of northern Spain, Scotland, Ireland, Scandinavia, and Russia

+ **diet:** Algae from grazing; will eat small crustaceans like barnacles

+ **name origin:** A combination of Latin *pina*, for a kind of mussel, and Danish *vincle*, for "snail shell"

The Common Periwinkle can live several days without water by closing its shell and keeping its gills moist inside. It moves by alternating the right and left sides of its single foot. Two antennae on its head are used to see and taste. The Common Periwinkle is considered to be a delicacy in Africa and Asia. Sometimes its soft parts are used as fish bait. Its predators include sea stars and several kinds of fish. It was accidentally introduced to the northwestern Atlantic and is now the common marine snail from New Jersey northward to Newfoundland. It is eaten like popcorn on the beaches of Ireland and Scotland.

Violet Sea Snail

Janthina janthina Height 1.5 in, width 1.6 in (38 by 41 mm)

The Violet Sea Snail is a small snail that is rarely eaten and floats atop the water by means of air bubbles and a very lightweight shell. Its colors range from pale to dark purple.

KEY FACTS

Its shell is paper thin. The family Janthinidae comprises five species, all colored purple.

+ range: Worldwide in tropical and temperate seas

+ diet: Other floating animals such as the Portuguese Man-of-War and the By-the-wind Sailor

+ name origin: Greek *snecko*, or "snail," and *snahhan*, "to creep"

The Violet Sea Snail floats along the ocean on a light shell with its point down and its darker flat part facing up, which makes it harder for predators to see it. Its shell, though light, is too heavy to support it, so it churns a raft of bubbles on which it rests. If it loses its raft of bubbles, it sinks to the ocean floor and dies. The snail has a very flexible neck and small eyes located at the base of its tentacles. It begins life as a male and changes to a female as it grows older. Males release sperm, which fertilize the female's eggs, and she gives birth to larvae that build their own raft of bubbles.

Black Turban Snail

Chlorostoma funebralis Diameter 1–1.25 in (25–32 mm), L 1.5 in (38 mm)

The medium-size Black Turban Snail is found primarily in kelp beds. Native Americans once hunted it, but it is not endangered because there is no market for it. It lives to be 25 years old.

KEY FACTS

Turban Snails are black or purple. Their sensors are on tentacles that can be withdrawn in case of a threatening situation.

+ range: West Coast from Vancouver Island to Baja California

+ diet: Algae and kelp from grazing

+ name origin: For the dark turbanlike whorl on its shell

The Black Turban Snail lives in rocky intertidal zones, where it attaches itself with thin suckers on its single foot to kelp plants on which it finds algae to eat. If it falls from the kelp, it ends up on the ocean floor where it often is prey to sea stars and bottom-feeding predators. Blind, it can recognize a predator by scent and by the touch of its tentacles, and tries to move away. If on a slope, it detaches itself from its shell and tries to roll away. Its predators are Sea Otters, Red Rock Crabs, octopuses, and predaceous snails. Hermit Crabs often take up housing in empty Black Turban Snail shells.

Atlantic Plate Limpet/Tortoiseshell Limpet

Tectura testudinalis L to 1 in (2.5 cm)

The Atlantic Plate Limpet, also called the Tortoiseshell Limpet, has a hard, conical shell that comes to an off-center peak. It attaches to rocks by strong suction from its single foot.

KEY FACTS

This limpet is light tan with darker brown blotches.

+ **range:** Atlantic shore from Massachusetts to the Arctic

+ **diet:** Algae scraped from rocks

+ **name origin:** For the resemblance of its markings and colors to a tortoiseshell; "limpet" from Medieval Latin *lampreda*, meaning "lamprey"

The Atlantic Plate Limpet is a snail, though its conical shell does not look like one. As do all limpets, this cold-water creature attaches its shell to a rock with a strong, muscular foot. The tight adherence protects its soft parts from predators and from waves that might otherwise wash it off the rock. Tiny projections on the edge of the shell resemble teeth, which are used to tear food apart. A mouth moves the food toward the stomach, then into an intestine, and finally the waste matter is expelled. The limpet never strays far from where it started, returning to the same spot when it has finished feeding.

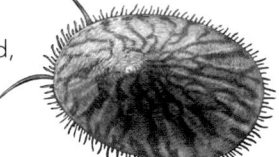

Frilled Dogwinkle

Nucella lamellosa up to 4 in (10 cm), usually 1–2 in (2.5–5 cm)

The Frilled Dogwinkle is a whelk that preys on other mollusks. This species is very common, and variable in appearance. The shell may be ridged or smooth, and may have dark bands.

KEY FACTS

The shell's color can be white, orange, brown, and rarely, purple.

+ **range:** Along West Coast from the Bering Strait to central California

+ **diet:** Mainly acorn barnacles and mussels

+ **name origin:** "Frilled" may refer to frills on the shell's ridges, which may be worn by waves. "Whelk" is from Danish *welc*.

The Frilled Dogwinkle is a medium-size predatory snail that lives in intertidal and subtidal rocky areas of the Pacific coast and is commonly found in mussel beds. To feed, the snail reaches the victim's soft parts by drilling a hole in its shell with a rasplike tongue; it may also inject a poison. The Frilled Dogwinkle then uses its tongue to scrape out the meat. It is preyed upon by the Red Rock Crab. Its shell is unpolished and has a well-developed spire. It sometimes congregates in large groups in the spring to breed. Each female can lay up to 1,000 eggs a year, but only one percent of the larvae survive.

Atlantic Auger
Terebra dislocata L 2.75 in (7 cm)

The Atlantic Auger is a carnivore that uses a poisonous barb to stun its victims. Also called Auger Shell and Auger Snail, this slender, pointed sea snail resembles a drill or a spike.

KEY FACTS

The auger is white with brown markings.

+ **range:** Offshore waters on sand flats and on beaches of the eastern Americas

+ **diet:** Diverse marine worms

+ **name origin:** For its resemblance to a drill

The Atlantic Auger is a predatory snail living in shallow sand flats offshore and on beaches. It subdues its marine worm prey in an unusual manner. After locating a victim, it stuns it with a poisonous barb, which immobilizes it. Then the Atlantic Auger eats at its leisure. This auger is easy to recognize because it is extremely long and thin with numerous whorls that come to a fine point at one end, like an auger or screw. It has a circular opening at the large end. The shell, which has a glossy exterior, is a familiar find to beachcombers on southeastern Atlantic shores. There are 313 auger species worldwide.

Lettered Olive
Oliva sayana L to 2.5–3 in (6–7.6 cm)

The Lettered Olive is a large predatory warm-water snail with a shiny shell that colonists and Native Americans turned into jewelry. It is also called Olive Shell and Olive Snail.

KEY FACTS

The shell is creamy to grayish with reddish brown zigzag markings. The empty shell is often washed up on beaches.

+ **range:** Shallows from North Carolina to the Gulf of Mexico

+ **diet:** Bivalves and small crustaceans

+ **name origin:** For the dark markings on its shell that resemble letters

The Lettered Olive lives in shallow sand flats in nearshore waters. Its shell is sometimes washed onto the beach where it is easily recognizable for its shiny cylindrical shape and the gap that runs the length of the shell. The Lettered Olive immobilizes small bivalves and crustaceans with its powerful foot and then drags them deeper into the sand to eat and digest them. The foot, or mantle, covers the outside of the shell as the snail moves along the bottom in search of prey. Females lay floating egg capsules, which are round and can be found in ocean drift. It ranges southward off Mexico and may be found in Brazil.

Atlantic Slipper Shell

Crepidula fornicata L 0.75–2 in (2–5 cm)

The shell of the Atlantic Slipper Shell looks like a shoe. One shell attaches to another in a stack; the bottom shell adheres to a hard surface. Its soft parts live on a hidden deck on its underside.

KEY FACTS

The shell is highly arched and white to whitish gray, often with tan to brown markings.

+ **range:** Native to East Coast in tidal zones or to depths of 50 feet (15 m)

+ **diet:** Food particles in the water

+ **name origin:** For its shape, which looks as if a foot could slip into the deck under the shell

The Atlantic Slipper Shell is a filter-feeding hermaphrodite, which can change sex if need be. Once stacked, one shell atop the other, the top shells are males and send down an extension to the bottom shell, which is a female, and fertilize her eggs. The larvae then attach to additional shells or attach to a surface on their own. If the latter, the shell becomes female, releasing a chemical that attracts males. The Slipper Shell has accidentally been introduced to other parts of the world. In England's waters, it is destructive to oyster beds because the shells block out food before it reaches the oyster.

Common Marsh Snail

Melampus bidentatus L 0.5 in (13 mm)

The Common Marsh Snail is air-breathing and grazes on plant foods in saltmarsh stalks. An absence of predators (Blue Crabs and turtles) may increase the snails and endanger marshes.

KEY FACTS

Its color ranges from white to grayish to brown.

+ **range:** Abundant on the East Coast in salt marshes that serve as fish nurseries along the edge of the Atlantic Ocean

+ **diet:** Microalgae on cordgrass, and sometimes decaying cordgrass

+ **name origin:** Its abundance in salt marshes

Look more closely at salt marshes and you will see many Common Marsh Snails. The whorled snail has lungs and is air-breathing and climbs to the tops of the grass to avoid high tides, but it can undergo submersion for two days if necessary, withdrawing into its shell and closing it up with a tight cap. The snail can be lured out of its shell by contact with warm human flesh. Although a larger snail, *Littorina irrorata*, probably disrupts the Common Marsh Snail's eggs by grazing over the same stems, it likely doesn't eat them. Marsh destruction was previously blamed on salt and poor soil.

Knobbed Whelk

Busycon carica L up to 12 in (30 cm)

The Knobbed Whelk is a predaceous sea snail that resembles a conch and divides its time between deep and shallow water. If the shell is held in front of you, the opening is to the right.

KEY FACTS

The shell is ivory and pale gray on the outside and orange inside.

+ range: Cape Cod to Florida on the Atlantic coast; the state shell of New Jersey and Georgia

+ diet: Oysters, clams, and other marine bivalves

+ name origin: For the knoblike projections on the widest part of its shell

The Knobbed Whelk is used much like its look-alike the conch shell. With a hole made in the empty spire, the part that comes to a point, it can become a bugle. The whelk spends summer and winter in deep water down to 40 feet (12 m), and in shallow water in temperate spring and fall where it preys on smaller bivalves. Its shell is thick and hard, and the inside of the opening is orange. The widest part of the shell has several knobby protrusions. The whelk's spent egg cases often wash onto the Atlantic beaches. Before the young hatch, the female plants the thin end of the case in the sand so it won't wash ashore.

Florida Fighting Conch

Strombus alatus L up to 4 in (10 cm)

The Florida Fighting Conch is a vegetarian that is very gentle but defends itself with a spur on its foot. Smaller than the Knobbed Whelk, it lives in sea grass beds, sand, and gravel shallows.

KEY FACTS

The animal appears in many colors including white, ivory, yellow, brown, gray, orange and green.

+ range: Warm ocean waters from North Carolina through the Caribbean Sea

+ diet: Algae and detritus from grazing; also dead animals

+ name origin: "Conch" comes from Greek *konche*

The Florida Fighting Conch is a popular aquarium dweller because it cleans up algae in the tank. An omnivore, it also preys on smaller snails if no algae are present. It "hops" if it is in danger, and defends itself with a spur on its single foot. A long snout something like that of an anteater can extend if the conch finds food; its rough tongue scrapes the food into its mouth. The conch burrows into the sand when it is not feeding. Its name distinguishes it from the West Indian Fighting Conch, and it is called fighting conch because it resists predators and males will fight each other over territory.

Atlantic Long-finned Squid

Loligo pealei L up to 20 in (51 cm)

A ubiquitous carnivore on the continental shelf, the Atlantic Long-finned Squid has commercial value as food; it spawns year-round, and those born in warm water grow most rapidly.

KEY FACTS

Naturally colored a reddish hue, it can vary its color to a soft pink.

+ **range:** Western Atlantic Ocean from Newfoundland to Gulf of Venezuela

+ **diet:** Crustaceans and small fish

+ **name origin:** For the length of its fins; "squid" from its habit of squirting ink when in danger

The Atlantic Long-finned Squid never grows very large, but it is an important food for other fish and controls populations of other marine life. Originally it was caught only as bait, but recently it has a market of its own for human food. The squid holds its prey in its tentacles and moves it toward its beak. It reproduces by a male and female copulating, the male injecting sperm into the female and the female then producing fertilized eggs. Many larvae are eaten by larger fish. It is differentiated from the Short-finned Squid in that its fins are longer, extending nearly half the length of its cylindrical body.

Common Octopus

Octopus vulgaris body to 4.3 ft (1.3 m), arms nearly four times as long

The Common Octopus is an intelligent eight-armed predator with a large head and a venomous beak. Its arms have suckers that clutch its prey, and a poison in its saliva paralyzes other animals.

KEY FACTS

The Common Octopus can change colors almost instantly to blend into its surroundings. Most males are bigger than females.

+ **range:** Temperate or warm waters of the eastern Atlantic

+ **diet:** Crabs, crayfish, and two-shelled mollusks

+ **name origin:** Greek *oktopous*, which means "eight arms"

The sea's most intelligent invertebrate has eight arms and a bulbous head. The Common Octopus can be trained to open a screw-top jar, and it has been known to raid lobster traps. This octopus can evade predators such as sharks or dolphins by blending in with its surroundings or exuding black ink, which dulls the predator's sense of smell while the octopus swims away. It can squeeze into cracks that the enemy cannot enter. The octopus can also lose an arm, which will grow back. Its beak can deliver a nasty bite. Its color may change according to its mood: If terrified, it turns white, and blushes if angry.

East Pacific Red Octopus

Octopus rubescens body to 9 in (23 cm)

The East Pacific Red Octopus is smaller than the Common Octopus but displays many of the same characteristics, including intelligence and the ability to vary its colors.

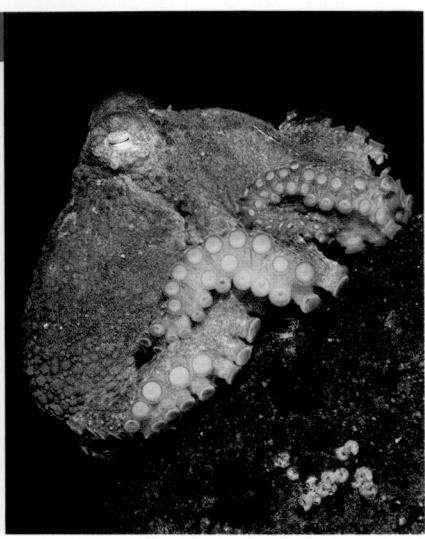

KEY FACTS

Although it can change colors readily, it is a deep brick red when relaxed.

+ **range:** Colder waters than the Common Octopus; much of the North American West Coast

+ **diet:** Clams, scallops, crabs, and barnacles, among other marine life

+ **name origin:** For its range and its red color

Like its southern cousin, the Common Octopus, the East Pacific Red Octopus can blend into its surroundings, changing into various colors and even textures. It is highly intelligent and can be trained in captivity. Two Red Octopuses may seem to be shaking hands, but they are a male and female breeding by attaching a special arm that extrudes and receives semen. Once eggs are released, the Red Octopus guards the eggs until they hatch. Many of the larvae will be eaten by fish. If individual characteristics are a sign of intelligence, then the octopus is highly intelligent, for it demonstrates individualism.

Giant Pacific Octopus

Enteroctopus dofleini record: 30 ft (9 m) across body, with weight more than 600 pounds (270 kg)

Largest of its kind, the Giant Pacific Octopus is found in deeper water than other octopuses, though it can live in shallows. It hunts at night, whereas the Common Octopus hunts at dusk.

KEY FACTS

Usually reddish brown, the animal changes its colors to blend with its surroundings.

+ **range:** Northern Pacific including coasts from Alaska to Baja California; also in deep ocean waters

+ **diet:** Primarily shrimp, clams, lobsters, and fish

+ **name origin:** For its large size and its range

The fearsome Giant Pacific Octopus is a creature to be avoided, but fortunately it lives in waters generally too cold for divers. The record size is measured in the hundreds of pounds, but most individuals are about half that size in body length and about 110 pounds (50 kg). It lives to approximately four years, which is twice as long a life as the Common Octopus. The female tends to the eggs fastidiously but does not eat during this period, dying soon after. Like other octopus species, it enlarges or contracts pigment cells to change colors and evade predators such as Harbor Seals, Sea Otters, and Sperm Whales.

Atlantic Horseshoe Crab

Limulus polyphemus up to 24 in (61 cm)

The Atlantic Horseshoe Crab is not a crab; it is more closely related to spiders, ticks, and scorpions. It is sometimes called a living fossil because it has changed little in millions of years.

KEY FACTS

Its colors range from greenish gray to dark brown.

+ **range:** Along the East Coast from the Gulf of Mexico to northeastern waters

+ **diet:** Mollusks, marine worms, other bottom-dwelling invertebrates, and bits of dead fish

+ **name origin:** For its shape, like a horseshoe

Medical researchers are quick to point out this strange-looking crab's value. Its blood has qualities that are important to researchers, and it is used to test for bacterial diseases. Shorebirds such as Red Knots depend on its eggs during the spring migration, and the Atlantic Loggerhead Turtle relies on it for food. The crab has nine eyes scattered throughout its body and six appendages for feeding and locomotion. Horseshoe Crabs have no jaws; they tear up their food with their legs to eat it. They have gizzards to further grind edibles. Atlantic Horseshoe Crabs undergo 17 molts and live 20 to 40 years.

Atlantic Lobster

Homarus americanus up to 25 in (64 cm)

The American Lobster, also called Maine Lobster, is the world's largest crustacean. Its meat is delectable. It lives at depths of 13 to 160 feet (4–49 m) but can be found at 1,500 feet (460 m).

KEY FACTS

It is bluish green or greenish brown. Rare ones are other colors. When cooked, it turns red.

+ **range:** Cold waters from Cape Hatteras to Labrador; uncommon south of New Jersey

+ **diet:** Mostly mussels; also other crustaceans and brittle stars

+ **name origin:** Old English *loppestre*, from "loppe spider"

The American Lobster can grow to great size, the biggest on record weighing 44.4 pounds (20.1 kg). The first of its eight legs are huge claws, one of which has notches for grasping and the other of which has sharp edges for tearing food. The female carries fertilized eggs under her abdomen for 9 to 11 months until she shakes her tail and the larvae emerge. The larvae swim near the surface, feeding on zooplankton until their fifth of 25 to 27 molts, when they sink to the bottom of the ocean and eat adult food. The adult lives for 40 to 60 years. Its closest relative is the European Lobster (*H. gammarus*).

Spiny Lobster/Rock Lobster/Langouste/Sea Crayfish
Genus Panulirus Various sizes

"The Spiny Lobster" is a general name given to lobsters resembling an American Lobster without big claws. It hides under coral reef ledges during the day and hunts at night.

KEY FACTS

This lobster has a striped body, brown-gray in color with yellow spots on its segmented tail.

+ **range:** Warm-water seas such as the Caribbean and Gulf of Mexico

+ **diet:** Sea snails and slugs, clams, crabs, sea urchins, and carrion

+ **name origin:** For its long antennae, which are spiny

The Spiny Lobster is the warm-water lobster that looks like the American Lobster, except that it lacks the large claws, which are the first two of its legs. The tail is good to eat and is a principal export of the Bahamas. A social creature, the Spiny Lobster tends to stay in groups but moves away if one is diseased. The lobster migrates in groups of 50 or so, keeping in touch by its antennae. Predators may be deterred from eating it by its sounds—a loud screech made by rubbing its antennae against the smooth part of its exoskeleton, or by rubbing the base of the antennae against a nearby "file," producing a raspy noise.

Dungeness Crab
Metacarcinus magister carapace width up to 7.9 in (20 cm)

The Dungeness Crab is an armored five-legged crab whose meat is tasty. It lives in eelgrass beds and muddy to sandy bottoms in excess of 600 feet (180 m). It can bury itself if threatened.

KEY FACTS

The shell is brownish, and the claws are white-tipped.

+ **range:** Off West Coast from the Aleutians to Point Conception, California

+ **diet:** Clams, other crustaceans, small fish—and each other

+ **name origin:** Port of Dungeness, Washington, where it was first found in abundance

The Dungeness Crab is a popular delicacy, especially in the western states and provinces. All five sets of legs are armored like the shell and end in a claw. The foremost pair is used in defense and for tearing apart food, and the smaller legs are used to pass food to its mouth. Once in the stomach, digestion is completed by toothlike appendages. The crab molts several times. Reproduction is accomplished by the female releasing a pheromone after which the male attaches to the female and fertilizes her eggs. The eggs are released several months later, and they remain attached to the female's abdomen until they hatch.

Yellow Shore Crab

Hemigrapsus oregonensis width 1.2–1.4 in (3–3.5 cm)

The Yellow Shore Crab is small and lives on mudflats. Its first two legs are large claws. It is also called the Hairy Shore Crab, Green Shore Crab, Mudflat Crab, and Oregon Shore Crab.

KEY FACTS

It often has a gray, green, or yellow carapace with small blue-black spots on its lighter-colored legs.

+ **range:** Found in great numbers in San Francisco Bay and on shore areas of Washington and Oregon

+ **diet:** Mostly algae, but also meat

+ **name origin:** For its color and its shoreline habitat

The Yellow Shore Crab has short eyestalks, short antennae, powerful front claws, and five pairs of hairy legs. It often lives in muddy, sandy, and gravelly habitats. The crab digs burrows for protection, often under rocks. If you turn over rocks and see it, it is a very fast digger and will disappear quickly. It eats mostly algae but will prey on invertebrates if they are available, and will scavenge when possible. It can also filter-feed for tiny organisms when necessary. It is preyed upon by shorebirds and an invasive species, the Common Littoral Crab. The Yellow Shore Crab is too small to be popular food for humans.

Pacific Rock Crab

Romaleon antennarium carapace width up to 6 in (15 cm)

The Pacific Rock Crab, also known as the California Rock Crab, can live either on the ocean floor or near land. It seeks out rocky bottoms where it hides under ledges and in crevices.

KEY FACTS

The upper side is usually red or brown, varying to shades of orange or gray.

+ **range:** West Coast from Alaska to Baja California, in intertidal zones up to 138 feet (42 m)

+ **diet:** Snails and crustaceans such as hermit crabs; also dead animals

+ **name origin:** For the rocks in which it hides

Many people associate the rock crab with Florida coasts, but the West Coast has the Pacific Rock Crab. It is fished recreationally and commercially, and it has a taste similar to that of the Dungeness Crab but is less preferred. Otters, sharks, octopuses, and large sea bass prey upon the Pacific Rock Crab, which is most vulnerable after it has molted and has a soft shell. In reproduction, the male places a sperm packet within the female, which blocks another male from fertilizing her eggs and is used for several spawnings. Fertilized eggs exit the female's body, and the larvae hatch after seven or eight weeks.

Green Crab

Carcinus maenas carapace width up to 4 in (10 cm)

The Green Crab is a medium-size crab at home in waters ranging from warm to cold; for that reason, the European native is spreading around the world and displacing native species.

KEY FACTS

Its primary color is green, though it can be brown, gray, or red.

+ **range:** From Europe to Atlantic and Pacific coasts of North America, Australia, South Africa, and South America, preferring soft ocean bottoms

+ **diet:** Mollusks, marine worms, and small crustaceans

+ **name origin:** For its prominent color

The Green Crab first came to the U.S. in the mid-1800s, spreading to Maine about a century later. It may have traveled by ballast water in ships, or by kelp wrapped around lobsters or bait worms. Less than 50 years after it had first reached the West Coast and colonized there, it has not proliferated because of control measures. Where it has become established, it impacts the populations of smaller shore crabs, clams, and oysters. It eats oyster larvae, and it can dig 6 inches (15 cm) for clams. The Green Crab prefers protected or semiprotected marine or estuarine habitats, including those with mud, sand, or rock.

Atlantic Blue Crab

Callinectes sapidus carapace width 9 in (23 cm)

The Atlantic Blue Crab is Maryland's state crustacean, and a commercial fishery is active in the Chesapeake Bay because of the restaurant market. Crab-eating festivals are sometimes held.

KEY FACTS

This crab can be all blue but is usually brown with blue on upper front legs.

+ **range:** Native to western Atlantic, Gulf of Mexico, and Pacific coast of Central America

+ **diet:** Its own species; also plants, small fish, clams, oysters, mussels, snails, worms, and insects

+ **name origin:** For its blue color

The Atlantic Blue Crab's scientific name is derived from the Greek *calli*—"beautiful"—and *nectes*—"swimmer," and the Latin *sapidus*—"savory." The crab is a popular meal, especially to East Coasters who use wooden hammers to break its shell and extract the meat. Males prefer fresher water, whereas females when spawning prefer more saline waters, which accounts for most males being caught in the upper bay while more females are caught in the lower. The Blue Crab stays near the mouth of the bay in its larval stage, and as it grows moves back into the Chesapeake Bay where it stays in marshes or sea grass beds to avoid predators.

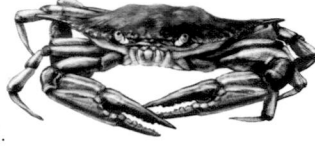

Ghost Crab

Ocypode quadrata up to 2 in (5 cm)

If you see a crab walking sideways, it is probably a Ghost Crab, known for its speedy locomotion. A striking feature is that one claw is larger than the other. It is also called a Sand Crab.

KEY FACTS

It operates at night and is mostly a pale color, usually yellow.

+ range: Tropical and subtropical shores worldwide

+ diet: Other crabs, clams, insects, and vegetation

+ name origin: Common name for its pale color; scientific name from Greek *ocy*, meaning "fast," and *podos*, meaning "foot"

The adult Ghost Crab digs a burrow with chambers at the end, in which it spends the hottest part of the day and the coldest part of the winter. It emerges at night to feed. When walking, it uses four pairs of legs, holding the larger clawed legs upward and occasionally alternating which side leads. At higher speeds, it raises one pair of legs, and at the highest speeds it uses only two sets of legs. Its survival success is due to a low number of land predators, its camouflage against sand, and its ability to endure starvation for long periods. It needs to wet its gills, using fine hairs on its legs to wick up water.

Mud Fiddler Crab

Uca pugnax 1 in (2.5 cm)

The Mud Fiddler Crab is a small species that lives in a burrow. It has one claw nearly the width of its body. If the claw is lost in a fight, the crab can grow another, but on the opposite side.

KEY FACTS

The crab's carapace is bluish green, but its large claw is different, usually yellow.

+ range: Most of the Atlantic coast

+ diet: Decaying detritus, and edible algae, microbes, and fungus it filters

+ name origin: For the claw, which is held across the body as if the crab were playing a fiddle

A Mud Fiddler Crab is easy to recognize because it is blue, almost black, and lives in a burrow in the mud. The crab will retreat to the burrow, which is up to 3 feet (0.9 m) long, if danger is present or the tide comes in. When the latter happens, the entrance or entrances are plugged with mud, and the crab opens them again when the tide recedes. This crab can live underwater because it has gills as well as a primitive lung for breathing air. The big claw seems to have no purpose other than attracting mates and fighting other males. Despite its pugnacious nature, it travels in groups when feeding.

Hairy Hermit Crab

Pagurus hirsutiusculus L to 2.8 in (7 cm)

The Hairy Hermit Crab is a small crustacean that carries another animal's shell on its back for protection. It has hair all over its body and varies in color from olive green to brown to black.

KEY FACTS

This species' walking legs have white and sometimes blue bands.

+ **range:** The Bering Strait south to California and Japan, from the intertidal zone to 360 feet (110 m)

+ **diet:** Detritus and seaweed

+ **name origin:** From the fact that it lives alone and for the small hairs that cover it

Like most crabs, the Hairy Hermit Crab has a hard exoskeleton, but a soft abdomen. To protect its abdomen, the cold-water crab finds a vacant shell of the right size and crawls inside, twisting its abdomen in the direction that will give it the most protection. Small pads keep the animal secure within the shell, although it may leave its protective quarters if a predator confronts it. The crab is covered with small hairs, except for the claws, one of which is bigger than the other and used as a weapon or shredder. It feeds itself with the smaller claw. The Hermit Crab picks a new shell when it outgrows the old one.

Brown Shrimp

Penaeus aztecus L to 9.3 in (23.6 cm)

The Brown Shrimp is commercially valuable in U.S. fisheries in the Gulf of Mexico. It burrows into the mud bottom to escape predators, and when the water is cold. It is an active swimmer.

KEY FACTS

The shrimp is brown when alive, and turns pink when cooked.

+ **range:** Atlantic coast and the Gulf of Mexico.

+ **diet:** Marine worms, algae, microscopic animals, and organic debris

+ **name origin:** For its color; "shrimp" from German *schrimpfen*, "to contract," and Norse *skorpna*, "to shrivel up"

The Brown Shrimp feeds at night and enters its burrow by day to escape predators. Females grow larger than males. They propel themselves with five pairs of swimming legs and their lobsterlike tail. They are generally found at depths from less than 180 feet (55 m) up to 360 feet (110 m), but they move to deeper water as they grow older. These shrimp prefer muddy bottoms that are rich in organic matter and decaying vegetation. They live less than two years. The larvae move into estuaries, where they eat plankton and then move into deeper water as they mature. They have antennae slightly longer than their bodies.

Northern Prawn

Pandalus borealis L to 6.5 in (16.5 cm)

The Northern Prawn is an edible shrimp found in colder waters and at greater depths than the Brown Shrimp. It is also known as Pink Shrimp, Deepwater Prawn, and Northern Shrimp.

KEY FACTS

This shrimp is generally pink, and larger and longer-lived than the Brown Shrimp.

+ **range:** Cold water with mud bottoms of Atlantic and Pacific coasts, and coasts worldwide

+ **diet:** Zooplankton

+ **name origin:** For northern cold waters in which it is found; "prawn" from Middle English *prane*

The Northern Prawn looks and acts like a shrimp, but it is larger and lives in cold waters instead of warm. It begins its life as a male; at the age of three years it changes sex and is a female for the next five years. It lives over soft, muddy bottoms at depths varying from 65 to nearly 4,300 feet (20–1,300 m). It feeds on the bottom by day and rises to the surface to feed at night. The female carries the eggs around on her abdomen until they hatch in April or May. Although it is heavily fished and popularly enjoyed as an appetizer, this species does not seem to be depleting, because of its huge population.

Mole Crab

Emerita talpoida 1.4 in (3.5 cm)

The Mole Crab, also called a Sand Crab or Sand Fiddler, is a small crab that stays in wave wash so it can easily burrow into sand as predators approach. It uses feathery antenna to filter-feed.

KEY FACTS

The gray or sand-colored, barrel-shaped Mole Crab has legs but no claws, and is among the smallest of the crabs.

+ **range:** Around the world, including beaches of Atlantic and Pacific coasts

+ **diet:** Plankton filtered from beach sand, brought in by waves

+ **name origin:** Its habit of burrowing into the sand

The Mole Crab may be small, but it is interesting. You may see it if you walk as far up as the waves go when they wash in and then recede. If it is exposed, the Mole Crab quickly burrows in the water-soaked sand, rapidly disappearing. It belongs to the genus *Emerita*, of which there are many members. Sometimes called a Sand Flea, it is not a flea at all, but a crab. As with many crabs, the female carries orange-colored eggs on her abdomen until they hatch, at which time the young are on their own. As an adult, it is used as bait for some kinds of fish, particularly pompano and redfish, but it is too small to be eaten by humans.

Common Barnacle

Semibalanus balanoides up to 0.8 in (2 cm)

The ubiquitous Common Barnacle is a small crustacean that adheres to an object and stays there for the remainder of its life. Ships' hulls are treated so that barnacles will not adhere to them.

KEY FACTS

A species of acorn barnacle, it is whitish in color. Larvae resemble tiny crabs.

+ **range:** Atlantic and Pacific coasts and seacoasts worldwide

+ **diet:** Plankton filtered from seawater

+ **name origin:** Alternate name "acorn barnacle" for its conical shape; *balanos* in Greek, meaning "acorn"

The Common Barnacle has both male and female reproductive organs, and the eggs can be laid and fertilized by one barnacle. If barnacles are clustered together in colonies, they receive sperm from a neighbor. Eggs are released into the water to become plankton, but most are eaten by other animals. Survivors attach themselves with a kind of cement to ships' hulls, piers, and other animals—literally anything that seawater touches. They impede a ship and must be scraped off. The Common Barnacle attaches itself with its head and sweeps plankton from the water with its feet. It may live up to seven years.

Gooseneck Barnacle

Pollicipes polymerus 3 in (8 cm)

The Gooseneck Barnacle attaches itself to an object with a stalk that resembles a goose's neck. It can be found attached to ships' hulls, piers, wharves, and shelled animals worldwide.

KEY FACTS

The Gooseneck is whitish with black cirri, or appendages that sweep food into its mouth.

+ **range:** Most common on West Coast

+ **diet:** Plankton filtered from the water

+ **name origin:** For the gooselike appearance of its stalk, which it fastens to objects that spend part of the day underwater

The ancients thought the Barnacle Goose came from Gooseneck Barnacles because the stalk by which they attach themselves to an object resembles a goose's neck. Monks successfully argued that since the goose was a crustacean, they could eat it during Lent when they were supposed to forsake things born of flesh. The Gooseneck Barnacle filters plankton by waving its extended cirri to catch food. It may live 20 years, longer than an acorn barnacle. Although not threatened, it is susceptible to pollution and recovers slowly from disturbances. Humans in Spain and Portugal eat the Gooseneck Barnacle.

Green Sea Urchin

Strongylocentrotus droebachiensis width 1.2–3.9 in (3–10 cm)

The Green Sea Urchin is a spine-covered creature that is round on top and flat on the bottom and moves with water-filled feet. It camouflages itself by attaching algae or detritus to its spine.

KEY FACTS

It is colored from greenish to black and shades of brown, purple, and red.

+ range: All oceans, from shallow water to 16,000 feet (4,877 m)

+ diet: Algae; also sea cucumbers and invertebrates such as mussels and sponges

+ name origin: "Urchin," an archaic word for "hedgehog," referring to the spiny hedgehog it resembles

The Green Sea Urchin looks like a hedgehog in the water, and its sharp spines, which are mostly to deter predators, may poke into flesh if they are touched. They are not poisonous. Touch a spine, and other spines may group around it. It has no visible eyes or legs but can move freely over hard surfaces, using small feet that have adhesives. It eats mostly algae, which it scrapes from its host by its mouth, which is equipped with teeth that can cut through stone. The Green Sea Urchin has two sexes, and the eggs on the female are fertilized by free-flowing sperm, which the male emits into the water.

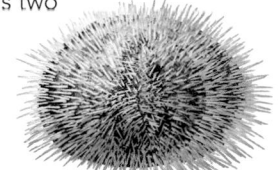

Common Sand Dollar

Echinarachnius parma diameter 3 in (7.6 cm)

The Common Sand Dollar is a circumpolar cold-water bottom-dweller covered with fine hairs that help it move. It is also called Sand Cake, Cake Urchin, Pansy Flower, and Sand Cookie.

KEY FACTS

The Sand Dollar is bleached white when it dies and washes onto the shore.

+ range: Isolated areas on sandy bottoms from New Jersey north, occurring in Alaska, British Columbia, Siberia, and Japan

+ diet: Detritus and other organic matter

+ name origin: For its disk shape, like a silver dollar

People think the Sand Dollar is white and calciferous when they find it on the beach, but actually it is covered with fine hair and is purplish brown when it is alive on the sea bottom. The Sand Dollar is round, in the same family as sea stars, but it lacks the arms. When feeding, it burrows at an angle in the substrate, moving food detritus and other food particles toward its mouth with the fine hairs that cover it. It grinds the food particles with large, triangular teeth that ring its mouth. There are two sexes, although they are indistinguishable to humans by appearance. A head has never been developed.

Inflated Sea Biscuit

Clypeaster rosaceus diameter 5 in (13 cm)

Of the same family as the Sand Dollar, the Inflated Sea Biscuit, also called the Caribbean Sand Dollar, is larger and hairier than its cousin, though it has some of the same characteristics.

KEY FACTS

The Sea Biscuit ranges from yellow to reddish to brown.

+ range: Generally a warm-water creature, from South Carolina to Venezuela on the Atlantic coast

+ diet: Algae, small copepods, crustacean larvae, detritus, and other organic matter

+ name origin: For its thick, biscuit-shaped appearance

The Inflated Sea Biscuit is related to the Common Sand Dollar. Its body is oval, but it rises to a blunt peak at the center and is thicker than the Sand Dollar. Like the Common Sand Dollar, it has five markings radiating from its center, with a mouth on the bottom and an anus on the top of its body from which it eliminates waste. It is found most often in shallow water off the Florida Keys. The Sea Biscuit has short spines all over its body that are covered with smaller hairs, which help it to move along the ocean bottom to find food. The creature is often camouflaged among shells on turtle grass beds.

Purple Sea Star

Pisaster ochraceus 4–10 in (10–25 cm)

The Purple Sea Star is a bottom-dweller with five stout rays branching out from a central disk. It is also called the Ochre Sea Star and Ochre Starfish. There is no parental care for the young.

KEY FACTS

Most are purple, but can be orange, orange-ochre, yellow, reddish, or brown.

+ range: Mussel beds and tidal pools on the Northwest coast, Alaska to California

+ diet: Mostly algae; also devouring mussels and other shelled animals

+ name origin: For its star shape, and because most are purple

A Purple Sea Star is easy to identify because of its star shape with five rays extending from a central disk. The cold-water creature can easily be seen in the Northwest, hugging rocks in tidal pools, but it also lives in depths down to 295 feet (90 m). Its rays have suckers, allowing the sea star to cling to objects that may be subjected to wave action. Small hairs on the animal's underside allow it to move slowly. If prey is too big for its mouth, it is able to extrude its stomach and digest the engulfed animal with enzymes. Male sea stars eject sperm, and females locate it and fertilize their eggs.

Daisy Brittle Star

Ophiopholis aculeata disk 0.8 in (2 cm) wide; arms 3.5 in (9 cm) long

The Daisy Brittle Star is related to sea stars, but it has thinner, more whiplike arms that are regenerative. Its other names include Painted Brittle Star and Ubiquitous Brittle Star.

KEY FACTS

It can be orange with red stripes or blotches, interspersed with brown and green. Some individuals are black, white, gray, or brown.

+ range: Circumpolar; in deep water and tide pools along our northwestern and northeastern coasts

+ diet: Detritus and other organic matter

+ name origin: For its brittle body

When handled, the cold-water Daisy Brittle Star may break off one of its arms, which can be regenerated. The loss is only a defensive measure. If a piece of the central disk is broken, it can result in another Brittle Star. The arms differentiate it from other sea stars because they are thinner and more limber. They help with locomotion, reaching out and pulling the animal along. That motion gives it one of its names, the Serpent Star. The arms also pull food toward its mouth, which is centrally located on its underside. The Daisy Brittle Star tends to occupy the ocean depths, down to 6,500 feet (1,981 m).

Orange-footed Sea Cucumber

Cucumaria frondosa L 8 in (20 cm)

The Orange-footed Sea Cucumber, largest of the sea cucumbers, is a bottom-dweller resembling a caterpillar or soft cucumber. It has ten branched tentacles, which bring food to its mouth.

KEY FACTS

Skin ranges in color from yellowish white to dark brown or black.

+ range: Most abundant in North Atlantic and the Barents Sea

+ diet: Phytoplankton in spring and summer when water is warm and plankton bloom is at its peak

+ name origin: For the color of its feet, and its resemblance to a cucumber

If you see something that looks like a small caterpillar inching with orange feet along on the bottom of the sea, it is likely an Orange-footed Sea Cucumber. When this sea cucumber is an adult it is prey to many fish, but when threatened it shoots out sticky threads to ensnare the animal. To feed the predator it may also jettison some internal organs from its anus, which it can regenerate. It is found in waters down to 100 feet (30 m). When feeding on the bottom, it pulls food items toward its mouth with oral tentacles. When reproducing, both sexes release eggs and sperm into the water, and the two combine.

Orb weaver spiders use their ornate webs to capture flying insects.

5 Insects & Spiders
The Little Animals That Rule the Earth

Insects, spiders, and their kin are absolutely essential for a healthy environment and for our own well-being. They pollinate crops, control pests and weeds, provide food for wildlife, recycle nutrients, and produce valuable products such as dyes, silk, honey, and wax. These benefits far outweigh the harm some species cause by spreading disease, destroying crops, infesting food stores, and damaging structures.

■ What Is an Arthropod?

Insects, spiders, and their kin belong to the animal phylum Arthropoda. Arthropods lack backbones and are characterized by having a tough outer skeleton, or exoskeleton. The exoskeleton comprises a series of ringlike segments that give arthropods their flexibility. Their mouthparts, antennae, and legs are also segmented. The phylum Arthropoda is divided into four living subphyla, whose members share similar features.

■ Insects

Ants, beetles, flies, silverfish, and other six-legged arthropods are classified in the subphylum Hexapoda. All insects have three major body regions: a head, thorax, and abdomen. The heads of most adult insects have mouthparts adapted for chewing, piercing-sucking, or lapping, one pair of antennae, and large compound eyes with multiple lenses. The thorax has six legs and one or two pairs of wings. The forewings are sometimes uniformly thick and leathery or membranous, or a combination of the two.

■ Arachnids

Spiders, scorpions, mites, ticks, and other eight-legged arachnids are placed in the subphylum Cheliceriformes. They have two major body regions, no antennae, and pincherlike jaws or fangs accompanied by a pair of leglike structures called pedipalps. In North America, spiders are distinguished from other arachnids by their unsegmented abdomens. Spiders and mites produce silk to capture food, build shelters, use in courtship, and protect their eggs.

■ Centipedes and Millipedes

Centipedes, millipedes, and their relatives all have heads with chewing mouthparts and a pair of antennae, followed by a long, segmented body trunk with up to 750 pairs of legs. Centipedes are fleet-footed predators with one pair of legs per body segment. The first pair of legs, or gnathopods, is used not for walking, but for defense, and capturing and killing prey. Millipedes are typically lumbering scavengers of plant tissues. Many species defend themselves by coiling their body around their head and exuding noxious fluids from pores along the sides of their body.

■ Crustaceans

Crayfish, beach hoppers, pill bugs, and sow bugs form a class of freshwater and terrestrial species within the subphylum Crustacea. Most crustaceans, such as lobsters, crabs, shrimp, and barnacles, are marine. All crustaceans have segmented heads with chewing mouthparts and two pairs of antennae, and two body regions with 5 to 22 pairs of appendages.

■ Development and Growth

Insects, spiders, and their kin typically begin their lives as fertilized eggs, although a few species reproduce asexually by parthenogenesis. In ants, bees, and wasps, unfertilized eggs always develop into males. Aphids alternate between winged males and females that reproduce sexually, and wingless females that produce genetic clones of themselves asexually.

In many arthropods, including primitive wingless insects such as silverfish, the young hatch from the egg looking very much like a small version of the adult, but are incapable of reproduction. In insects, this type of metamorphosis or development is called ametaboly. Unlike most other adult insects that stop molting when they reach adulthood, silverfish continue to molt for the rest of their life.

Winged insects, such as grasshoppers and dragonflies, develop by hemimetaboly, hatching from eggs as nymphs or naiads, respectively. They, too, resemble the adults, but lack wings. As they reach maturity, developing wings become evident on the outside of the body if they are present in the adult.

GLOSSARY

+ **alate:** possessing wings.

+ **antennal club:** gradually or abruptly expanded tip of the antenna.

+ **antennomere:** an article or "segment" of the antenna.

+ **elytra (singular elytron):** soft, leathery, or hard shell-like forewings of beetles.

+ **exoskeleton:** external skeleton of insects and other arthropods.

+ **femur (plural femurs):** the third of six leg segments in insects and the seven segments in spiders and kin.

+ **forewings:** front pair of wings, sometimes modified into tegmina or elytra.

+ **genal comb:** row of comblike spines on the head of fleas.

+ **hemimetabolous:** insects that hatch from eggs as nymphs or naiads that resemble adults, but lack wings; wing development evident on outside of body, if present in adult.

+ **hind wings:** the back pair of wings, modified into knoblike halteres in flies and mosquitoes.

+ **holometabolous:** insects with four distinct stages of development—egg, larva, pupa, and adult; wing development occurs internally during the larval stage, if present in adult.

+ **mandibles:** first pair of jaws in insects variously modified to bite, cut, chew, grind, or pierce plant and animal tissues.

+ **molt:** shedding of the arthropod exoskeleton.

+ **naiad:** aquatic nymph of dragonflies, damselflies, mayflies, and stoneflies.

+ **nymph:** the immature stage between egg and adult of hemimetabolous insects.

+ **öotheca (plural öothecae):** a protective coating surrounding a batch of eggs.

+ **osmeterium (plural osmeteria):** fleshy, eversible gland in some swallowtail caterpillars that produces a strong, repellent odor.

+ **ovipositor:** the egg-laying apparatus of some female insects that resembles a tube, needle, or blade; modified as a stinger in ants, bees, and wasps.

+ **pronotum:** the exoskeletal plate across the top of the thoracic segment behind the head, or prothorax.

+ **prothorax:** the first of three thoracic segments; the midsection of beetles, cockroaches, grasshoppers, and kin.

+ **pupa:** the third in four stages of development in holometabolous insects.

+ **scutellum:** small, often triangular thoracic plate, present at base of elytra.

+ **setae:** soft, bristly hairlike structures on the arthropod exoskeleton; sometimes flat, scalelike, forming patterns on moth and butterfly wings.

+ **tarsus (plural tarsi):** outermost segment of leg, or foot; attached to tibiae in insects, metatarsus in spiders and kin; consists of one or more articles, or tarsomeres.

+ **tegmina:** leathery forewings of grasshoppers and kin, cockroaches, mantises, and earwigs.

+ **tibia: (plural tibiae):** the fourth of six leg segments in hexapods and the fifth of seven segments in spiders and kin.

Glowing lightning bugs blanket a prairie of wild alfalfa.

Beetles, bees, and butterflies develop by holometaboly, undergoing four distinct stages of development—egg, larva, pupa, and adult. Although wing development occurs internally during the larval stage, the first external evidence of wings and other adult characteristics first appear in the pupa.

To grow, arthropods shed their exoskeleton in a series of molts that are hormonally controlled. The period between each molt is called an instar, the number of which varies among different groups. When the old, outgrown exoskeleton is shed, a new soft and pale one is revealed underneath. Over the next few hours or days, this new exoskeleton will undergo a chemical hardening and darkening process akin to tanning. Recently molted insects that are soft and pale are said to be teneral. The entire life cycle, from egg to adult, varies depending on species and conditions. Some species need only several days, while others may require years to complete their development.

|||||||||||||||||||||||||||||||||||||||||||||||||||||||||||||||||||

Among the tangled weeds of the roadside or in the grassroot jungles of your own backyard, you encounter strange and incredible forms of life.

—EDWIN WAY TEALE, *GRASSROOT JUNGLES*

|||||||||||||||||||||||||||||||||||||||||||||||||||||||||||||||||||

■ Identifying Insects and Spiders

The 160 species of arthropods presented in this chapter are but a fraction of the nearly 100,000 species known to occur in the United States and Canada. Most of the species were selected because they are conspicuous in appearance or behavior, and occupy a significant portion of North America north of Mexico. The arthropods featured include both native species and adventive (exotic) species that have become established in the region.

The species accounts are organized first by sub-phyla (Hexapoda, Cheliceriformes, Myriapoda, Crustacea), then by order and family. Each species account includes both scientific and common names, if available. Because of the sheer diversity of species, very few arthropods have widely accepted common names; however, they all have a universally accepted scientific name that serves as an access point to all of the information available for each species. The length (L) or wingspan (W) and a brief introduction that highlights essential or unusual features or behaviors are also presented. Length relates to the body and not the legs. Next is a brief list of basic diagnostic features that will help you to distinguish the species from its nearest relatives or other similar species. This information is followed by habitat preferences, distribution in the United States and Canada, and preferred foods of adults and immatures. The account ends with a brief essay that provides more information on behavior, life cycle, and similar species.

■ Where to Look

Wherever you live, you won't have far to go to see insects, spiders, and their kin. The following are some of the best places to look, but be sure you do not disturb these and other sensitive microhabitats so that they continue to attract and support species.
Flowers and Vegetation Flowers are incredibly attractive to nectar- and pollen-feeding insects, as well as their predators and parasites. Carefully examine fruits, seedpods, cones, stems, branches, and trunks on various plants, day and night.
Decaying Snags, Logs, and Stumps Decaying wood attracts the greatest diversity of arthropods. As wood decomposes, it will attract a succession of different species. Carefully pull back loose bark and check its underside and the exposed wood.
Shorelines of Streams, Rivers, and Lakes Many arthropods take refuge under piles of debris along the shore, or are found resting on nearby vegetation. Floating debris on flowing and standing waters may harbor animals swept from the shore by wind.
Freshwater Wetlands Search for aquatic and semiaquatic species in, on, and near bogs, swamps, ponds, and quiet pools along streams and rivers. Some glide over the surface; others swim through the water column or cling to various submerged objects and substrates.
Carrion, Dung, and Decaying Plants Look under dead animals and feces in various stages of decay for species that are unlikely to be found anywhere else. Rake through leaves and needles at the bases of trees and shrubs. Compost heaps and other piles of decomposing plants will attract all kinds of arthropods.
Lights Nocturnal insects are attracted to porch lights, storefronts, and well-lit gas stations, especially in less developed areas. The bluish glow of mercury vapor streetlights is the most attractive to insects. Some species will fly or crawl directly to the light, and others remain in the shadows. Predatory arachnids and centipedes are commonly attracted to insect activity around lights.

ACTIVITIES

Here are some things you can do to enhance your knowledge and enjoyment of insects, spiders, and their kin:

+ **Make** an insect collection.

+ **Conduct** a survey of the species in and around your home.

+ **Participate** in a BioBlitz, butterfly count, or other citizen science project.

+ **Start** an insect garden at a school or park.

+ **Observe** insects and spiders through close-focusing binoculars.

+ **Record** your insect and spider observations in a journal.

+ **Take** pictures of arthropods with your camera or cell phone.

+ **Plan** a day or evening trip to a natural area to see arthropods.

+ **Visit** an insect zoo, a butterfly house, or a museum or university with an entomology department.

+ **Contact** your local agricultural extension office to learn more about local species.

INSECTS & SPIDERS

Silverfish

Lepisma saccharina L 0.4–0.5 in (10–12 mm)

Silverfish are fleet-footed pests that thrive in warm, humid conditions in human dwellings, especially in attics and wall spaces, and under flooring. They are often trapped in sinks and tubs.

KEY FACTS

Silverfish are clothed in silvery gray scales; abdomen tipped with 3 long, threadlike appendages.

+ **habitat:** Homes, offices, libraries, warehouses

+ **range:** Throughout the U.S. and Canada

+ **food:** Any human food, paste, glue, paper, and fabrics, especially cotton, linen, and silk

Among the most primitive insects, Silverfish may live three or more years. Unlike most other insects, they continue to molt, or shed their exoskeletons, long after they have reached the adult stage. They are pests in libraries, damaging books, bindings, and paper documents. There are three domestic relatives of silverfish. The Gray Silverfish (*Ctenolepisma longicaudata*) inhabits the warmer Southeast and southern California, and is uniformly light or dark gray. The stouter, less shiny, tannish gray Four-lined Silverfish (*C. lineata*) has dark lines down its back. The Mottled Firebrat (*Thermobia domestica*) is distinctly mottled.

Silverfish damage

Giant Mayfly

Hexagenia limbata L 0.7–1.2 in (18–30 mm), without filaments

One of the largest mayflies, the Giant Mayfly is a familiar sight on summer nights at lights along mud-bottomed rivers in the East. Mayflies live just long enough to mate and reproduce.

KEY FACTS

The leading edge of the adult's forewings is dark purple; outer margins are edged with smoky gray; the body is pale to dark purplish brown; the tails are long.

+ **habitat:** Vicinity of lakes, streams, and rivers

+ **range:** Eastern U.S. and Canada

+ **food:** Adults do not feed; naiads eat plant materials.

In summer, Giant Mayflies engage in large, dense, nocturnal emergences that appear on weather radar. These and other mayflies are unique insects because they have two winged stages. Anglers know the first dull yellow or yellow-green winged stage, the subimago or subadult, as a dun. After leaving the water, duns rest on riparian vegetation and take up to three days to develop into a sexually mature adult, or imago. The aquatic naiad typically digs a U-shaped burrow in mud along stream and riverbanks. Four of the five *Hexagenia* species live in the East; the fifth is found in the Southwest.

Mass adult emergence

Common Green Darner

Anax junius L 2.8–3.3 in (70–84 mm)

With their plain green thorax, and blue and green abdomen, male Common Green Darners are common and familiar dragonflies patrolling ponds, lakes, and brackish waters.

KEY FACTS

The head has a ringed black spot, thorax is green, and abdomen is mostly blue and green (male), or brick red (female and immature male).

+ **habitat:** Vicinity of permanent, temporary, and brackish standing waters

+ **range:** U.S. and Canada

+ **food:** Adults and naiads eat insects.

In the Southwest, the male Common Green Darner and Giant Darner (*A. walsinghami*) are similar, but the Giant Darner appears much larger in the air and flies with its abdomen arched. The female is similar to both sexes of the southeastern Comet Darner (*A. longipes*), but her abdomen is duller than the male Comet's and less spotted than the female Comet's. Males search shores for females in spring and summer. Females lay eggs on floating or submerged vegetation, often with a male in tandem. Feeding swarms with other species may form late in the day. Large numbers often migrate along coasts and mountain ridges in fall.

Naiad

Twelve-spotted Skimmer

Libellula pulchella L 1.7–2.3 in (43–58 mm)

The attention-grabbing black-and-white wings of male Twelve-spotted Skimmers are mesmerizing to watch as they patrol and hover over the water. They perch on plants with wings held flat.

KEY FACTS

The wings have white patches in the mature male; abdomen is chalky gray (male), or brown with yellow side stripes (female and immature male).

+ **habitat:** Bogs, marshes, ponds, lakes, and other permanent standing water

+ **range:** U.S. and Canada

+ **food:** Insects plucked from tips of vegetation

Male and female Twelve-spotted Skimmers have three black or brown patches on each wing. The Western Eight-spotted Skimmer (*L. forensis*) is similar but lacks dark wing tips. Females resemble those of the Common Whitetail (*L. lydia*) but lack white zigzag stripes on the sides of the abdomen. Territorial males patrol over the water and engage in aerial acrobatics with rival males by flying vertical loops around them. They prefer to perch in open, sunny spots alongshore, especially on the top of tall, bare perches close to the shore or surrounded by water. This species forms migratory swarms along coasts.

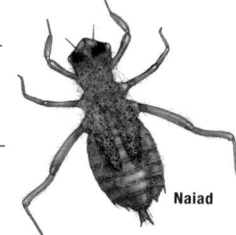

Naiad

Ebony Jewelwing
Calopteryx maculata L 1.5–2.2 in (39–57 mm)

The flight of an Ebony Jewelwing moving slowly along shaded streams in spring and summer is butterfly-like. Males hover over females as they lay eggs on floating or submerged vegetation.

KEY FACTS

This large metallic green damselfly has densely veined wings that are uniformly dark (male) or with a swollen white spot on the tips (female).

+ **habitat:** Adults perch on branches and reeds along freshwater streams and rivers in wooded areas.

+ **range:** Eastern U.S. and Canada

+ **food:** Small insects

Male and female Ebony Jewelwings fly in small groups along shady, sandy-bottomed streams. Individuals perched on sun-dappled shrubs often display by slowly opening and closing their wings. Receptive females fly along streams occupied by territorial males. After a brief courtship, the male transfers sperm to a sexual organ at the base of his abdomen. Inserting it into the female's genital opening, he scrapes out a previous suitor's sperm and supplants it with his own. To make sure that his sperm does not suffer the same fate, he defends his paternal investment by remaining with the female as she lays her eggs.

Naiad

Familiar Bluet
Enallagma civile L 1–1.6 in (25–40 mm)

Well-known and widespread, the Familiar Bluet is a common sight along the vegetated edges of swamps, ponds, and other standing water habitats. Mature females have blue and green forms.

KEY FACTS

Male is blue, with a black-striped thorax and an abdomen with 6 to 7 mostly black segments. Female is blue, greenish, or brownish, with an abdomen that is black on top.

+ **habitat:** Standing water, including ponds, reservoirs, irrigation ditches, and rice fields

+ **range:** Widespread in U.S. and Canada

+ **food:** Insects

In the East, male and female familiar bluets are nearly identical in color and pattern to the coastal Atlantic Bluets (*E. doubledayi*). Only males can be separated, best done by microscopic examination of the genitalia. This species resembles several western bluets, and females are hard to separate from those of the Arroyo (*E. praevarum*) and Tule (*E. carunculatum*) Bluets. Familiar Bluets begin flying in spring and continue through late fall. They often occur away from water, flying over lawns in gardens and parks. Tandem pairs seek emergent stems and rushes where the female will lay her eggs above the water.

Mating pair

German Cockroach

Blatella germanica L 0.5–0.6 in (13–16 mm)

Common household pests, German Cockroaches look for food and water at night, and hide during the day in walls and other tight places, especially around refrigerators, stoves, and sinks.

KEY FACTS

This species is tan or pale brown, male paler than female, with dark stripes on pronotum; abdomen tapered at tip (male) or broad (female).

+ **habitat:** Homes, restaurants, food preparation areas in hospitals, hotels, other facilities

+ **range:** U.S. and Canada

+ **food:** Plant and animal tissues

German Cockroaches seek narrow, secluded places where there is optimal heat and moisture, especially in kitchens. They will infest bathrooms if a common wall is shared with the kitchen. Seeing these normally nocturnal insects during the day may indicate a severe infestation. A female produces four or five purse-shaped egg cases, or oothecae, each containing up to 40 or more eggs. The female carries the öotheca on the tip of her abdomen for about a month, and then drops it anywhere a few days before the eggs hatch. The nymphs are mostly uniformly dark gray to black with a pale brown stripe down the thorax.

Nymph

American Cockroach

Periplaneta americana L 1.1–1.6 in (27–40 mm)

American Cockroaches are pests in basements and food preparation areas of large buildings and commercial facilities rather than in homes and apartments, and are abundant in sewer systems.

KEY FACTS

This cockroach is shiny, reddish brown; the pronotum is paler, with a central pair of reddish brown blotches; wings are brown, extending beyond the abdomen in the adult male.

+ **habitat:** Restaurants, grocery stores, hotels, hospitals, and sewers

+ **range:** Throughout U.S. and Canada

+ **food:** Plant and animal tissues

The American Cockroach, also known as the Palmetto Bug, is the most common of four *Periplaneta* species in the U.S. and Canada. Adults reluctantly take to the air and are weak fliers. A female produces up to 90 öothecae in her lifetime, each containing 14 to 16 eggs. The average life cycle from egg to adult is about 600 days, with adults living for about another 400 days. Nymphs are usually mostly reddish brown, but darkened posterior margins on the thoracic and abdominal segments give them a striped appearance. The Southern Brown Cockroach (*P. brunnea*) from Florida and Texas is similar, but the midsection is less distinctly marked.

Öotheca

Oriental Cockroach

Blatta orientalis L 0.7–1.2 in (18–30 mm)

The Oriental Cockroach, or "water bug," frequently enters homes and apartments through sewer pipes and occupies secluded places. Dense infestations have a pungent "roachy" odor.

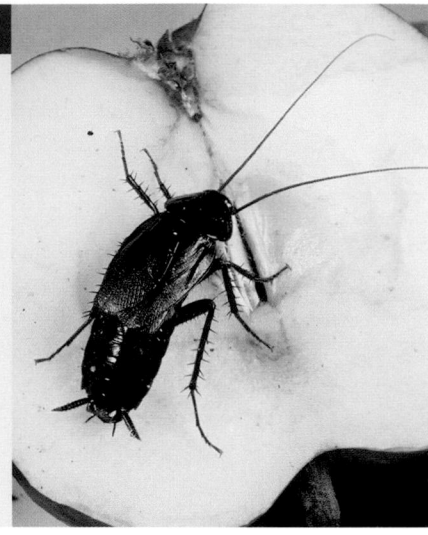

KEY FACTS

A shiny black cockroach; adult male has fully developed wings covering three-fourths of abdomen; female has short wing pads.

+ **habitat:** Crawl spaces, basements, cellars, buried water meter vaults in alleys

+ **range:** Throughout the U.S. and Canada

+ **food:** Plant and animal tissues

Originally from North Africa, the Oriental Cockroach is a significant year-round house pest. Although tolerant of cooler, drier conditions, nymphs and adults prefer warm, moist habitats. They are often found in basements and crawl spaces behind radiators and floors. Newly hatched nymphs are reddish and become darker as they age. Of all house pest cockroach species, the Oriental Cockroach appears to be the best adapted for living outdoors, especially in warm climates. There, they are commonly encountered in hollow stumps and tree holes, hiding in leaf litter or under stones, or on sidewalks on warm, humid evenings.

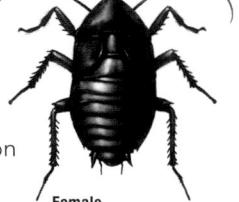

Female

Pacific Dampwood Termite

Zootermopsis angusticollis L winged reproductive 0.9–1 in (23–26 mm), including wings; soldier 0.6–0.8 in (15–20 mm)

Like all termites, Pacific Dampwood Termites are social, living in colonies of workers, soldiers, and reproductives. The worker caste consists of male and female adults and nymphs.

KEY FACTS

Workers are pale yellowish white; soldiers are yellowish to reddish brown, with well-developed mandibles; alates are darker with long wings.

+ **habitat:** Under bark of decaying conifer logs in coastal and montane forests

+ **range:** Coastal British Columbia to Baja California

+ **food:** Decaying wood

The Pacific Dampwood Termite is our largest termite, living in stumps, logs, and wooden structures in moist coastal and montane habitats. Soldiers use outsized mandibles to defend the colony from ants and other predatory arthropods. Winged reproductives (kings, queens), called alates, live in colonies year-round. They usually take flight after rains in summer and early fall, and are attracted to light. The wings detach soon after mating. The Nevada Dampwood Termite (*Z. nevadensis*) is smaller and darker, occurring from British Columbia to northern California east to the mountains of Montana and Nevada.

Winged reproductive, or alate

Eastern Subterranean Termite

Reticulitermes flavipes L winged reproductive 0.4 in (10 mm), including wings; soldier 0.2–0.3 in (5–7 mm)

Mature colonies of Eastern Subterranean Termites contain 20,000 to 5 million individuals. Although very destructive, they are essential to the breakdown of dead plant matter.

KEY FACTS

Soldiers and workers are pale; soldiers have long mandibles; reproductives are darker with grayish brown wings.

+ **habitat:** Disturbed and natural areas

+ **range:** Southern Ontario to Florida, west to southern Minnesota, Colorado, and Texas

+ **food:** Stumps, logs, lumber, wooden structures and products

The most common and important structural pest in the eastern United States and Canada, Eastern Subterranean Termites infest any type of wood. They build tubes of soil and bits of wood up tree trunks and across foundations to maintain a connection between their food supply and the soil. Winged reproductives (kings, queens), or alates, swarm in spring and summer, mate, and shed their wings, before establishing a new colony. Dark *(R. virginicus)* and Light *(R. hageni)* Southeastern Subterranean Termites are restricted to the Southeast. The Western Subterranean Termite *(R. hesperus)* occurs throughout the West.

Winged reproductive, or alate

Chinese Mantid

Tenodera aridifolia L 3.3–4..1 in (83–104 mm)

The oval egg cases of the Chinese Mantid, an opportunistic predator, are sold widely to gardeners and farmers for use as biological control agents to help keep insect pests in check.

KEY FACTS

An elongate, robust mantid, it is mostly green or brown, or green with brown wing margins; wings extend beyond abdomen in male; female heavy-bodied.

+ **habitat:** Gardens, parks, old fields, roadsides

+ **range:** Southern Ontario and Quebec, throughout eastern U.S., and in California

+ **food:** Insects

Introduced into the United States from China in 1896, the Chinese Mantid is a voracious predator that will eat any insect it can catch. The mantid will occasionally stake out bird feeders to capture and eat hummingbirds. After mating, adult females are especially common near flowers where they can capture and eat large numbers of insects to assure proper egg development. In late summer and fall, Chinese mantids lay up to 200 eggs within a single foamy öothecae that soon hardens into a protective case, which they attach to a branch or flat surface. The tiny nymphs will hatch the following spring.

Öotheca

European Earwig

Forficula auricularia L 0.4–0.7 in (9–17 mm)

Although they can be pests in gardens and greenhouses, and nuisances in homes, European Earwigs are sometimes considered beneficial because they prey on garden pests.

KEY FACTS

Earwigs are flattened, brown to blackish brown with paler appendages and tegmina; antennae with 14 to 15 antennomeres; males' forceps curved, female's crossed at tips.

+ **habitat:** Gardens, parks, fields, woods

+ **range:** Southern Canada, throughout U.S.

+ **food:** Omnivorous, eating plants and small insects

Typically nocturnal, European Earwigs spend the day in cool, dark, narrow spaces in flowers, buried in leaf litter, or under loose bark. Females dig a chamber in spring to lay eggs, brood them, and will stay with the nymphs for several days after they hatch. When earwig populations are dense and insect prey is scarce, earwigs will shift their diet to plant material. They can be destructive to flowers and are sometimes serious pests in greenhouses. Although fully winged, they seldom fly. Adults overwinter underground in well-drained soils or in aggregations under bark, hollow stems, and other frost-free places.

Female

Salmonflies

Pteronarcys species L 2.2–3.3 in (55–84 mm)

The largest of our stoneflies, Salmonflies emerge in great numbers along big rivers charged by snowmelt in spring or summer. They help anglers to identify the best spots for fly-fishing.

KEY FACTS

Adults are large, flat, sprawling; brown or gray, sometimes with bright markings; head has three simple eyes; first segment of foot is short.

+ **habitat:** Clean, cold streams and rivers

+ **range:** Throughout U.S. and Canada

+ **food:** Adults do not feed; naiads eat algae, vegetation, small invertebrates.

Nocturnal and short-lived, adult Salmonflies are attracted to light and spend the days resting on rocks and vegetation along streams and rivers in spring. Females lay eggs on the water, which hatch the following year. The naiads crawl on rocky stream bottoms in accumulations of leaves and other detritus. They avoid fish and other predators by releasing noxious fluids from their leg joints or by feigning death. Mature naiads leave the water and crawl onto boulders, logs, and tree trunks to molt into adults. The life cycle takes about three years. There are ten species of *Pteronarcys*, most of them in the East.

Naiad

Carolina Grasshopper
Dissosteira carolina L 1.5-1.7 in (37-42 mm)

Dark-winged Carolina Grasshoppers resemble butterflies as they fly over dirt roads in a fluttering zigzag pattern with distinct clicking sounds. They are seldom garden or crop pests.

KEY FACTS

This large grasshopper is yellowish gray to reddish brown, sometimes with a small dark spot; pronotum sharp with notched ridge; hind wings black with yellowish margins.

+ **habitat:** Unpaved roads, undisturbed fields, clearings

+ **range:** Southern Canada, throughout U.S.

+ **food:** Grasses and forbs

Easily recognized by its large size and the flash of its pale-margined flight wings, the Carolina Grasshopper is among the most common grasshoppers near human habitats. The species is sometimes found on streets of towns and fairly large cities. Adults are active from late spring through late fall in the southern part of the range, and in summer and early fall to the north. Although recorded as causing minor damage to tobacco, cereal, alfalfa, and turf, it does not generally have significant economic impact, even when it is numerous. Forty or more eggs overwinter in a pod, and the nymphs hatch the following spring.

With spread wings

Pictured Grasshopper
Dactylotum bicolor L 0.8-1.4 in (20-35 mm)

Their riot of bright and contrasting colors gives the flightless and lumbering Pictured Grasshopper several other common names, including Rainbow, Painted, and Barber-pole Grasshopper.

KEY FACTS

Adults are short-winged and marked blue-black, red, orange, yellow, white, and pinkish; males are smaller than females.

+ **habitat:** Prairies, desert grasslands, eroded hillsides; occasionally alfalfa fields

+ **range:** Saskatchewan to Arizona and Texas

+ **food:** Grasses and low-growing forbs

Somewhat clumsy and slow-moving, Pictured Grasshoppers are usually encountered during the day walking over bare, gravelly ground or feeding in clumps of vegetation. Despite their conspicuous coloration, they are sometimes hard to see against natural backgrounds because the "fractured" color pattern breaks up the outline of their body. Egg pods contain up to 25 eggs that hatch in late summer. Adults are active in late July through September and tend to be localized in distribution. Darker markings of northern prairie populations are blackish. Those in southern deserts are bluish and more brilliantly marked.

Mating pair

Differential Grasshopper
Melanoplus differentialis L 1.1–2 in (28–50 mm)

The Differential Grasshopper is a large, distinctive species of *Melanoplus,* which includes many agricultural pests and the recently extinct Rocky Mountain Locust *(M. spretus).*

KEY FACTS

This species is greenish or yellow with black markings, or is black; pronotum has black grooves; forewings unmarked; hind legs have black chevrons, yellow tibiae.

+ habitat: Developed and natural areas with lush vegetation

+ range: Central U.S., southeastern coastal plain

+ food: Many kinds of plants

The Differential Grasshopper is common in moist cultivated crop areas and is a significant crop pest in the Midwest. It eats grasses, but prefers broad-leaved plants. Males are generally smaller than females. Each fragile, curved egg pod deposited in the soil contains up to eight eggs. Nymphs hatch in spring and reach adulthood in about a month. One generation is produced annually. Nymphs and adults are often seen feeding during the day in groups, and such swarms can be highly destructive in gardens, agricultural crops, vineyards, and orchards. At night, individuals usually rest high up on various plants.

Female laying eggs

Eastern Lubber Grasshopper
Romalea microptera L 1.7–2.8 in (43–70 mm)

Well-known in the Southeast, the large and flightless Eastern Lubber Grasshopper has colorful monikers including Georgia thumper, devil's horse, and graveyard grasshopper.

KEY FACTS

Adults are large, short-winged, and black with yellow markings or yellow with black markings; nymphs are dark with red or yellow stripe down the back.

+ habitat: Nymphs in low moist, dense vegetation; adults more widespread

+ range: North Carolina to Florida, west to Texas

+ food: Low-growing forbs

Eastern Lubber Grasshoppers, or Southern Lubbers, are seen in gardens in the summer. Nymphs are gregarious and sometimes abundant enough to damage flowers, vegetables, citrus, ornamental plants, and fruit trees. Nymphs and adults dissuade predators with color patterns that serve as warnings of distastefulness. Adults further defend themselves by spreading their stubby reddish wings, hissing from abdominal spiracles, and exuding foul-smelling and foul-tasting secretions from the thorax. The Horse Lubber Grasshopper *(Taeniopoda eques)* is mostly black with yellow markings and lives in desert shrub and oak habitats in Arizona and New Mexico.

Nymph

Broad-winged Katydid

Microcentrum rhombifolium L 2–2.6 in (50–65 mm)

Broad-winged Katydids have visible "ears" with exposed tympana, or "eardrums," on their front legs at the bases of the tibiae. The female may answer calling males with audible ticks.

KEY FACTS

Forewings not strongly bulging, hind wings are longer; front pronotal margin has a tooth; hind femurs do not reach last fourth of wing tips.

+ habitat: Trees and shrubs in neighborhoods, parks, and woodlands

+ range: Southern and most of eastern U.S.

+ food: Leaves of broad-leaved trees and shrubs

Broad-winged Katydids are attracted to lights. Males produce a loud lisp every few seconds to attract females, and a series of ticks resembling a thumbnail run slowly over the teeth of a comb when approaching a ticking female. Flat, seedlike eggs are laid along leaf margins. One generation is produced annually. Adults' activity peaks in summer and fall, but occurs year-round in Florida. The Lesser Angle-wing Katydid (*M. retinerve*) is smaller, with brown in front of the sound-making area, and occurs in the Southeast. The California Angle-wing Katydid (*M. californicum*) lives in California and southern Arizona.

Close-up of eggs

Fall Field Cricket

Gryllus pennsylvanicus L 0.6–1 in (15–25 mm)

The slow chirps of male Fall Field Crickets attract females on late summer and warm fall evenings. The sound is soothing, except when it originates inside your home.

KEY FACTS

This robust cricket is mostly black; forewings nearly covering abdomen in male; female has long, needlelike ovipositor.

+ habitat: Mainly in grassy areas or in leaf litter

+ range: Southern Canada, most of U.S. except southeastern coastal plain

+ food: Variety of plant and animal tissues

The Fall Field Cricket overwinters in the egg stage. The nymphs hatch the following May and June and mature in late July and August. The omnivorous adults live on the ground In burrows or among plant debris. They sometimes damage vegetable crops, but they will also consume the eggs and pupae of insect pests. Until 1957, all *Gryllus* in the United States and Canada were thought to be one species. Careful examination of their calls and life cycles revealed ten species in the East and at least twice as many in the West, most of which are new to science and await formal scientific description by entomologists.

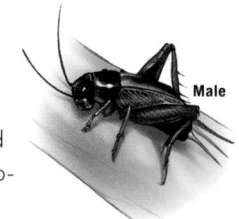

Male

Jerusalem Crickets
Stenopelmatus species   L 0.8–2.7 in (21–69 mm)

Also called "potato bug" or *niña de la tierra* (child of the earth), Jerusalem Crickets resemble a cross between Jiminy Cricket and a Cootie toy, and needlessly inspire fear.

KEY FACTS

This antlike cricket is brown; head is large, smooth, shiny; abdomen is large, soft, banded; legs are stout, spiny.

+ **habitat:** Under objects or in burrows in coastal plain, valleys, foothills, mountains, deserts

+ **range:** Mostly western U.S., especially California

+ **food:** Plant materials and invertebrates

Adult Jerusalem Crickets communicate by drumming their abdomen on the soil. They deliver a painful nip with their powerful mandibles. Mating pairs resemble wrestlers grappling, after which the female may kill and eat the male. Eggs are probably laid in small clutches in the soil. They hatch in spring and reach maturity in two to five years. Dead adults in or near pools and streams are often parasitized by horsehair worms. The name Jerusalem Cricket is also applied to species in the coastal and dune-dwelling *Ammopelmatus* and the Mexican *Viscainopelmatus*. At least 50 species in these three genera are new to science.

Close-up
of head

Northern Stick Insect
Diapheromera femorata   L 2.2–4 in (55–101 mm)

Northern Stick Insects are sometimes common but are seldom seen in large numbers because of their cryptic colors and habits. They rarely damage fruit trees and other broad-leaved trees.

KEY FACTS

The head is slightly longer than wide; antennae are long; legs have swollen and banded middle femur, and hind femur has spine underneath.

+ **habitat:** Deciduous woods and mixed forests

+ **range:** Throughout the East

+ **food:** Leaves of hardwoods, especially oak, hickory, hazelnut, wild cherry, and black locust

Adult Northern Stick Insects move slowly as they feed in shrubs and trees in late summer and early fall. They sometimes remain motionless for short periods, stretched out along a twig with front legs and antennae extended forward. They are easiest to spot among leaves and branches at night with a flashlight. Males grasp females with abdominal claspers during copulation. About 100 shiny black-and-gray striped eggs are dropped haphazardly onto leaf litter. Nymphs hatch in spring, feed on understory plants, and molt five times before reaching maturity. Seven other *Diapheromera* species occur in the U.S. and Canada.

Mating
pair

California Timema

Timema californicum L 0.5–1 in (12–25 mm)

California Timemas are relatively small and robust stick insects restricted to mountainous regions of the western U.S. They are excellent subjects for studying ecological adaptations.

KEY FACTS

This timema is stout and short-legged; green or pink; males are smaller, abdomen is tipped with asymmetrical genitalia.

+ **habitat:** Coastal chaparral and mountains

+ **range:** California

+ **food:** Oak, fir, chamise, California lilac, mountain mahogany, silk tassel bush, toyon, and chaparral pea

California Timemas are most abundant in spring and early summer. They rest on branches and leaves of their host plants during the day and feed on leaves at night. Cryptic colors and behavior help to protect them from predators. Timemas run quickly when disturbed and sometimes release an unpleasant odor. Although they cannot change their colors, three species of timemas have multiple color forms that are genetically determined and match their food plants. Most of the 21 species in the genus *Timema* are best identified by male genitalia. At least three species reproduce asexually, and their males are unknown.

Mating pair

Head Louse

Pediculus humanus L 0.08–0.1 in (2–3 mm)

The bane of students and parents, Head Lice appear identical to Body Lice, but the two are ecologically distinct. They began to diverge when people began wearing clothing.

KEY FACTS

Head Lice are longer than wide, pale grayish, with reddish gut contents clearly visible through the translucent body. The legs extend away from the body, and all are approximately of equal thickness.

+ **habitat:** Hair and scalp of human heads

+ **range:** Wherever humans live

+ **food:** Human blood

Adults and nymphs of the Head Louse spend their lives crawling and feeding on humans, especially children in schools. In dense infestations, they may occupy other hairy parts of the body. Bites usually cause formation of small, itchy, and raised pustules. Eggs, or nits, are glued to hair shafts. The Body Louse (subspecies *P. h. humanus*) is slightly smaller and can survive for longer periods without feeding. Unlike the Head Louse, the Body Louse is a vector of deadly diseases. Head Lice are spread via physical contact; Body Lice are transmitted via infested clothing and unsanitary conditions. Pets do not transmit human lice.

Louse egg, or nit

Giant Water Bug
Lethocerus americanus L 1.6–2.4 in (40–62 mm)

Also called "electric light bugs" or "toe biters," Giant Water Bugs are attracted to lights. They can deliver a painful bite if carelessly handled, and they can be pests in fish hatcheries.

KEY FACTS

Flat and brown, this bug's front legs are modified to grab prey. The middle and hind legs are adapted for swimming.

+ **habitat:** Muddy bottoms or vegetated margins of standing, clear, shallow water

+ **range:** Southern Canada, northern two-thirds of U.S.

+ **food:** Insects, tadpoles, small fish

Adult Giant Water Bugs are active from spring through fall. They inject prey with a potent mixture of paralytic compounds and digestive enzymes, and are able to attack and subdue animals much larger than themselves, including tadpoles and small fish. After mating, the female glues about 70 brown-striped eggs in batches on plant stems just above the water. The eggs are kept moist and guarded by the male until they hatch in about two weeks. Development time from nymph to adult takes about a month. When threatened, Giant Water Bugs will feign death. Four species of *Lethocerus* occur in the U.S. and Canada.

Eggs

Water Boatman
Hesperocorixa vulgaris L 0.4 in (10 mm)

Water Boatmen, especially males, can produce chirping sounds by rubbing a group of bristle-like setae on the bases of their forelegs against sharp margins on the sides of their head.

KEY FACTS

Head and legs are pale; pronotum more than half as long as wide, with ten dark bands; forewings have narrow yellow bands.

+ **habitat:** Lakes, ponds, and slow-moving streams

+ **range:** Southern Canada and the U.S.

+ **food:** Fluids and small particles of algae, protozoa, and aquatic insect larvae

The Water Boatman propels itself through the water with oarlike hind legs much like the Backswimmer, only right side up. When feeding, it uses its scooplike front legs to stir up bottom ooze to find small plants and animals. Fluids and very small particles are consumed with the aid of its short and highly modified beaklike mouthparts, which can also deliver a painful bite. Eggs are laid on any submerged surface. Nymphs hatch in a few weeks and shed their exoskeletons five times before reaching adulthood. The winter is passed in the adult stage. There are 19 species of *Hesperocorixa* in the U.S. and Canada.

Close-up of reticulate pattern

Backswimmer

Notonecta kirbyi L 0.5–0.6 in (12–16 mm)

The Backswimmer is a jerky swimmer, constantly working to maintain its position in ponds and streams when not clinging to an object or resting at the surface.

KEY FACTS

Head is two-thirds as wide as pronotum, eyes are separate; scutellum is black; wings are red-orange with two small light spots at tip.

+ habitat: Ponds, spring-fed pools, and shallow lake margins with emergent vegetation

+ range: Western U.S. to Dakotas

+ food: Small aquatic insects and crustaceans

The Backswimmer is deep-bodied and swims upside down with long, flattened, fringed oarlike hind legs. It typically rests by clinging to submerged vegetation or by hanging upside down from the surface film. It darts after prey on the surface or swims in open water and grasps the prey with its front and middle legs. The Backswimmer replenishes its air supply by breaking the surface film with stiff setae on the tip of the abdomen to draw a silvery bubble down the back. Eggs are typically laid on aquatic vegetation or on submerged rocks and debris. There are 19 additional species of *Notonecta* in the U.S. and Canada.

Fringed, oarlike hind legs

Water Strider

Gerris remigis L 0.5–0.6 in (13–15 mm)

Water Striders glide over the water on layers of air trapped under their feet. They move forward when each leg stroke creates vortices that push against the surface surrounding each foot.

KEY FACTS

Long and slender, the strider is dark brown with reddish markings; pronotum with silky setae; middle and hind legs long.

+ habitat: Slow-moving waters in clear permanent streams with coarse substrates; also ponds, lakes, and stock tanks

+ range: Canada and U.S.

+ food: Small aquatic insects and crustaceans

The biology, behavior, and ecology of the Water Strider, also called pond skater and Jesus bug, have been studied intensively. It preys mostly on or scavenges insects trapped on the water's surface using piercing-sucking mouthparts. Individuals sometimes rest on nearby steeply sloping rocks and boards. Females lay eggs just below the waterline that hatch in about two weeks; nymphs reach adulthood in less than two months. Adults exhibit varying degrees of wing development within a population, including wingless, reduced-winged, and fully winged forms. There are 18 species of *Gerris* in the U.S. and Canada.

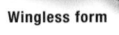

Wingless form

Bed Bug

Cimex lenticularius L 0.16–0.24 in (4–6 mm)

With resurgent Bed Bug activity, people should be more vigilant when traveling, visiting theaters and laundries, or buying used items at garage sales to prevent infestations at home.

KEY FACTS

The Bed Bug is flat and wingless, rusty red, and clothed in short, erect setae; pronotum is broadly notched in front; sides are broadly expanded and fringed.

+ **habitat:** Homes, hotels, theaters, laundries, office buildings, taxis, and furniture

+ **range:** Throughout U.S. and Canada

+ **food:** Human blood

Bed Bugs typically congregate in dark recesses of beds, furniture, and clothing. They may crawl long distances from hiding places, seeking sleeping humans to obtain a blood meal. Most bite victims have allergic reactions to the saliva, including bumps and swelling. Avoid bites in hotels, motels, and hostels by isolating luggage rather than unpacking, and by carefully inspecting beds and furniture. Don't bring home used furniture, especially pieces found outdoors. Despite blood-feeding habits, the Bed Bug has not been associated with outbreaks of disease. There are six other species of *Cimex* in North America.

Bed bug bites

Wheel Bug

Arilus cristatus L 1.1–1.4 in (28–36 mm)

One of our largest and most distinctive assassin bugs, the Wheel Bug is so named because of the eye-catching semicircular crest on the pronotum bearing 8 to 12 short, thick spines.

KEY FACTS

This species is dark brown, densely clothed with short, silky, gray setae; pronotum has a conspicuous spiny ridge down the middle; front legs are enlarged.

+ **habitat:** Gardens, parks, and woodlands

+ **range:** Ontario to Florida, and west to Iowa, Kansas, Oklahoma, and Arizona

+ **food:** Insects

Wheel Bugs are large, robust predators that hunt especially for bees and caterpillars on trees and shrubs. They can deliver a painful bite if carelessly handled. Females lay more than 100 bottle-shaped eggs in clutches on twigs and tree trunks. Young nymphs are black with their bright red abdomen held high, and are antlike in appearance. Older nymphs more closely resemble adults but lack fully developed wings. Although a few adults are found in midsummer, most are encountered in late summer and early fall. Of five species in the mostly neotropical genus *Arilus*, only one occurs in the U.S. and Canada.

Nymph

Western Conenose

Triatoma protracta L 0.6–0.8 in (15–20 mm)

Also known as big bed bug or cross bug, the Western Conenose belongs to a group known as kissing bugs because some species bite sleeping people on their thin-skinned lips.

KEY FACTS

The Western Conenose is dark brown to blackish; antennae with seven antennomeres; a straight, three-segmented beak; oval abdomen; flared sides that extend past wings.

+ **habitat:** Open, arid areas, especially near pack rat nests

+ **range:** California east to Colorado and Texas

+ **food:** Vertebrate blood

The Western Conenose is usually found during the warm months in rodent nests but often occurs near human habitation. Nocturnal nymphs and adults suck blood from Bed Bugs and other blood-feeding insects, and from mammals, including humans. Human bites typically occur in homes where light attracts adults on warm evenings. Although a vector of the protozoan parasite *Trypanosoma cruzi*, or Chagas disease, the Western Conenose doesn't transmit that pathogen to humans. Some people may be sensitized to its saliva and go into anaphylactic shock from bites. There are 11 additional species of *Triatoma* in the U.S. and Canada.

Nymph

Large Milkweed Bug

Oncopeltus fasciatus L 0.5–0.7 in (13–18 mm)

The colors of feeding aggregations of Large Milkweed Bugs on milkweed seedpods warn predators of their bitter taste, which comes from chemical defenses acquired from milkweed sap.

KEY FACTS

Mostly dark brown or black, this bug has an orange Y-shaped mark on the head, pronotal margins, and two broad bands across the forewings.

+ **habitat:** Variety of environments from low to middle elevations

+ **range:** Southeastern Canada and throughout the U.S.

+ **food:** Seedpods of milkweeds

The Large Milkweed Bug is active during warm months in disturbed and natural habitats, where it is found on seedpods and flowers of milkweeds and their relatives. It has evolved adaptations for dispersal and increased reproduction in response to season and available food sources. Adults tend to develop flight muscles in response to lack of food, but when food is abundant, females reduce their muscles and become flightless. Because it is hardy and easy to maintain in captivity, this bug is used for physiological research. There are six, possibly seven, more species of *Oncopeltus* in the U.S. and Canada.

Nymph

Squash Bug

Anasa tristis L 0.5–1 in (13–24 mm)

The Squash Bug is a serious agricultural pest, and its feeding can lead to anasa wilt. Nymphs and adults use their piercing-sucking mouthparts to draw sap from stems, leaves, and fruits.

KEY FACTS

This bug is robust, broad-shouldered, dull yellow to grayish brown to dark brown, and densely covered with coarse punctures; young nymphs are green, older ones are gray.

+ **habitat:** Gardens and agricultural fields

+ **range:** Southern Canada and the U.S.

+ **food:** Sap of melons, pumpkins, and squash

The adult Squash Bug emerges from its over-winter hiding place in early summer to feed and mate. The female lays coppery eggs in masses along veins on the underside of leaves, where the young nymphs hatch and feed. Older nymphs have distinct wing pads. Microorganisms invade and foul wounds where the bugs have fed, quickly destroying the surrounding plant tissues. Only one generation is produced in the northern reaches of the range, and two or three are produced in the South. Adults that develop late in the season do not lay eggs and eventually seek shelter for the winter. Six *Anasa* species occur in the U.S. and Canada.

Nymphs, eggs

Boxelder Bug

Boisea trivittata L 0.4–0.6 in (11–14 mm)

A familiar sight on sunny spring and fall days on tree trunks, fences, and buildings, masses of red-and-black Boxelder Bugs cause unnecessary alarm. They are seldom more than a nuisance.

KEY FACTS

The Boxelder Bug is blackish, with red eyes and three red lines on the pronotum; margins of the thickened portion of the forewings are red; the abdomen is bright red.

+ **habitat:** Gardens, parks, and wooded areas

+ **range:** Eastern U.S. and Canada

+ **food:** Sap of box elder, also maple, oak, and ailanthus

Boxelder Bugs emerge in spring to lay batches of rusty brown eggs near box elder, maples, and sometimes ash. The young red nymphs feed on seeds and other low, herbaceous growth, even dead insects. Mature nymphs have black wing pads. Nymphs and adults may feed on developing fruits, causing puckered "cat-face" injuries. Another generation is produced in late summer. In fall, adults seek shelter to overwinter and can become a nuisance when they congregate on walls and porches, or invade homes through windows, ducts, and other gaps. The similar Western Boxelder Bug (*B. rubrolineata*) occurs in the West.

Mass of adults/ nymphs on wall

Brown Marmorated Stink Bug

Halyomorpha halys L 0.6–0.7 in (14–17 mm)

The Brown Marmorated Stink Bug inflicts considerable damage to ornamental, fruit, and vegetable crops. In fall, it becomes a nuisance to homeowners by taking shelter inside.

KEY FACTS

This frequent pest is mottled, with a blunt head; the antennae and abdominal edges have broad, alternating light and dark markings; each tibia has a broad, pale ring.

+ **habitat:** Gardens, parks, and agricultural fields

+ **range:** Asia; now established across the U.S.

+ **food:** Juices of fruits

The adult Brown Marmorated Stink Bug emerges in spring. In summer, the female lays batches of 20 to 30 light green eggs on the undersides of leaves, and the eggs hatch in about a month. When squashed, these bugs release a disagreeable odor. Up to five generations are produced annually in the South. The feeding activities of the nymphs and adults result in pitting and scarring of fruit and introduce microorganisms that cause decay. As the weather cools, adult bugs seek shelter, sometimes by the hundreds, indoors via gaps in vents, windows, and doors. The Brown Marmorated Stink Bug does not bite people or pets, nor does it spread disease.

Nymph

Harlequin Bug

Murgantia histrionica L 0.3–0.5 in (8–12 mm)

Clad in red, orange, yellow, white, and black, the Harlequin Bug is common on plants in the mustard and caper families in gardens, parks, vacant lots, and fields in late spring and summer.

KEY FACTS

Shield-shaped, this distinctive true bug has bright, variegated marks ranging from mostly orange to nearly all black.

+ **habitat:** Urban, rural, agricultural, and other areas

+ **range:** Southern half of the U.S.

+ **food:** Mostly sap of plants in the mustard family, especially cabbage and related crops

The Harlequin Bug, also called the calico bug or fire bug, is a major pest of cabbage, broccoli, radishes, and related crops. Using piercing-sucking mouthparts to suck sap, it causes plants to wilt, brown, and die. Males and females are often seen copulating end to end. The female lays two rows of black-and-white, barrel-shaped eggs on food plants. The eggs hatch within a month, depending on temperature. Nymphs take two to three months to mature. The Harlequin Bug migrates in spring and summer, and sometimes occurs as far north as New England, but its northern limit varies depending on the severity of the winter.

Eggs and nymph

Seventeen-year Cicada

Magicicada septendecim L 1.1–1.2 (27–30 mm)

The mass emergence of black-and-orange Seventeen-year Cicadas, heralded by the intense rising and falling chorus of thousands of males seeking a mate, is one of nature's great phenomena.

KEY FACTS

This cicada is mostly black; orange mark behind red eye; orange and black wing veins; orange bands under abdomen.

+ **habitat:** Hardwood forests, mixed woodlands

+ **range:** Connecticut to Georgia, west to Wisconsin, Nebraska, and Kansas

+ **food:** Adults do not feed; nymphs suck sap from tree roots.

Female Seventeen-year Cicadas slit living twigs with their reproductive organs and lay eggs inside. In the process, they may cause minor damage in nurseries. Nymphs burrow into the soil, seeking roots to tap with their piercing-sucking mouthparts. Their subterranean feeding activities rarely harm healthy, mature trees. Each emergence is called a brood. Each brood was assigned a Roman numeral by entomologist C. L. Marlatt in 1907 so entomologists can track the broods over time. There are seven species of *Magicicada*. Three emerge as adults every 17 years. The other four are on a 13-year cycle. Several broods have become extinct.

Adult emerging from "shell"

Oak Treehopper

Platycotis vittata L 0.3–0.4 in (8–10 mm), without horn

Oak Treehoppers are gregarious, living on oak branches in groups of 50 to 100 individuals, but their sap-feeding activities, even when in large numbers, seldom cause damage to the tree.

KEY FACTS

This treehopper is triangular and heavy bodied. It is striped or mottled with red; the pronotum has a variable hornlike crest present or absent; hind feet are relatively short.

+ **habitat:** Hardwood forests and mixed woodlands

+ **range:** Most of the U.S.; scarce in Midwest

+ **food:** Sap of oak

The Oak Treehopper feeds only on deciduous and evergreen oaks, but adults are occasionally found on other trees. The mature male and female are variable in color, have bright red eyes, and are either horned or hornless. After undergoing their last molt they are contrastingly light turquoise with brilliant red stripes, but soon become more somberly colored as their exoskeletons harden. The female mates multiple times, but only the sperm of the last male fertilizes most of her eggs. She guards her eggs and stays with the nymphs almost until they reach adulthood. Spring and fall generations are produced annually.

Nymph

Meadow Spittlebug
Philaenus spumarius L 0.3–0.4 in (8–10 mm)

The sudden release of muscle energy stored in elastic pads of protein at the bases of the hind legs enables the Meadow Spittlebug to jump more than 100 body lengths in a single bound.

KEY FACTS

This bug is grayish green, yellowish, or brown with pale spots, sometimes black; head of pale form has two black spots; wings have raised veins.

+ **habitat:** Meadows, woodland edges

+ **range:** Widespread in U.S. and Canada, except in the Southwest

+ **food:** Sap of alfalfa, clover, and weedy plants

The Meadow Spittlebug, also called froghopper, is sometimes a serious pest of alfalfa, red clover, wheat, oats, corn, and strawberries. Females lay masses of 50 or more eggs in late summer and fall on plant stems near the ground. Nymphs hatch the following spring and surround themselves in a white, frothy mass of bubbles uniform in size, called spittle. It is produced from a mixture of waste and sticky glandular secretions that protects them from predators, temperature fluctuations, and dehydration. Scientists studying spittle are applying the data to develop stronger and more efficient foam insulation.

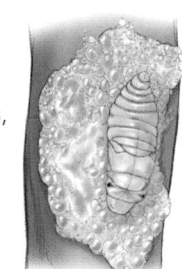

Nymph in frothy spittle

Broad-headed Sharpshooter
Oncometopia orbona L 0.4–0.5 in (11–13 mm)

The female Broad-headed Sharpshooter has brochosomes, small chalky lumps of microscopic protein granules on the forewings, which are applied to their eggs as a protective coating.

KEY FACTS

The head, front of the pronotum, scutellum, and appendages are orange-yellow; remainder of the pronotum and forewings are blue, all with irregular black markings.

+ **habitat:** Gardens, parks, fields, vineyards, and orchards

+ **range:** Eastern U.S. to North Dakota and Texas

+ **food:** Plant sap

The sharpshooter is a solitary or gregarious feeder and a vector of plant viruses that cause phony peach and Pierce's diseases, and oleander leaf scorch. Females generally lay eggs on herbaceous vegetation. Nymphs reach maturity in about two months, and adults live for about another two months. "Sharpshooter" is applied to many kinds of leafhoppers that feed via piercing-sucking mouthparts and filter large quantities of fluids laden with nutrients. Most of the fluid consists of water and carbohydrates, which are forcibly excreted from the anus as droplets in a fine stream, earning them the moniker "sharpshooter."

Brochosome/close-up of particles

Oleander Aphid

Aphis nerii L 0.08–0.1 in (2–3 mm)

The Oleander, or Milkweed, Aphid is commonly found on oleander and milkweeds. Females reproduce asexually, giving birth to nymphs that are clones of their mother. Males are unknown.

KEY FACTS

Plump, soft-bodied, and leggy, this aphid has piercing-sucking mouthparts; bright yellow with black antennae, legs, and cornicles; the female is winged or wingless.

+ **habitat:** Gardens, parks, fields

+ **range:** Temperate and tropical regions of the world

+ **food:** Sap of oleander, other milkweed species

Infestations of Oleander Aphids build quickly because the aphids reproduce asexually by parthenogenesis and have high reproduction rates and short development times. Nymphs undergo five stages of development and, under ideal conditions, reach maturity in a few weeks. Adult females are generally wingless, but winged forms can migrate to new areas when host plants become crowded or begin to die. Light brown husklike aphid "mummies" have had their internal organs eaten by wasp parasitoids. Larvae of flies, lacewings, and lady beetles prey on aphids and often live in or near the aphids' colonies.

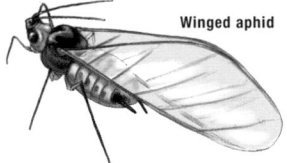

Winged aphid

Cochineal Insects

Dactylopius species female L 0.08–0.1 in (2–3 mm)

Female Cochineal Insects are little more than ovary-filled bags with reduced mouthparts and legs. They form masses of sticky wax on Beavertail Cacti and can harm the plants when abundant.

KEY FACTS

Wingless adult female Cochineal Insects are deep red, covered with tangled strands of white wax, with reduced legs; the male has two long wings.

+ **habitat:** Open developed and natural habitats that include cacti

+ **range:** Saskatchewan, central and southern U.S.

+ **food:** Cactus sap

Native Americans in the Southwest used Cochineal Insects as a source of dye. It gained popularity in Europe where the Spanish introduced it during their conquest of the New World. Soon, Cochineal Insects were raised on cultivated cacti in several parts of the world, harvested, dried, crushed, and boiled to extract the crimson dye. Their importance waned with the advent of synthetic dyes, but recent interest in natural coloring has revived the cochineal dye industry. Today, the dye provides color for candy, drinks, cosmetics, and textiles. There are four species of *Dactylopius* in Canada and the United States.

Adult male

Eastern Dobsonfly

Corydalus cornutus L up to 2 in (50 mm), not including male's mandibles

In spite of their long and fearsome mandibles, male Eastern Dobsonflies cannot bite. The female and the larva, or hellgrammite, can deliver a painful nip with their short mandibles.

KEY FACTS

Adult is large, elongate; translucent gray and white-spotted wings folded flat over body; male's mandibles are very long.

+ **habitat:** Along streams and rivers

+ **range:** Eastern U.S. and Canada

+ **food:** Adults do not feed; larvae prey on aquatic insects and other invertebrates.

Nocturnal adult Eastern Dobsonflies are often attracted to lights near streams and rivers. They spend the day hidden among leaves on the ground or under bridges. Females lay chalky white, bottle cap–size masses of eggs on rocks along flowing water. Hellgrammites are nocturnal, preying on anything they can catch. They respire through eight almost leglike gills protruding from the sides of the abdomen. Mature larvae move on shore to build chambers under rocks and logs, where they pupate. Adults emerge in about a week. Three more species of *Corydalus* occur in the U.S., all with limited distributions in the Southwest.

Hellgrammite

Green Lacewings

Chrysopa and *Chrysoperla* species L 0.5–0.6 in (12–14 mm)

Green Lacewings larvae are dubbed "aphid lions" because they hunt and eat plant pests. Their eggs are sold to gardeners and farmers as biological controls of aphids and other pests.

KEY FACTS

Adults are bright green and soft-bodied, with long, threadlike antennae; clear, green-veined wings folded over body; and long, cylindrical abdomen.

+ **habitat:** Foliage in gardens, parks, fields, woodlands

+ **range:** Throughout U.S. and Canada

+ **food:** Adults eat aphids, honeydew, and pollen; larvae eat aphids.

Active in spring and summer, nocturnal Green Lacewings are attracted to lights and are a common sight on porches in cities and suburbs. The slow-flying adults can hear ultrasonic signals of bats through earlike organs at their wing bases and can release a foul-smelling fluid when attacked. Females lay batches of eggs on leaf surfaces affixed to slender silken stalks, presumably to keep them out of reach of their cannibalistic siblings. There are nine species of *Chrysopa* and seven species of *Chrysoperla* in Canada and the United States. Several other genera and species of Green Lacewings are not green.

Larva

Antlions
Myrmeleon species L to 1.6 in (40 mm)

Antlions are so named for the predatory feeding habits of the larvae, known as doodlebugs. They leave sand scribbles in their wake as they search for sites to make their pit traps.

KEY FACTS

The Antlion is brownish or grayish with darker markings; short, thick antennae; and long wings with finely branched veins.

+ **habitat:** Sandy soils under branches, rock faces, overhangs

+ **range:** British Columbia, throughout U.S.

+ **food:** Adults, nectar, pollen, and insects; larvae eat insects.

Adult Antlions are attracted to lights in spring and summer. The female drops her eggs from branches and overhangs. Using their abdomen as a plow, doodlebugs push backward through sand in ever tighter spirals to construct a conical pit. With a flat shovel-like head, they flip sand out of the pit before settling out of sight at the bottom. With tonglike mandibles, the larvae pump digestive enzymes into insect prey that tumble into the pit. Then the Antlions suck out the liquefied tissues and organs. Only 12 species of *Myrmeleon* dig pits; other Antlion species hunt for prey on branches or in leaf litter.

Larva, or doodlebug

Whirligig Beetles
Gyrinus species L 0.1–0.3 in (3–8 mm)

With a streamlined body, paddlelike legs, and compound eyes for seeing above and below the waterline, the Whirligig Beetle is supremely adapted for life on the water's surface.

KEY FACTS

The shiny black Whirligig has distinctly divided eyes; visible scutellum; and elytra with 11 rows of punctures, without setae on sides.

+ **habitat:** Aquatic margins with vegetation; stream pools with stony or sandy substrates

+ **range:** Throughout Canada and U.S.

+ **food:** Insects on water surface

Adult Whirligigs sometimes crawl out of the water to rest on emergent stems and other objects. When attacked, they rely on speed and chemistry. It can dart across the surface in bursts up to 39 inches (100 cm) per second, and can release a foul-tasting milky secretion from the tip of the abdomen. Species of *Gyrinus* are most reliably identified by careful examination of the male genitalia. Two more genera of whirligigs are commonly encountered in North America. One is the much larger *Dineutus*, which has relatively smooth elytra and no visible scutellum. It is found mainly in the East and the Southwest.

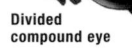

Divided compound eye

Fiery Searcher

Calosoma scrutator L 1–1.4 in (25–36 mm)

One of the largest and handsomest of our ground beetles, the Fiery Searcher, also called the Caterpillar Hunter, is widespread. It often climbs vegetation in search of moth larvae.

KEY FACTS

Elytra are shiny green, purplish along edges, with rows of punctures; legs reddish brown with faint metallic green or bluish highlights.

+ habitat: Woodlands and woodland edges in lowlands and mountains

+ range: Southern Canada, most of the U.S.

+ food: Caterpillars, insect larvae, earthworms

Fiery Searchers overwinter as adults and are active in spring and summer. They often climb trees and shrubs at night in search of cankerworms, tent caterpillars, and armyworms. They are sometimes attracted to lights in large numbers. Their days are spent hiding under loose bark, rocks, boards, and other debris. Like other ground beetles, this one emits a pungent and burning defensive fluid from the anus when alarmed. The life cycle from egg to adult takes one year, with adults living another two or more years. Ground-dwelling and predatory larvae help to control larvae of Gypsy Moths and tent caterpillars.

Larva

Bombardier Beetles

Brachinus species L 0.16–0.24 in (4–6 mm)

Equipped with an explosive defense system, Bombardier Beetles can direct a spray of noxious chemicals at an attacker through their turret-like anus with a "pop" and puff of "smoke."

KEY FACTS

Body and legs are amber, reddish; elytra are wide, squared-off at tips, and iridescent blue, green, or black, without short, hairlike setae.

+ habitat: Moist spots under rocks, bark, logs

+ range: U.S. and Canada

+ food: Adults eat insects; larva parasitizes immature stages of aquatic beetles.

The Bombardier Beetles' defense system consists of two abdominal glands that produce hydroquinones and hydrogen peroxide, which are stored in separate chambers. When threatened, the chemicals are released into a third chamber, where enzymes trigger a violent reaction that forces boiling fluids out through the anus. It can burn human skin for an instant and leave a persistent stain. The sound and smoke of the reaction are startling. Eggs are laid singly in mud balls attached to rocks. After hatching, the larvae search for beetle larvae or pupae to parasitize. Fifty species of *Brachinus* live in Canada and the U.S.

Defensive spray

Predaceous Diving Beetles

Dytiscus species L 0.9–1.6 in (22–40 mm)

Predaceous Diving Beetles, in the genus *Dytiscus*, are among the largest of our aquatic beetles. The male uses broad front feet to grasp the slippery female while mating under the water's surface.

KEY FACTS

This beetle is large, oval, streamlined with upper and lower surfaces moderately convex, and broadest behind the middle; elytra of female often have deep grooves.

+ **habitat:** Permanent ponds and marshes

+ **range:** U.S., Canada, and Eurasia

+ **food:** Aquatic insects, other invertebrates, and small vertebrates

Adult Predaceous Diving Beetles overwinter and will remain active in ponds that are not completely frozen. Mating takes place in spring and fall. The female lays eggs on submerged leaves and stems, and the eggs hatch within two weeks. The predatory larvae reach maturity in about four to six weeks. They leave the water to pupate in chambers dug in damp soil or rotten logs near the shore. Adults emerge in late summer, and they may live up to 18 months in the wild. Large numbers of this beetle are sometimes attracted to lights in late summer and fall. Twelve species of *Dytiscus* occur in the U.S. and Canada.

Larva

Giant Water Scavenger Beetle

Hydrophilus triangularis L 1.3–1.6 in (32–40 mm)

The Giant Water Scavenger Beetle is a large, smooth beetle that is distinguished from *Dytiscus* by its body form. It is convex on top and flattened with a distinct keel underneath.

KEY FACTS

This species is elongate, oval, and dark olive-black; the front tip of the spinelike keel fits into a pocket behind the head.

+ **habitat:** Large, deep, weedy ponds and temporary pools

+ **range:** Across southern Canada and the U.S.

+ **food:** Animal and plant tissues

Adults of this species are usually encountered at lights during early spring and fall dispersal flights. Eggs are laid in a silken case with a mastlike protrusion attached to floating bits of vegetation and debris. The aquatic larvae seize prey, mostly fly larvae and snails, with forceps-like mandibles. They also capture and eat small vertebrates, such as tadpoles and fish. Mature larvae leave the water to dig spherical pupal cells near the shore. A similar species (*H. ovatus*), living only in eastern U.S. and Canada, is more oval and convex, and notched underneath with the keel extending forward to the head.

Larva

Burying Beetle

Nicrophorus marginatus L 0.6–1 in (14–25 mm)

The Burying Beetle exhibits some of the most advanced parental care behavior known in beetles. Pairs bury small bird and mammal carcasses as food for their progeny.

KEY FACTS

Antennal clubs and sides are orange; elytra have two orange bands often connected along sides.

+ **habitat:** Open fields, mountain meadows, prairies, and desert woodlands

+ **range:** Southern Canada, and U.S. except Florida

+ **food:** Adults eat carrion and fly larvae; larvae eat carrion provided by adults.

Overwintering adult Burying Beetles become active in spring and will mate and reproduce throughout the summer. It is not clear whether they are primarily diurnal or nocturnal, although some individuals are sometimes attracted to lights at night. The beetles locate bird or mammal carcasses by smell, quickly bury them to eliminate competition with other scavengers, and treat them with a fungicide in their saliva to retard decomposition. There are 15 species of *Nicrophorus* in the United States and Canada. One, the American Burying Beetle (*N. americanus*), is designated as a U.S. endangered species.

Beetles on carcass

Hairy Rove Beetle

Creophilus maxillosus L 0.4–0.9 in (11–23 mm)

Forensic entomologists use insects, including the Hairy Rove Beetle, found on human remains to develop legal and medical evidence to determine the time and place of death.

KEY FACTS

This is a large beetle; elytra short, exposing at least five abdominal segments; black with bands of grayish white setae on elytra and abdomen.

+ **habitat:** Gardens, parks, woods, strandlines of lakes and rivers

+ **range:** U.S. and Canada

+ **food:** Carrion, maggots

The adult Hairy Rove Beetle hunts for prey mostly on fungus, carrion, and dung from spring through fall. It is a fast flier and runs rapidly over the ground with the tip of its abdomen raised in a menacing manner. It is incapable of stinging but not defenseless. It has glands beneath the abdomen that produce a smelly and irritating fluid that is particularly repellent to ants. The predatory habits of the adults and larvae help to keep pest fly populations in check on cattle feedlots. Each female lays about 500 eggs in her lifetime, and it takes just over a month for them to hatch and reach adulthood.

Larva

Giant Stag Beetle

Lucanus elaphus L (including mandibles) 1.1–2.4 in (28–60 mm)

Male Giant Stag Beetles, with broad heads and curved, antlerlike mandibles, are among our most impressive beetles. Their mandibles are used against rival males over females.

KEY FACTS

This is a chestnut brown beetle, with dark brown or blackish legs; male's mandibles longer than head, female's shorter.

+ habitat: Deciduous and mixed woodlands

+ range: Southern Ontario to Florida Panhandle, west to Minnesota and eastern Texas

+ food: Adults eat honeydew and plant sap; larvae eat moist, rotten wood.

This species is the largest stag beetle in the United States and Canada, and is the only species of *Lucanus* in the region with males that have mandibles longer than the head and midsection combined. When threatened, males raise their body up with their forelegs and spread their mandibles menacingly. Adults and larvae live in oak stumps and decaying logs. Females lay their eggs in the crevices of stumps and logs. The nocturnal adults are active mostly in June and July and are attracted to lights. There are three other species of *Lucanus* in the United States, all of which have males with shorter mandibles.

Female

Horned Passalus

Odontotaenius disjunctus L 1.1–1.5 in (28–37 mm)

The "Bess Beetle" lives year-round under loose bark on rotting stumps and logs. Adults and larvae eat decaying wood and digest it with the aid of microorganisms living in their gut.

KEY FACTS

This large, shiny black, straight-sided beetle has a short curved horn on its head, a midsection (prothorax) that is loosely attached to the rest of the body, and deeply grooved elytra.

+ habitat: Deciduous and mixed woodlands

+ range: Most of the eastern U.S.

+ food: Adults and larvae eat decaying wood.

The Horned Passalus is also known as the Bess Beetle or Patent Leather Beetle. Adults tunnel along the grain of decaying hardwood logs to feed, mate, and reproduce. Eggs, larvae, and pupae typically inhabit the tunnels in the summer. Adults communicate with larvae by rubbing filelike patches on their abdomen against hardened veins that support their flight wings. Larvae respond by scraping their pawlike hind legs over ridges at the bases of the middle legs. Larvae depend on adults for pre-chewed wood as food and for feces laced with microorganisms, which they need to digest cellulose.

Larva

Rainbow Dung Scarab

Phanaeus vindex L 0.4–0.9 in (11–22 mm)

The Rainbow Dung Scarab is beneficial, breaking down and burying animal feces to feed itself and its young, recycling nutrients, and eliminating breeding sites of flies and other pests.

KEY FACTS

This scarab is shiny metallic green, with coppery red and gold reflections; male with or without a horn of variable length.

+ **habitat:** Fields, grasslands, sand hills, forests

+ **range:** Massachusetts to Florida, west to Montana, southeastern Arizona

+ **food:** Vertebrate dung, occasionally carrion

Diurnal adults fly in spring and summer in a zigzag pattern as they search for fresh feces. Upon finding a mass, they land and construct a burrow beside it, with an entrance marked by a distinct "push-up" of excavated earth. Burrows are built either for adult food storage or for burying brood balls. Each brood ball contains one egg and is covered with a thin layer of clay to prevent desiccation. Horn development is likely the result of larval nutrition. Small horned males are called "minors"; individuals with larger horns are called "majors." Six additional species of *Phanaeus* occur in the United States.

Female

Ten-lined June Beetle

Polyphylla decemlineata L 0.7–1.2 in (18–31 mm)

The most common *Polyphylla* in the West, the Ten-lined June Beetle was first described in 1824 by the "father of American entomology," Thomas Say, from specimens destroyed by pests.

KEY FACTS

Elytra are striped with different colored scales; male has 7-segmented fanlike antennal club, female's club smaller, 5-segmented

+ **habitat:** Chaparral, grasslands, desert scrub, oak–juniper woodlands, forests

+ **range:** Western U.S. and Canada

+ **food:** Adults eat conifer needles and leaves; larvae eat roots

The C-shaped grubs can be pests in agricultural crops, orchards, and pine plantations because they eat roots of wild and cultivated plants. The life cycle may take two to five years, depending on local conditions. Mature larvae pupate deep in the soil to avoid freezing and move back to the surface in spring to feed and complete their development. The pupal stage may last up to five weeks. Adults begin emerging in late spring and are active through the summer. They fly at dusk and in the evening, with males more strongly attracted to lights than females. Thirty-two species of *Polyphylla* are recognized north of Mexico.

Male antennae

Little Bear

Paracotalpa ursina L 0.4–0.9 in (10–23 mm)

The Little Bear, so named because of its robust, hairy body, often appears in large numbers on warm spring days flying low over valley grasslands or clambering over chaparral vegetation.

KEY FACTS

Head and pronotum are coarsely punctured, black, steel blue, or metallic green; elytra brick red, yellowish, or black.

+ **habitat:** Grasslands, chaparral, and desert scrublands

+ **range:** Central and southern California

+ **food:** Adults eat chamise and sage leaves; larvae eat roots.

The Little Bear is likely a complex of similar species. Adults are found on chamise (*Adenostoma fasciculatum*) and sage (*Artemisia* species). Three more *Paracotalpa* species occur in western North America. The black *P. deserta* lives along the western edge of the Colorado Desert in California. The metallic green and brick red *P. granicollis* occupies juniper woodlands of central eastern California to Washington and Idaho. *P. puncticollis* is metallic green with yellow and thin black-striped elytra, and inhabits juniper belts west of the Colorado Desert, mountains of the eastern Mojave Desert, and western Arizona.

Green form

Grapevine Beetle

Pelidnota punctata L 0.7–1.1 in (18–27 mm)

The Grapevine Beetle, or Spotted Grape Beetle, commonly found in the East, is distinguished from other large scarabs by its uniform color above, and is often flanked by dark spots.

KEY FACTS

This beetle is yellowish to reddish brown above; variable black spots sometimes absent; underside yellowish brown to black.

+ **habitat:** Deciduous and mixed woodlands

+ **range:** Eastern U.S. and Canada

+ **food:** Adults eat wild and cultivated grape leaves; larvae eat decaying logs and stumps.

The adult Grapevine Beetle flies in early summer and is sometimes found clinging to stems and the undersides of leaves. It begins flying at dusk in search of food and mates, and it is commonly attracted to lights. The females lays their eggs in moist soil at the base of stumps or under logs. Mature grubs construct pupal cells from wood fragments and other bits of nearby vegetation. The entire life cycle takes two years, and the adults live for about a month. The only other species of *Pelidnota* in the U.S. and Canada is *P. lugubris*, smaller, completely black, and restricted to southeastern Arizona and southwestern New Mexico.

Larva

Pupa

Glorious Jewel Scarab

Chrysina gloriosa L 1–1.1 in (25–28 mm)

The Glorious Jewel Scarab is among our most spectacular beetles, yet its brilliant and conspicuous body is well camouflaged as it feeds and mates among the foliage of junipers.

KEY FACTS

This robust scarab is bright, shiny, iridescent emerald green, and it has four broad metallic silver or silvery-gold stripes on each elytron.

+ **habitat:** Juniper and oak–juniper woodlands

+ **range:** Western Texas to southeastern Arizona, and south to Mexico

+ **food:** Adults eat juniper foliage; larvae eat roots.

In summer, adults begin to fly at dusk and are often attracted to lights. The silvery stripes are cryptic, visually breaking up the body's outline and reflecting light to resemble drops of resin. The life cycle takes one or two years to complete. Although *Chrysina* is mainly tropical, three species occur in the United States. The dark green *C. lecontei* feeds on pine needles in Arizona and New Mexico. The larger apple green and lavender-legged *C. beyeri* eats oak leaves in southeastern Arizona, southwestern New Mexico, and Mexico. *Chrysina woodi* has metallic yellow legs, eats walnut leaves, and lives in western Texas.

In flight

Ox Beetle

Strategus aloeus L 1.2–2.4 in (31–61 mm)

The large Ox Beetle has few predators, though immature stages are susceptible to microbial attacks. It is the most widespread, abundant, and variable species in the genus *Strategus*.

KEY FACTS

This large beetle is mahogany to black; major males are equipped with well-developed horns; minor males have only short, rounded knobs; females are unarmed.

+ **habitat:** Woodlands and wooded canyons

+ **range:** South Carolina to Florida, west to southeastern Arizona

+ **food:** Adults eat roots and leaves; larvae eat decaying wood.

Little information is available on the life history of the Ox Beetle in the United States, although there is a record of adults feeding on the roots of date palms in Arizona. Both the adults and the larvae are sometimes minor pests in palm groves in Mexico, Central America, and South America. The adults are mostly nocturnal and are commonly encountered at lights in summer. The grubs feed and develop in rotten hardwood logs, such as ash (*Fraxinus*) and live oak (*Quercus*) trees; they are also found under boards on the ground. Four additional species of *Strategus* inhabit the southern United States.

Female

Eastern Hercules Beetle
Dynastes tityus L 1.6–2.4 in (40–60 mm)

Among our largest beetles, Eastern Hercules Beetles lay their eggs in ground-level tree holes, where their grubs eat the crumbling dead heartwood, but do not harm living trees.

KEY FACTS

A large species, the Hercules Beetle is olive or gray-green and mottled with irregular black spots; the underside and legs are black. The male has horns, but the female is unarmed.

+ **habitat:** Hardwood forests

+ **range:** Eastern U.S.

+ **food:** Adults eat sap of ash; grubs eat decaying wood.

The adult male chews a hole in the limb of an ash tree and guards the sapping wound until a female arrives. Rival males use their horns like forceps to dislodge one another from the wound. Females lay their eggs in hardwoods, especially oaks. Grubs take about two years to reach adulthood. They pupate in a protective chamber fashioned from their own waste mixed with soil. Adults emerge from the pupa in fall but remain in the chamber until the following summer. They are nocturnal and attracted to lights at night. The Southwestern Hercules Beetle *(D. grantii)* inhabits wooded canyons in Arizona, New Mexico, Utah, and Mexico.

Female

Green June Beetle
Cotinis nitida L 0.6–1.1 in (15–27 mm)

Green June Beetles are a familiar sight and sound on summer days, flying in zigzag patterns with a loud buzz over lawns, gardens, fields, and decomposing plant materials.

KEY FACTS

This beetle is dull green and yellowish brown above, shiny underneath; pronotum with central lobe extending backward.

+ **habitat:** Gardens, parks, agricultural fields

+ **range:** Connecticut to Florida, west to Nebraska and Texas

+ **food:** Adults eat plant tissues and sap; larvae eat decaying plant materials and roots.

Adults emerge from pupal chambers in summer and eat fruits and foliage of trees and shrubs. They are sometimes pests in gardens and orchards, attacking vegetables and ripening fruit. Large numbers sometimes fly low over lawns, grass clippings, compost heaps, and manure piles on warm days. Females lay eggs in these and other accumulations of organic matter. The subterranean grubs, which usually crawl on their back, occasionally damage roots of turf, vegetables, and ornamental plants. Five species of *Cotinis* occur north of Mexico. The Green Fig Beetle *(C. mutabilis)* is the most common species in the West.

Larva

Large Flathead Heartwood Borer

Chalcophora virginiensis L 0.9–1.3 in (23–33 mm)

Large Flathead Heartwood Borers are commonly found sunning themselves on bark or patches of wood on pine snags and logs in spring and summer. They take to the air with a loud buzz.

KEY FACTS

A dull black or bronze borer, it has elytra with dark or shiny elevations around patches of rough depressions; underside brassy.

+ **habitat:** Pine forests, mixed woodlands

+ **range:** Southern Canada, northern and eastern U.S.

+ **food:** Pine buds, needles (adults); pine sapwood, heartwood (larvae)

The female Large Flathead Heartwood Borer lays her eggs around fire scars on living trees or in cracks or holes of bark on downed logs and stumps. Full-grown larvae are large and have a distinct Y-shaped mark on the broad thoracic plate. Living trees are sometimes severely damaged by their tunneling and feeding activities. Logs cut for timber are sometimes destroyed by the larvae if left outdoors too long. Three additional species of *Chalcophora* are found in the East: *C. fortis*, *C. liberta*, and *C. georgiana*. Another species, *C. angulicollis*, is restricted to mountainous pine forests of western North America.

Larva

Eyed Click Beetle

Alaus oculatus L 1–1.7 in (24–44 mm)

Large, mottled with small patches of short and hairlike black-and-white setae, the Eyed Click Beetle has eyelike spots that are not visual organs and are unlikely to frighten predators.

KEY FACTS

Pronotum has a distinct pair of round or elliptical velvety black eyelike spots ringed with a narrow band of white scales and variable white patches on elytra.

+ **habitat:** Deciduous woodlands and mixed forests

+ **range:** Eastern U.S. and Canada

+ **food:** Larvae of wood-boring insects

Adult Eyed Click Beetles are usually found during spring and summer days on decaying hardwood logs and stumps infested with wood-boring insects. They are attracted to odors of solvents and freshly painted surfaces. The long, slender, yet tough larvae have a dark head and a shiny yellowish or orange-brown body. They crawl in and out of burrows in rotten wood in search of prey. Four additional species of *Alaus* occur north of Mexico. *A. myops* also occurs in the East, and *A. melanops* is found only the West. *A. lusciosus* inhabits the south-central United States, and *A. zunianus* is restricted to Arizona.

Larva

Big Dipper Firefly
Photinis pyralis L 0.4–0.6 in (9–15 mm)

The male Big Dipper Firefly makes repeated dips in flight as it flashes along hedgerows and woodland edges to locate a sedentary female. If she is receptive, she responds with her own light.

KEY FACTS

The head is covered by a hoodlike pinkish pronotum with a large black spot; elytra are dark with narrow, pale margins; light-producing organs are under the abdomen.

+ **habitat:** Gardens, parks, and woodland edges

+ **range:** Eastern U.S.

+ **food:** Adults do not feed; larvae prey on worms and slugs.

Also called lightning bug, the Big Dipper Firefly and its relatives are neither flies nor true bugs, but are soft-bodied beetles. Different species have distinctive flash patterns so that males and females can recognize and locate one another. Bioluminescence also deters predators, warning them of the firefly's bad taste. Fireflies are bioluminescent from egg to adult, although some species lose the ability to produce light just hours or days after becoming adults. Reliably identifying firefly species requires observing the flash patterns of the male and microscopic examination of their genitalia.

With light-producing organs

Brown Leatherwing Beetle
Pacificanthia consors L 0.5–0.8 in (13–20 mm)

A large beetle known only from California, the Brown Leatherwing often flies to porch lights in late spring and early summer. It is a predator of the Citrus Mealybug *(Pseudococcis citri).*

KEY FACTS

Reddish brown with partly dark legs, this species' head is visible, not narrow behind eyes; elytra are soft, leathery, and velvety gray.

+ **habitat:** Suburbs, coastal chaparral, oak woodlands

+ **range:** California

+ **food:** Adults eat sap-feeding insects; larvae probably eat small insects.

Although common and easily recognized, very little is known about the natural history of the Brown Leatherwing. Nocturnal adults spend the day under loose bark and give off a strong and unpleasant odor when handled roughly or crushed. The nocturnal larvae probably develop under bark or in damp areas beneath rocks and logs. Some species of soft-bodied soldier beetles resemble fireflies, but are easily distinguished because their head is not covered by the pronotum and is visible from above. These and other species of soldier beetles have bright, contrasting patterns that warn predators of their bad taste.

Larva

Goldenrod Soldier Beetle

Chauliognathus pennsylvanicus L 0.4–0.5 in (9–12 mm)

Individuals or clusters of mating pairs of contrastingly colored Goldenrod Soldier Beetles are familiar sights during late summer and early fall on goldenrods and other flowers.

KEY FACTS

Elongate and soft-bodied, this beetle's head is black; pronotum is wider than long; leathery elytra are tipped with dark spots.

+ habitat: Gardens, parks, fields, woodland edges

+ range: Maritime Provinces to Florida, west to Ontario, Colorado, Texas

+ food: Adults, pollen; larvae probably insects.

A fascinating species for research, the Goldenrod Soldier Beetle has aided scientists in studies of mating behavior, variation in color pattern, dispersal, and genetics. Females deposit their eggs in leaf litter where the larvae develop and probably prey on small insects and their eggs. Adults found dead with their wings spread on flowers and branch tips are victims of lethal fungi that are harmless to vertebrates. The Margined Leatherwing (*C. marginatus*) has a similar range, but is active in spring and early summer. It has an orange head with a black, V-shaped mark and a pronotum that is longer than wide.

Mating pair

Red Flat Bark Beetle

Cucujus clavipes L 0.4–0.7 in (10–17 mm)

Overwintering larvae of Red Flat Bark Beetles produce antifreeze and proteins that protect them in the winter. They are studied by scientists researching adaptations to cold climates.

KEY FACTS

Flat and thin, this species is bright red with black eyes, antennae, tibiae, and feet; their head is broad and triangular, bulging behind eyes.

+ habitat: Deciduous woodlands and mixed forests

+ range: U.S. and Canada

+ food: Adults eat insects, fungi, and plants; larvae eat insects.

Little is known about the biology of the Red Flat Bark Beetle. Adults search in narrow spaces under bark of recently dead trees for prey in summer and fall. In winter, the amber-colored and flat-bodied larvae are found more often under bark than the adults. They are adapted for surviving low winter temperatures by substituting up to 40 percent of their extracellular body fluids with chemicals that reduce the freezing point of water. This works in concert with proteins that induce ice formation between, rather than inside, cells, thus protecting tissues from being destroyed by razor-sharp ice crystals.

Larva

Convergent Lady Beetle

Hippodamia convergens L 0.16–0.28 in (4–7 mm)

Adult Convergent Lady Beetles are collected in California and sold to organic farmers and gardeners to control aphids and other pests, but the larvae are actually more effective predators.

KEY FACTS

The somewhat oval, hemispherical pronotum of this beetle is black with white bars converging toward the rear; elytra are orange, each usually with six black spots.

+ **habitat:** Wide variety of habitats

+ **range:** Throughout the U.S. and Canada

+ **food:** Adults eat pollen, other plant materials, and aphids; larvae eat aphids.

The Convergent Lady Beetle is one of our most common lady beetles. Females lay batches of 15 to 30 bright yellow eggs on plants in spring and summer. The larvae molt four times in about a month before pupating; the pupal stage lasts about a week. In the West, adults escape hot, dry conditions by flying up cooler river canyons into the mountains in search of food. They eventually overwinter in large aggregations and return to their valley feeding grounds in spring. This species is the official state insect of South Dakota. Most species of lady beetles, or ladybugs, are neither red nor spotted.

Larva

Stink Beetle

Eleodes obscurus L 1–1.3 in (25–34 mm)

The Stink Beetle is adapted for desert life. Its thick, fused elytra protect it from abrasion and desiccation, and a cavity under the elytra helps as insulation from extreme temperatures.

KEY FACTS

This large beetle is dull or weakly shiny black or reddish brown; elytra are grooved.

+ **habitat:** Grasslands, mesquite and oak woodlands

+ **range:** Southern British Columbia to northern Mexico, west to Colorado and Texas

+ **food:** Adults eat plant and animal detritus; larvae probably eat plant detritus

Stink Beetles are commonly encountered at dusk walking along trails, roads, and in other open habitats in late spring and summer. When disturbed, this and other species of *Eleodes* assume a defensive posture by standing on their head and releasing a smelly dark fluid from the tip of their abdomen, which contains hydrocarbons that repel predators, especially other arthropods. As a result, they are sometimes called acrobat or clown beetles. There are about 130 species of *Eleodes* in the U.S. and Canada, most living in the West. Incapable of inflicting harm, these large and long-lived beetles make interesting pets.

In defensive position

Horned Fungus Beetle

Bolitotherus cornutus L 0.3–0.5 in (7–12 mm)

Horn development in the male Horned Fungus Beetle is a function of the quality of larval nutrition, which is determined by the fungal quality of the egg-laying site selected by the mother.

KEY FACTS

This beetle is rough, rectangular, dull brownish black to reddish brown; males armed with pairs of small bumps to large curved and fringed horns.

+ **habitat:** Deciduous woodlands, mixed forests

+ **range:** Eastern U.S. and Canada

+ **food:** Bracket fungi on decaying hardwood logs and stumps

The adult Horned Fungus Beetle is found on bracket fungi on warm summer evenings and is sometimes attracted to lights. Well-armed males use their horns in battles with rival males over females, including those already copulating. They nudge or butt other males forward and upward with their head, sometimes with their mandibles open. In the absence of larger males, less-endowed males can still locate and mate with a female. Eggs are laid singly on fungal surfaces. Larvae tunnel inside to develop and pupate. Adults live at least two years, typically overwintering inside fungus or under loose bark.

Female

Magisterial Blister Beetle

Lytta magister L 0.6–1.3 in (16–33 mm)

The Magisterial Blister Beetle has blister-forming cantharidin in its blood. It oozes from leg joints when a beetle is threatened or disturbed. Contrasting colors warn predators of the threat.

KEY FACTS

This large beetle's head is antlike; head, midsection, and legs are mostly reddish orange; elytra are rough, leathery, and black.

+ **habitat:** Mojave and Sonoran Deserts

+ **range:** Southeastern Utah and southern California to Sonora

+ **food:** Adults eat flowers and leaves of brittlebush and other desert plants

The Magisterial Blister Beetle is most abundant in spring. It appears singly or sometimes in thousands, and the scraping of the elytra against stones and dried vegetation creates a loud rasping sound as it awkwardly crawls about. This lumbering, heavy-bodied insect laboriously takes flight with a loud buzzing sound. The first-stage larvae have been described, but later stages and their habits are unknown. Based on the natural history of their nearest relatives, the larvae probably feed on pollen, larvae, and pupae of ground-nesting bees. This species is also known as the Arizona or Desert Blister Beetle.

Mating pair

Striped Blister Beetle

Epicauta vittata L 0.4–0.6 in (9–15 mm)

The Striped Blister Beetle is economically damaging because it often gathers in large mating swarms that consume fruits, flowers, and leaves of vegetable and forage crops.

KEY FACTS

Distinct, antlike head has two black spots; leathery elytra have two or three black stripes.

+ **habitat:** Gardens, parks, and fields

+ **range:** Southern Ontario and Quebec, eastern U.S.

+ **food:** Adults are partial to plants in amaranth, pea, and nightshade families; larvae consume grasshopper eggs.

Striped Blister Beetles in the Northeast tend to be darker, with two narrowly separated stripes on each elytron. In the Southeast, especially on the coastal plain, they tend to have three stripes as a result of more distinctly separated middle and outer stripes. Adults are attracted to lights at night, sometimes in large numbers. Like other blister beetles, this species produces caustic yellowish droplets from the leg joints, which blister the skin. For food, the leggy and active first-stage larvae seek pods of grasshopper eggs buried in the soil. Once settled, the larvae develop into legless, sedentary grubs.

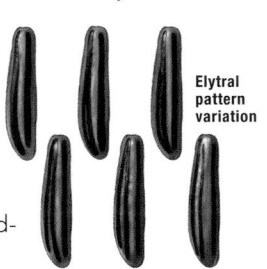

Elytral pattern variation

Pine Sawyer

Ergates spiculatus L 1.6–2.6 in (40–65 mm)

Pine Sawyer larvae typically chew large, meandering galleries in the sapwood before tunneling into the heartwood of trees that have been killed by fire or bark beetles.

KEY FACTS

The Pine Sawyer is large and brown, with sometimes darker head, prothorax, and legs; pronotum has three large calluses, fringed with many sharp spines.

+ **habitat:** Montane coniferous forests

+ **range:** Western U.S. and Canada

+ **food:** Adults do not feed; larvae develop in Douglas fir and other conifers.

The adult Pine Sawyer is encountered during the summer, flying at dusk or emerging from broad, flat tunnels at the bases of stumps. The male has long antennae about two-thirds the length of the body. It is often attracted to light. The heavier-bodied female has antennae that are only half as long as the body, and is seldom drawn to light. The adult is also known as ponderous borer and spiny wood borer. Loggers refer to the larvae as timber worms. A subspecies, *E. s. neomexicanus*, ranges from the Rockies to Arizona and New Mexico and is pale reddish brown with pale makings on the elytra.

Larva

Tile-horned Prionus

Prionus imbricornis L 1.3–2 in (34–52 mm)

Isolated mature trees in recreational areas, as well as trees that are sick, injured, or affected by drought, are particularly susceptible to damage by Tile-horned Prionus larvae.

KEY FACTS

This prionus is large, robust, dark brown to black; antennae have 15 to 18 segments, thick and reaching middle of elytra (male), or not (female).

+ **habitat:** Deciduous woodlands

+ **range:** Southern New England to Florida, west to Nebraska and Texas

+ **food:** Adults do not feed; larvae eat roots of hardwoods.

Nocturnal Tile-horned Prionus beetles are active in summer and spend the day hiding under bark or in leaf litter at the base of trees. Males are attracted to lights, but the larger, heavy-bodied females do not fly. They lay batches of 100 to 200 eggs at the base of deciduous trees and shrubs, especially oaks. Mature larvae burrow up near the soil surface to pupate. The larvae feed exclusively on living roots of all sizes. Four more species of *Prionus* occur in the East, all with three stout spines on the sides of the prothorax. The California Prionus (*P. californicus*) is widespread east of the Rocky Mountains.

Female

Locust Borer

Megacyllene robiniae L 0.4–1.1 in (11–28 mm)

Native to the East, the colorful and wasplike Locust Borer has expanded its range westward because of widespread planting of its larval host plant, Black Locust, along streets.

KEY FACTS

Adult is black with broad yellow bands; bands at basal third of elytra form the letter W; legs are reddish; and underside is mostly yellow.

+ **habitat:** Meadows, uncultivated fields, other open habitats

+ **range:** Eastern U.S. to Washington, Colorado, Arizona

+ **food:** Adults eat pollen; larvae develop in black locust.

Adult Locust Borers appear in late summer and early fall and are frequently found feeding on goldenrod flowers. Females scurry up and down tree trunks both day and night in search of cracks, crevices, and scars in which to lay their eggs. Young larvae feed under the bark before tunneling into the sapwood and heartwood of limbs and trunks. Small trees and saplings riddled with larval tunnels and adult exit holes are especially prone to breakage at ground level. Trees growing in nutrient-poor soils, poorly planted on eroded sites, and subjected to fire and droughts are especially susceptible to damage.

Mating pair

Argus Tortoise Beetle
Chelymorpha cassidea L 0.3–0.4 in (8–11 mm)

The red and black-spotted Argus Tortoise Beetle is sometimes mistaken for a very large lady beetle. It feeds only on plants in the Morning Glory family, including Sweetpotato and false bindweed.

KEY FACTS

This beetle is oval, convex, and dark red or yellowish with black spots; pronotum has 4 to 6 small spots; elytra has 13 spots, including a shared basal spot.

+ **habitat:** Gardens, meadows, old fields, other open areas

+ **range:** Southern Canada and the U.S.

+ **food:** Morning Glory species, including Sweetpotato

The Argus Tortoise Beetle is one of the largest leaf beetles in North America. Adults overwinter in dry places under bark or in leaf litter and emerge in spring to feed, mate, and reproduce. Females lay slender, stalked yellow eggs singly or in batches on undersides of leaves. The spiny, flattened larvae protect themselves with an umbrella-like shield consisting of their own cast exoskeletons and feces. Adults and larvae damage crops by chewing holes in the leaves. Although it feeds only on relatives of morning glories, this species is sometimes known by the inappropriate name of "milkweed tortoise beetle."

Larva

Colorado Potato Beetle
Leptinotarsa decemlineata L 0.2–0.4 in (5–11 mm)

The Colorado Potato Beetle became a serious pest of potatoes in about 1840, and is one of the few North American insects to become established as a pest in Europe, Russia, and Asia.

KEY FACTS

These beetles are round and humpbacked; their pronotum is a black "U" flanked by six small spots; elytra are yellowish, each with five black stripes.

+ **habitat:** Gardens, parks, and old fields

+ **range:** U.S. and Canada

+ **food:** Leaves of potatoes, eggplants, and their relatives

Overwintering adult Colorado Potato Beetles emerge in spring to find food and mates. Females lay masses of 20 to 60 eggs on the undersides of leaves. The humpbacked, orange-red or reddish larvae have a double row of black spots on each side. Younger larvae feed in groups, while older ones are solitary. Mature larvae burrow into the soil to pupate. This species is strongly similar in form and habit to another eastern species, the False Potato Beetle (*L. juncta*), but it has elytra with single, not double, rows of pits and only three or four stripes that usually meet at both ends on each elytron.

Larva

Dogbane Leaf Beetle

Chrysochus auratus L 0.3–0.5 in (8–13 mm)

Iridescent colors of the Dogbane Leaf Beetle are created by small plates in the outer exoskeletal layers that reflect light at various wavelengths when observed from different angles.

KEY FACTS

This oblong, convex beetle is iridescent green with bluish, golden, or coppery luster; appendages and underside bluish black.

+ **habitat:** Gardens, parks, old fields, grass-lands, and prairies

+ **range:** Mostly east of Rockies

+ **food:** Leaves (adults) and roots (lar-vae) of dogbane and Indian Hemp

The Dogbane Leaf Beetle emerges in summer and is defended from predators by distaste-ful and toxic chemicals sequestered in its tissues from ingested plant tissues. After hatching from eggs laid at the base of food plants, larvae bur-row into soil to feed, develop, and pupate. In the Pacific Northwest, the Dogbane Leaf Beetle hybridizes with the western Blue Milkweed Beetle (*C. cobaltinus*). This species consumes leaves and roots of both milkweeds, *Asclepias* and *Apocynum*. Adults sequester little if any toxins in milkweed tissues, but will apply a protective coating of feces laced with toxin to their eggs.

Mating pair

Sisal Weevil

Scyphophorus acupunctatus L 0.3–1 in (8–24 mm)

Native to the southern United States, the Sisal Weevil is a serious pest elsewhere in the world on sisal plantations where it is harvested commercially for use as twine and floor covering.

KEY FACTS

A reddish black to black species, Its beak is nearly straight; antennal club spongy and concave; elytra shallowly punctate between grooves.

+ **habitat:** Wild and developed areas with food plants

+ **range:** South Caro-lina to Florida, west to Kansas and California

+ **food:** Agave, sotol, and yucca

The adult Sisal Weevil is typically found in the Southwest on stalks or trunks of host plants. Larvae bore into the stems and roots. The species was introduced into the Southeast via ornamental plants, and sometimes infests greenhouses. This weevil was inadvertently transported in agave to Hawaii, Australia, and Africa, where it is a pest. The Yucca Weevil (*S. yuccae*), a similar species found on Chaparral Yucca in California, has flat and ridged tips of antennal clubs. *Cactophagus spinolae* is a larger weevil with a curved beak associated with Saguaro and Beavertail Cactus in southern California and Arizona.

Larva

Cat Flea

Ctenocephalides felis L 0.08–0.2 in (2–4 mm)

The bane of cats, dogs, and people, the Cat Flea not only causes great discomfort with its bite but also serves as an intermediate host for tapeworm. Adults are most abundant in summer.

KEY FACTS

A reddish brown flea, has a low and sloping head outline; the head (genal) combs have 7 to 8 black teeth; inner surface of hind femur with 7 to 10 bristles.

+ **habitat:** Wherever there are domesticated animals and humans

+ **range:** Cosmopolitan

+ **food:** Adults eat blood; larvae eat bits of organic matter.

The Adult Cat Flea is a free-living, blood-feeding insect that normally attacks cats, but dogs, people, and other wildlife are also hosts. People sensitive to the bites develop large, red, itchy welts, especially around the ankles, that may take weeks to heal. Females lay hundreds of eggs in their lifetime. In homes infested with fleas, the wormlike larvae are found in cat bedding. They don't bite, but instead eat bits of dried blood passed by adult fleas, cat dander, and bits of organic debris. Mature Cat Flea larvae pupate in a loose silk cocoon. Large infestations of adults can cause severe anemia in domestic animals.

Close-up of head

Giant Eastern Crane Fly

Pedicia albivitta L 0.8–1.5 in (20–38 mm)

Crane flies such as Giant Eastern Crane Flies are sometimes called "giant mosquitoes" or "mosquito hawks," but they do not prey on mosquitoes, nor do they feed on blood.

KEY FACTS

This crane fly is brown with white markings; dark-banded wings have a short band at an angle nearly reaching the rear margin; legs are long and slender.

+ **habitat:** Near woodland pools and slow-moving rivers

+ **range:** Northeastern U.S.

+ **food:** Adults do not feed; larvae eat small adult and larval insects.

Adult Giant Eastern Crane Fly is commonly encountered near wetlands. The predatory larvae are aquatic and breathe air through a pair of openings, or spiracles, on a disk located at the tip of the abdomen. There are 60 species of *Pedicia* in North America, all very large, with a broad, bold triangle pattern with one side on the leading edge of the wing, and "hairy" eyes. Crane flies are often mistaken for mosquitoes, but they lack piercing-sucking mouthparts and are not clothed in scales. The thick-skinned larvae of some species, known as leatherjackets, are root feeders and some are considered pests.

Larva

Asian Tiger Mosquito

Aedes albopictus L 0.2–0.3 in (5–7 mm)

The Asian Tiger Mosquito is a vector of important diseases of humans and domestic animals, including dengue fever, dog heartworm, eastern equine encephalitis, and West Nile virus.

KEY FACTS

This mosquito is black with white stripes and bands on body and legs; male's antennae are bushier than female's; abdomen is pointed at the tip.

+ habitat: Gardens, parks, and woodlands

+ range: Asia; established in the eastern U.S. and in other countries

+ food: Blood of birds and mammals

A daytime biter in spring and summer, this mosquito was first reported from North America in Texas in 1985, apparently accidentally imported in scrap tires from Asia. Only females bite to obtain nutrients from blood for egg development. Both sexes drink nectar for their own nourishment. Eggs are laid individually in tree holes or any water-filled container. The aquatic larvae, or wrigglers, hatch in a few days, breathe air from the surface through a siphon at the tip of their abdomen, and eat bits of submerged vegetation. Floating, comma-shaped pupae, or tumblers, obtain oxygen through thoracic respiratory tubes.

Larva

Big Black Horse Fly

Tabanus punctifer L 0.8–1.1 in (20–28 mm)

To obtain blood-based proteins needed for proper egg development, a hungry female Big Black Horse Fly inflicts her painful bite on livestock, zoo animals, and occasionally humans.

KEY FACTS

Male's eyes touch in this large black fly; female's thorax is covered in short, dense, creamy setae; front legs are black and gray.

+ habitat: Open areas near fresh water

+ range: Western U.S. and Canada

+ food: Adults eat nectar and fruit juice; larvae eat slow-moving aquatic invertebrates.

The Big Black Horse Fly reaches peak activity in summer. The female bites cattle and horses, especially in the late morning and afternoon. She will sip sweet fluids to supplement her blood meals, and the male relies exclusively on these fluids. Pyramid-shaped egg masses containing 200 to 800 eggs are attached to cattails, spike rushes, and other plants overhanging water along edges of ditches, sloughs, ponds, and seepage areas. The larvae hatch in about two weeks and burrow in mud to hunt for shore-dwelling insects and snails. After about two years, the mature larvae crawl on shore to pupate in damp soil or debris.

Female laying eggs

Black Soldier Fly
Hermetia illucens L 0.6–0.8 in (15–20 mm)

Sluggish, wasplike Black Soldier Flies gather at decaying plants to mate and lay eggs. Larvae are scavengers and are sold commercially to break down compost and sanitize waste.

KEY FACTS

This species is black, with pale tibiae and feet; its smoky wings are folded flat over its back when at rest; the abdomen has a pair of translucent patches at its base.

+ **habitat:** Gardens, parks, and farms

+ **range:** Most of the U.S.

+ **food:** Larvae eat decaying plant and animal tissues.

The Black Soldier Fly resembles the Organ-pipe Mud Dauber wasp. Translucent patches at the base of the abdomen inspired the scientific name *illucens* and the common name Window Fly. Adults frequent the vicinity of compost heaps, where their somewhat flattened, tough, and leathery larvae feed and develop. Larvae are sometimes found at human remains and are useful as evidence for the time of death. They are available commercially to feed captive amphibians, reptiles, and tropical fish. Some farmers use them in manure management programs to consume animal waste and destroy the eggs of pestiferous house flies and blowflies.

Larva

Mydas Fly
Mydas clavatus L 1–1.2 in (25–30 mm)

One of the largest and most conspicuous flies in the East, the Mydas Fly is a spider wasp mimic. Mydas Flies are often seen on hot summer days visiting flowers.

KEY FACTS

A large and robust fly, it is mostly dull velvety black; antennae are thick; wings are brownish with a bluish sheen; abdomen has a reddish orange or yellow second segment.

+ **habitat:** Parks, gardens, meadows, shrubby margins of woods

+ **range:** Southern Ontario, eastern U.S.

+ **food:** Flower nectar

The Mydas Fly is the largest fly in the U.S. and Canada, but little is known about its biology. Males and females resemble spider wasps in the genus *Anoplius* as they cruise over open ground and flowering plants on sunny days. It was once thought to be an insect predator, but morphological studies and field observations have established that it drinks flower nectar, especially that of milkweed. Females deposit their eggs singly on chunks of wood embedded in soil and stumps at the soil line. The larvae are sometimes found under bark on decaying hardwood stumps, where they presumably develop and prey on beetle larvae.

Anoplius wasp model

Drone Fly

Eristalis tenax L 0.5–0.6 in (12–15 mm)

The big-eyed, beelike Drone Fly resembles the male European Honey Bee *(Apis mellifera).*
The flies produce a beelike buzz when captured, but they do not deliver a sting or bite.

KEY FACTS

With Honey Bee–like shape, color, and behavior, this fly has a head that is broader than its thorax, two wings, and a dark abdomen with brownish triangles at the base.

+ **habitat:** Various open, disturbed, and agricultural areas

+ **range:** Europe; throughout Canada and the U.S.

+ **food:** Flower nectar

The Adult Drone Fly commonly visits flowers, especially those in the Aster family, in late summer. Females lay up to 200 eggs near stagnant water or in liquid excrement in stockyards, outdoor toilets, and septic tanks. Sausage-shaped larvae feed on decaying animal waste and other organic matter in fetid pools with little dissolved oxygen. They are called "rat-tailed maggots" because they breathe air through a long, tapering, retractable tube extending from the tip of the abdomen. Drone Flies are distinguished from other beelike and native species of *Eristalis* by two narrow bands of setae across each compound eye.

Rat-tailed maggot

House Fly

Musca domestica L 0.2–0.3 in (5–8 mm)

The House Fly is associated with filth and the spread of disease and parasites. It liquefies foodstuffs with salivary secretions and sops them up with a fleshy, spongelike mouthpart.

KEY FACTS

This fly has reddish eyes; a gray thorax with four black stripes; clear wings held at an angle; a yellowish and gray abdomen; and hairy and clawed legs.

+ **habitat:** Neighborhoods, parks, farms, and garbage dumps

+ **range:** Cosmopolitan

+ **food:** Wide variety of plant and animal-based foods

A native of the Old World, the nonbiting House Fly was probably introduced to the New World by European colonists. The larvae develop in manure, decaying lawn clippings, compost heaps, and other accumulations of wet organic matter associated with farms and feedlots. They breathe through a pair of back-to-back D-shaped spiracles characterized by sinuous slits. The entire life cycle from egg to adult takes as little as eight days. These flies breed continuously across the southern United States. Adults are potential carriers of pathogens that cause diarrhea, typhoid, cholera, intestinal worms, and eye infections.

Larva

Green Bottle Fly
Lucilia sericata L 0.2–0.4 in (6–10 mm)

The shiny and brassy Green Bottle Fly is a common sight around dog feces, garbage cans, and carrion. Adults feed on animal fluids, and their maggots feed and develop in rotting flesh.

KEY FACTS

A metallic yellow- or coppery-green fly, it has reddish eyes; mostly clear wings, yellowish at the base with brown veins; and a shiny or dull abdomen.

+ **habitat:** Most common near human habitations

+ **range:** Throughout Canada and the U.S.

+ **food:** Adults eat plant and animal fluids; larvae eat decaying flesh.

The female Green Bottle Fly is among the first insects to arrive at a carcass, and the stage of maggot development at a human corpse may provide evidence assisting investigators in establishing time of death. This fly sometimes lays her eggs on open wounds of live animals. Maggots are used to treat diabetic ulcers and other deep-tissue wounds in humans when surgery and antibiotics are undesirable or likely to be ineffective. Sterile larvae introduced into the wounds eat only dead, infected tissue and produce antibacterial secretions that hasten the healing process, especially in patients infected with methicillin-resistant *Staphylococcus aureus* (MRSA).

Larva

Flesh Fly
Sarcophaga pernix L 0.4–0.6 in (10–14 mm)

Resembling an oversize House Fly, the female Flesh Fly broods her eggs internally, then deposits maggots that are ready to feed on dead insects, decaying carcasses, and excrement.

KEY FACTS

The Flesh Fly has brick-red eyes; a thorax with three stripes; clear wings; faintly striped abdomen with gray-black checkerboard pattern and a reddish tip in males.

+ **habitat:** Sunny, open areas near carcasses and animal waste

+ **range:** U.S. and southern Canada

+ **food:** Adults eat animal fluids; larvae eat decaying flesh.

The scientific name *Sarcophaga pernix* is an appropriate combination of Greek and Latin meaning "nimble flesh-eaters." The Flesh Fly is quick to find dead animals, even human remains, and immediately begins laying its larvae on the decomposing body. Thousands of wriggling maggots will occupy a carcass and, under ideal conditions, complete their development in a few days before pupating in the soil. The maggots can also develop in animal waste and are sometimes attracted indoors by fecal odors. Flesh Flies are similar to one another, and many species can be reliably identified only by examination of a male's genitalia.

Larva

Beelike Tachinid Fly

Hystricia abrupta L 0.4–0.5 in (10–13 mm)

Large, yellow, and bristling with spines, the Beelike Tachinid Fly is one of our most striking tachinid flies. Adults are usually seen visiting flowers or resting on vegetation.

KEY FACTS

These are robust, wide-bodied, yellow flies with brick red eyes; a yellow to black thorax; an orange abdomen with three bands of black bristles; and orange legs.

+ habitat: Meadows, fields, open areas near woodlands

+ range: U.S. and Canada

+ food: Adults visit flowers; larvae attack woolly bear caterpillars.

The Adult Beelike Tachinid Fly is active in late spring and summer. Like many other tachinid flies, this species is not a vector of disease, but it is a parasitoid of caterpillars. Unlike parasites that usually do not harm their hosts, parasitoids kill their host and, thus, are important biological controls of pestiferous species. After mating, the female Beelike Tachinid Fly lays one or two eggs on or near a woolly bear caterpillar. The larvae immediately latch on to their host and feed on their internal tissues and organs, keeping their doomed host alive long enough to complete their own development.

Milkweed woolly bear host

Pipevine Swallowtail

Battus philenor W 2.8–3.4 in (70–86 mm)

Male Pipevine Swallowtails focus on searching for females, drinking nectar, or gathering at puddles. Females concentrate on locating food plants for their eggs and visiting flowers.

KEY FACTS

Males' hind wings are iridescent blue-green with white spots; underwings are iridescent with orange spots and cream chevrons.

+ habitat: Woodlands, meadows, riparian areas

+ range: Southern Ontario, eastern U.S., to Arizona, Oregon, northern California

+ food: Adults sip nectar; larvae eat pipevine leaves.

Pipevine Swallowtails flutter their wings as they imbibe nectar or visit rain puddles or moist sand along streams and rivers during mid-morning and late afternoon. As they feed, their velvety black or red and tentacled caterpillars sequester toxins that make them as well as the adults unpalatable. The larger Spicebush Swallowtail (*Papilio troilus*) is distinguished by cream crescents on its forewings and hind wings. The Black Swallowtail (*Papilio polyxenes*) has an orange eyespot on the hind wings. The Red-spotted Purple (*Limenitis arthemis astyanax*) is also similar but lacks tails on its hind wings.

Larva

Western Tiger Swallowtail

Papilio rutulus W 2.8–3.9 in (70–98 mm)

In wooded watercourses along canyon bottoms, the large and brilliant yellow and black Western Tiger Swallowtail is also a familiar sight in urban and suburban neighborhoods and parks.

KEY FACTS

Wings are boldly marked yellow and black; hind wings each have a single tail and broad blue band underneath near the margin.

+ **habitat:** Gardens, parks, riparian woodlands

+ **range:** Mostly west of Rockies

+ **food:** Adults sip nectar; larvae eat pipevine leaves.

The Western Tiger Swallowtail overwinters as a pupa and emerges in spring to feed, mate, and reproduce. The female seeks riparian hardwoods on which to lay her spherical green eggs. The mature caterpillar is mostly green, with two conspicuous eyespots on its swollen thorax followed by a yellow band. It will rest in a shelter made from leaves drawn together with silk. The dark brown pupa is attached at its bottom to a button of silk affixed to a vertical stem or surface. It is supported in an upright position by a heavy strand of silk. Depending on elevation, up to three generations may be produced annually.

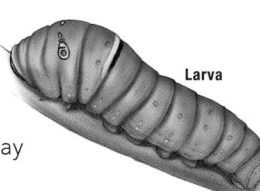

Larva

Black Swallowtail

Papilio polyxenes W 2.6–3.5 in (67–89 mm)

The Black Swallowtail is common in gardens. Its black-banded and yellow-spotted larva, or parsley worm, is sometimes a nuisance to gardeners growing carrots, dill, fennel, and parsley.

KEY FACTS

This species is black with yellow spots on wing margins; hind wings have a red eyespot, more metallic blue in females; and abdomen has yellow spots.

+ **habitat:** Gardens, parks, open and riparian areas

+ **range:** Eastern U.S. and Canada

+ **food:** Adults sip nectar; larvae eat leaves related to carrots and citrus.

The Black Swallowtail flies from spring through summer and drinks nectar from clovers, milkweeds, thistles, ironweeds, Joe-pye Weed, and other flowers. Yellow eggs are laid singly on the leaf tips of food plants. Young black-and-white larvae resemble bird droppings; mature larvae are mostly green with black bands and yellow spots. To repel birds, larvae extend a forked, orange thoracic gland that releases a noxious odor suggesting spicy vomit. Two broods are produced annually, and sometimes a third in warmer regions. This is one of six eastern species of butterflies that mimic the toxic Pipevine Swallowtail.

Larva with forked gland, or osmeterium

Cabbage White

Pieris rapae W 1.3–1.9 in (32–48 mm)

A European species first reported in North America in Quebec in 1860, the Cabbage White has become very common and widespread. It is one of the first butterflies to emerge in spring.

KEY FACTS

The wings above are white, with a black tip, and one spot (male) or two spots (female); underside of the hind wings is unmarked greenish yellow.

+ **habitat:** Gardens, parks, fields, and other disturbed areas

+ **range:** U.S. and Canada

+ **food:** Adults sip nectar; larvae eat plants in the mustard family.

Male and female Cabbage Whites fly low over the ground searching for flowers, nectar, and egg-laying sites. Eggs are laid singly on young growth of mustards, radishes, cabbage, broccoli, cauliflower, and turnips. The green caterpillars have thin, yellow stripes down the back and sides, and are clothed in short, fine, hairlike setae. The green or tan chrysalis is strongly tapered on both ends, with projections on both sides. From the tip of the abdomen, it attaches a button of silk to a leaf or other surface and suspends upright by a strand of silk. Multiple generations are produced from spring through summer.

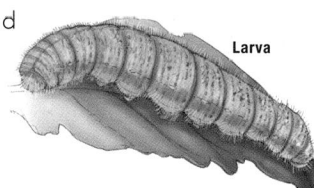

Larva

Cloudless Sulphur

Phoebis sennae W 2.1–2.8 in (54–70 mm)

The Cloudless Sulphur is a fast and powerful flier. It uses its very long proboscis to sip nectar from deep, tubular flowers that are inaccessible to other butterflies.

KEY FACTS

Wings are lemon yellow with broken dark margins; female has dark spot on upper forewing. Male's underside is greenish yellow; female's is yellow with pinkish brown marks and silver spots.

+ **habitat:** Open, disturbed areas

+ **range:** Southern half of the U.S.

+ **food:** Adults sip nectar; larvae eat *Cassia* leaves.

The Cloudless Sulphur is migratory and cold intolerant, but regularly strays north to Washington, New England, and the upper Midwest. In agricultural fields, along roadsides, and in other open sites it flies with Clouded and Orange Sulphurs, which have black wing margins. Caterpillars range from green to yellow, usually with a yellowish or greenish stripe over the spiracles and multiple bands of shiny black bumps on their back. The chrysalis is tan or green, pointed at both ends, and has a large, rounded "fin" extending down the back. Two generations are produced in most of the range, with more on the Gulf Coast.

Larva

Great Purple Hairstreak

Atlides halesus W 1.3–1.5 in (32–38 mm)

When the Great Purple Hairstreak rubs its hind wings together, the wings' "tails" and spots become a "pseudohead." If predators attack the false head, the butterfly has a chance to escape.

KEY FACTS

Wings are iridescent blue with crimson marks at base below and a blue-and-orange abdomen; male has a blue patch and female does not.

+ **habitat:** Oak woods, pine flats, riparian areas

+ **range:** Southeastern U.S. to New York; southwest to California and Oregon

+ **food:** Adults sip nectar; larvae eat mistletoe.

The male stakes out hilltops and other high perches to see a potential mate flying through the area, a behavior call "hilltopping." Rival males are challenged to a spiraling contest high into the air. The acrobatics are staged not for specific females but for control of the perch and its surroundings. Green eggs are laid singly on mistletoes parasitizing oaks, sycamores, junipers, and Red Maples. The mature caterpillar is oval, dark green, and clothed with short setae and has a dark diamond-shaped pattern behind the head, with occasional yellow stripes on the sides. Multiple generations are produced in spring and summer.

Larva

Gray Hairstreak

Strymon melinus W 1–1.3 in (25–32 mm)

The male Gray Hairstreak is distinguished from the female by his orange abdomen. The male perches low on herbaceous growth and shrubs to await a female and will chase rival males away.

KEY FACTS

The wings are slate gray with reddish orange eyespots on tailed hind wings; underside has narrow white, black, and orange bands; female's abdomen is gray.

+ **habitat:** Parks, fields, chaparral, woods, roadways

+ **range:** U.S., southern Canada

+ **food:** Adults sip nectar; larvae eat pea and mallow plants.

The adult Gray Hairstreak flies in spring through fall, visiting all kinds of flowers in gardens and parks. Hairlike tails on the hind wings are associated with bright eyespots and presumably direct attacks by predators away from the more vulnerable body. Pale green eggs are laid singly on young leaves and flower buds, especially clover, milk vetch, lupine, and mallow. The caterpillar excretes honeydew and is tended and defended by ants. The mature caterpillar is bright green, straw, reddish brown, or pink, marked with variable white, cream, or light purple stripes on the sides. Multiple generations are produced annually.

Larva

Gulf Fritillary

Agraulis vanillae W 2.5–3 in (64–76 mm)

The Gulf Fritillary resembles *Speyeria* fritillaries but has more elongate forewings. Gulf Fritillaries fly year-round in southern California, and southern Texas and Florida.

KEY FACTS

Wings are bright orange; forewings have dark margins, each with three white spots ringed in black; underside is brownish with long silvery spots.

+ **habitat:** Gardens, thorn scrub, and open woodlands

+ **range:** Southern third of the U.S.

+ **food:** Adults sip nectar; larvae eat leaves of Passionflower.

The Gulf Fritillary is particularly fond of red or white flowers as it searches for nectar. It migrates northward during the summer, especially in California, the Mississippi River Valley, and along the Atlantic coast, but is intolerant of cold winters. One or more may roost together on vines and vertical branches at night. The yellow eggs are laid singly on leaves, stems, tendrils, and buds. The mature caterpillar is blackish, dark brown, or purplish with six rows of branched spines, an orange stripe down the back, and orange and white stripes along the sides. Multiple generations are produced annually.

Larva

Mourning Cloak

Nymphalis antiopa W 2.9–3.4 in (73–86 mm)

The Mourning Cloak is among the first butterflies to appear in late winter and early spring, taking advantage of fleeting periods of warmth because it overwinters as a winged adult.

KEY FACTS

Wings are unmistakably dark brown with yellow margins and blue spots above; underside resembles dead leaves.

+ **habitat:** Riparian woodlands, forest edges, and along roads

+ **range:** New World; also Eurasia

+ **food:** Adults sip nectar and sap; larvae eat leaves of elm and other hardwoods.

The adult Mourning Cloak lives up to ten months, and it spends most of its life avoiding the summer heat. On cold days it basks in the sun using its dark wings as solar collectors to gather heat and transfer it to its flight muscles. To avoid freezing, it seeks shelter under loose bark, or in woodpiles and sheds. In spring, clusters of pale green eggs laid on twigs soon fade to white and then darken just before hatching. The very distinctive mature larvae are black with pairs of red spots down their back and armed with sharp, branched spines. One or two broods are produced annually.

Larva

Red Admiral
Vanessa atalanta W 1.7–2.2 in (44–57 mm)

With its forewings divided by a bold orange-red band, the Red Admiral is one of the most distinctive butterflies in its wide range. It is more attracted to dung and sap than to flowers.

KEY FACTS

Wings are mostly dark brownish black above; forewings have a reddish orange crossband and white spots; hind wings have a spotted, broad orangish margin.

+ **habitat:** Gardens, parks, woodlands, forest edges

+ **range:** Widespread in Northern Hemisphere

+ **food:** Adults sip mostly excrement and tree sap; larvae eat leaves of nettles.

The male Red Admiral stakes out hilltops, shrubs, sheds, open ground, and other landmarks to await the arrival of a female. Pale green eggs are laid on the upper surfaces of leaves. Young and gregarious larvae build silk nests that tie up shoot tips. Older larvae feed alone and chew through a petiole, attach the edges of the drooping leaf with silk, and begin feeding at the leaf tip from the inside. The mature caterpillar ranges from almost white to yellow-green or black with pale flecking and has a dark head and pale, branched spines. Some individuals have creamy blotches on sides that coalesce into a broad stripe.

Larva

American Lady
Vanessa virginiensis W 1.7–2.1 in (44–54 mm)

The female American Painted Lady butterfly darts erratically over the ground, searching for an egg-laying site. Territorial males tend to stake out bare patches, hilltops, roadsides, and mud puddles.

KEY FACTS

The forewing tips are extended and rounded, orangish with black and white markings above; hind wings below have two large bluish eyespots.

+ **habitat:** Parks, fields, and other open areas

+ **range:** U.S. and Canada

+ **food:** Adults sip nectar and mud; larvae eat leaves of plants in sunflower family.

Eggs of the American Lady are green, ribbed, and barrel shaped. Mature, spiny caterpillars vary in color but typically have thin, creamy bands on the front of each body segment and red spots at bases of branched spines along the sides. They tie up bunches of leaves with silk and feed from within the shelter. Several generations are produced annually. Two similar and widespread *Vanessa* species have hind wings with multiple eyespots underneath. The West Coast Lady (*V. annabella*) has extended forewings, but tips appear clipped. The Painted Lady (*V. cardui*) has rounded forewing tips that are extended only slightly.

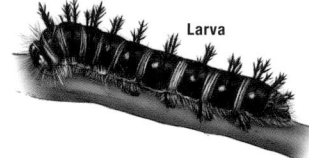

Larva

Common Buckeye

Junonia coenia W 2–2.5 in (51–63 mm)

The Common Buckeye is not a cold-tolerant species and does not breed in the north. Adults emerging in spring have grayish hind wings, but those of the fall brood are reddish brown.

KEY FACTS

The brownish fore-wings have pairs of orange bars, a broad white bar, and a large eyespot; hind wings each have large and small eyespots.

+ **habitat:** Gardens, parks, and other open areas

+ **range:** Eastern, southern U.S.; Pacific coast to Oregon

+ **food:** Adults sip nectar; larvae eat leaves of many plants.

A resident of the southern United States, the Common Buckeye regularly migrates north-ward to Oregon, New England, and southeastern Canada. The mature caterpillar is highly vari-able, usually dark above with pale and/or orange sides. The head is orange on top, and the back is defended by branched, metallic blue-black spines. Two to four generations are produced annually. Two similar buckeyes occur in the region. The Mangrove Buckeye in Florida (*J. evarete*) has a large eyespot on the fore-wing surrounded by orange. The Tropical Buckeye in southern Texas (*J. genoveva*) has a narrow band on the underside of the hind wing.

Larva

Viceroy

Limenitis archippus W 2.6–3 in (67–76 mm)

The Viceroy mimics the equally bad-tasting Monarch in coloration. Unrelated distasteful species that bear similar bold markings as a warning to predators are examples of Müllerian mimicry.

KEY FACTS

The rusty reddish orange wings are black-veined, with broad black-and-white spotted margins; a narrow black band crosses the middle of the hind wings.

+ **habitat:** Meadows, riparian areas

+ **range:** U.S. east of Pacific states

+ **food:** Adults sip nectar, dung, carrion, and decaying fruit, mud; larvae eat mostly willow.

The male Viceroy is territorial and patrols small areas, often returning to the same low perch after each sortie. While copulating on the wing, the female carries the male. Pale green eggs are laid singly on leaf tips of host plants. The mature caterpillar has two long thoracic horns behind its head and is mottled white with brown or olive, resembling a bird dropping. In southern parts of the range, viceroys apparently mimic the Queen butterfly, *Danaus gilippus*. Florida viceroys tend to be more reddish. Populations in the Southwest have less-pronounced wing veins, and bands across the hind wings are faint or lacking.

Larva

American Snout

Libytheana carinenta W 1.6–1.9 in (41–48 mm)

The American Snout looks like a dead leaf when resting on stems, its long mouthpart resembling the leaf stem. Massive, yet unpredictable, summer migrations occur in Arizona and Texas.

KEY FACTS

It has a long, snoutlike mouthpart; brown and orange forewings with white spots, extended and squared tips, and a mottled or plain underside.

+ **habitat:** Thorn scrub, open woodlands, and clearings

+ **range:** Southern and eastern U.S.

+ **food:** Adults sip nectar and sap; larvae eat hackberry leaves.

In addition to visiting flowers and sap flows on trees and shrubs, the American Snout also sips water and dissolved minerals from wet spots and mud puddles. Its erratic flight pattern alternates between bouncing and gliding. The males perches on or near larval host plants in anticipation of the arrival of females. Eggs are laid on hackberry leaves. The mature caterpillar is yellow-green with black stripes down the back and sides. Suspended upside down under a leaf, the green chrysalis has a distinctive oblique ridge. Two or three broods are produced in the northern portions of the range in late summer.

Larva

Monarch

Danaus plexippus W 3.5–4 in (89–102 mm)

The large and beautiful Monarch is one of our most familiar butterflies. It has been studied intensively because it is the only insect known that regularly undertakes a two-way migration.

KEY FACTS

The wings are burnt orange with black veins and white-spotted margins above, paler underneath; hind wings of male have a dark scent pouch.

+ **habitat:** Gardens, parks, fields, other natural and open areas

+ **range:** Southern Canada and throughout U.S.

+ **food:** Adults sip nectar; larvae eat milkweeds.

A tropical species breeding in a temperate climate, Monarchs in the last annual generation seek mild temperatures in late summer and fall to avoid freezing. Those that breed east of the Rockies fly to mountains in Mexico; those in the West migrate to coastal California and Baja California. The migrants fly north and east in spring and repopulate North America in three to five generations. Co-opting the chemical defenses of milkweed as they feed, the caterpillars advertise their distastefulness with bands of black, white, and yellow. Boldly and contrastingly colored adults also employ this defensive strategy.

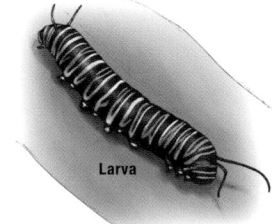

Larva

Silver-spotted Skipper

Epargyreus clarus W 1.7–2.6 in (43–67 mm)

One of our largest skippers, the chunky Silver-spotted Skipper darts among various garden flowers to feed while flashing big, conspicuous silvery patches underneath its hind wings.

KEY FACTS

It has chocolate brown forewings with dull, irregular bars above, and fringed hind wings with a narrow, checked pattern above and a large silvery spot below.

+ habitat: Gardens, parks, prairies, woodland edges

+ range: Southern Canada, most of U.S.

+ food: Adults sip nectar; larvae eat relatives of peas.

The male Silver-spotted Skipper perches low on twigs or leaves and chases any large insects that fly nearby. The female deposits her green eggs singly on leaves of larval food plants. Newly hatched larvae build shelters by cutting a section of a leaf and rolling it with the upper surface inside. The caterpillar makes larger shelters as it grows, which eventually consist of several leaves. The mature caterpillar has an oversize, brownish red head with a pair of orange spots, and a yellow-green body with thin, dark green bands. One or two broods are produced annually, with the last generation overwintering as a pupa.

Larva

Regal Moth

Citheronia regalis W 3.7–6.1 in (95–155 mm)

The giant caterpillar of the Regal Moth, called the Hickory Horned Devil, has long, reddish orange and black-tipped thoracic horns. It looks fearsome, but is harmless to humans and pets.

KEY FACTS

This large, stout moth has an orangish ground color, with broad gray stripes and yellowish spots and smaller, spotless hind wings with more orange than gray.

+ habitat: Deciduous and mixed forests

+ range: New England to Florida, west to Missouri and Texas

+ food: Adults do not feed; caterpillars eat hardwood leaves.

The Regal Moth emerges from the cocoon on late summer evenings. The male is a strong flier and begins seeking females at dusk; copulation continues through the next day. The female begins laying eggs singly or in small batches on upper and lower surfaces of leaves at dusk. Mouthparts of both sexes are nonfunctional. Males are more strongly attracted to lights than females. The Pine-devil Moth (*C. sepulcralis*) is smaller, mostly brownish gray, and more narrowly distributed along the East Coast. *C. splendens sinaloensis* in southeastern Arizona and Mexico is similar to *C. regalis* in size, but its ground color is reddish.

Larva

Imperial Moth

Eacles imperialis W 3.1–6.9 in (80–174 mm)

Occurring mostly east of the Great Plains, the Imperial Moth is commonly encountered at night in wooded areas, but it has largely disappeared from many urbanized areas in the Northeast.

KEY FACTS

This large, stout moth's wings are yellow, marked with pinkish, orangish, or purplish brown spots; yellow males are more heavily marked on outer forewings.

+ **habitat:** Woodlands, forests

+ **range:** Eastern U.S. to Nebraska and Texas

+ **food:** Adults do not feed; larvae eat needles, leaves of trees.

The adult Imperial Moth emerges from the cocoon on summer mornings at dawn. The male is a strong flier and takes wing at night to track females that release pheromones. Copulating pairs remain together through the following day, and females begin laying eggs singly or in small batches on upper and lower surfaces of leaves at dusk. Mature larvae pupate in loose soil without a cocoon, like *Citheronia*. Subspecies *E. i. pini* is restricted to pine forests across the northern Great Lakes Basin. Oslar's Eeacles *(E. oslari),* occurring only in southeastern Arizona and Mexico, have a bold, brown line across each wing.

Larva

Io Moth

Automeris io W 2–3.1 in (50–80 mm)

The plain forewings of the Io Moth help to conceal it when at rest. When disturbed, it suddenly opens its wings, revealing large mammal-like eyespots to intimidate would-be predators.

KEY FACTS

The Io Moth's forewings are yellow (male), or brownish, reddish, or purplish brown (female); hind wings have black eyespots and blue-ringed white centers.

+ **habitat:** Woodlands, forests

+ **range:** Eastern Canada and U.S. to Manitoba, Colorado

+ **food:** Adults do not feed; caterpillars eat tree and shrub leaves.

The adult Io Moth emerges in late morning or early afternoon in summer and remains still until evening. The female releases a sexual attractant pheromone late in the evening, and mating pairs remain in copula for most of the next day, separating at dusk. The female lays eggs singly or in small batches on leaf surfaces and stems. Male and female are attracted to lights. The mature caterpillar is bright green, with stinging spines. It spins a papery cocoon on the ground in leaf litter or in a crevice. The number of generations per year varies, from one in the North to three or four in southern Florida and Texas.

Larva

Polyphemus Moth

Antheraea polyphemus W 4–6 in (100–150 mm)

The Polyphemus Moth is our most widely distributed species of giant silk moth. It is sometimes common in late spring and summer, especially in forests with large stands of oak.

KEY FACTS

Wings are brown, each with large, black eyespots ringed in yellow; eyespots on hind wings are ringed in black and blue.

+ **habitat:** Parks, forests, woodlands

+ **range:** Southern Canada and U.S. states, except Arizona and Nevada

+ **food:** Adults do not feed; caterpillars eat leaves of trees and shrubs.

The Polyphemus Moth emerges from a cocoon in late afternoon from late May through the summer. Mating begins late in the evening, and mated pairs stay together through the next day, parting at dusk. Both sexes are attracted to light. The female lays off-white banded eggs singly or in short rows of two or three on leaves. Mature green larvae construct cocoons that often incorporate leaves of the host plant and sometimes have a functional peduncle. The related *A. oculea* has wings with eyespots ringed in orange, and extensive blue and black scaling. It is known only from mountains of Arizona and western New Mexico.

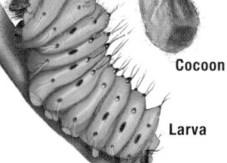

Cocoon

Larva

Luna Moth

Actias luna W 3–4.1 in (75–105 mm)

The spectacular Luna Moth was the first giant silk moth to appear in scientific literature. First reported in Maryland in 1700, it was not formally described as a species until 1758.

KEY FACTS

A large, robust moth, it has antennae that are featherlike in the male, less so in the female; wings are green or blue-green, with long tails on hind wings.

+ **habitat:** Suburbs, parks, forests, woodlands

+ **range:** Eastern U.S. and Canada

+ **food:** Adults do not feed; caterpillars eat tree leaves.

The Luna Moth is the most common giant silk moth in the East, especially in the southern United States. Eggs are laid singly or in small batches on the food plant. Larvae are solitary and somewhat sedentary feeders on white birch in the north and on hickories, walnuts, sumacs, persimmon, and sweet gum in the south. The mature caterpillar spins its irregularly shaped papery brown cocoon, which usually incorporates leaves at the base of its host plant. The adult usually emerges in the morning in late spring and summer. There is one generation in the north and up to three generations in the south.

Larva

Cecropia Moth

Hyalophora cecropia W 4.3–6 in (110–150 mm)

Long popular in natural history literature because of its large size and coloration, the Cecropia Moth is typically found at night. Its caterpillars are encountered eating ornamental plants.

KEY FACTS

This large moth's wings have white, reddish, and tan wing bands, large black-lined crescents, and forewings with blue crescent in each black eyespot.

+ **habitat:** Developed, undeveloped woodlands and forests

+ **range:** Eastern U.S. to Montana

+ **food:** Adults do not feed; larvae eat deciduous tree leaves.

The Cecropia Moth emerges mid-morning in spring and summer. The male seeks a female before dawn, with the female generally mating only once. Two to six eggs are laid in rows on upper and lower surfaces of leaves. Mature larvae leave the food plant to pupate. They seek dark, protected areas to construct a slender or bag-type cocoon that is pointed on both ends and affixed along its entire length to a stem. Four additional species of *Hyalophora* occur in the region. *H. columbia gloveri* occurs in eastern California, Nevada, and the Rockies. *Hyalophora euryalus* inhabits Canadian provinces and U.S. states along the Pacific coast.

Cocoon

Larva

Tobacco Hornworm Moth

Manduca sexta W 4.1–4.7 in (105–120 mm)

The Tomato Hornworm Moth is the most widely distributed hawk moth. It is an economic pest because the larvae regularly attack tomatoes, tobacco, potatoes, and peppers.

KEY FACTS

This large, robust species has a brown thorax, forewing with brownish ground color, and a serrated white band on most of the outer margin.

+ **habitat:** Riparian areas, agricultural fields

+ **range:** Southern Canada, eastern U.S. to California

+ **food:** Adults sip nectar; larvae eat tomatoes and other nightshades.

Flight of the Tobacco Hornworm Moth, also called the Carolina Sphinx, begins at dusk. It frequents many flowers, especially mesquites, daturas, Moonflower, Japanese Honeysuckle, and Evening Primrose. It is commonly attracted to light. Eggs are laid singly on upper and lower leaf surfaces of plants mostly in the nightshade family. Mature larvae are green or brown with seven oblique lines passing over the spiracles, and a reddish orange horn. They sometimes bear egglike cocoons of a wasp parasitoid (*Cotesia congregata*). One generation is produced throughout most of the range, possibly more in the Deep South.

Larva

Achemon Sphinx

Eumorpha achemon W 3.4–3.8 in (87–96 mm)

The speckled green or brown caterpillar of the widespread Achemon Sphinx has three-part whitish or yellowish diagonal bands along its sides. It feeds on grapevines.

KEY FACTS

Large and robust, this moth is mostly tan to brown with rusty triangles on thorax, forewings with dark brown patches at the tips and middle, and pinkish hind wings.

+ **habitat:** Fencerows, woodland edges

+ **range:** U.S. except Northwest

+ **food:** Adults sip nectar; larvae eat grape and Virginia Creeper leaves.

The Achemon Sphinx emerges in late afternoon, visits flowers at dusk, and is commonly seen at lights. Eggs are laid singly on the underside of grape leaves or on tendrils. When threatened or disturbed, the larger horned caterpillars can expand their last thoracic segment so the head may be withdrawn inside. Mature larvae dig chambers several inches underground before transforming into a dull reddish brown pupa. One generation is produced annually in the north and possibly two in the southern part of the range. Eleven more species of *Eumorpha* occur here, but none have mostly pink hind wings.

Larva

White-lined Sphinx

Hyles lineata W 2.5–3.5 in (63–90 mm)

Population explosions of White-lined Sphinxes sometimes occur in the West. Thousands of caterpillars cross desert highways, and hundreds of adults swarm to outdoor events at night.

KEY FACTS

It has a striped head and thorax; olive-brown forewings with a broad tan stripe along entire length, and white veins; and black hind wings with a pink band.

+ **habitat:** Gardens, fields, deserts, grasslands, forest edges

+ **range:** U.S. and southern Canada

+ **food:** Adults sip nectar; larvae eat many plants.

Mostly nocturnal, the adult White-lined Sphinx is sometimes seen at flowers in the afternoon. Mating takes place at dusk in spring and summer. Adults occur year-round in the southern U.S. Northern populations produce one generation, and southern populations produce multiple generations. A related species, *H. gallii,* is smaller, chunkier, lacks white veins on brownish forewings, and occurs farther north. A third species, *H. euphorbiae,* resembles *H. gallii,* but *euphorbiae* has more greenish forewings. *Euphorbiae* was introduced into the upper Midwest and Great Lakes region to control invasive leafy and cypress spurges.

Larva

Banded Woollybear Moth

Pyrrharctica isabella W 1.6–2.1 in (40–53 mm)

According to legend, the light-colored middle of the Banded Woollybear caterpillar can forecast the coming winter's severity, but the colors vary greatly and have no predictive value.

KEY FACTS

Forewings are pale orangish with brownish markings; hind wings are orangish (male) or rose (female), with spots sometimes obscure; abdomen has rows of spots.

+ **habitat:** Fields, pastures, bottomlands, woodlands

+ **range:** Widespread in U.S., Canada

+ **food:** Adults sip sugary bait; larvae eat forbs and shrubs.

The Banded Woollybear is also known as the Isabella Tiger Moth. The larvae, or woolly bears, are typically orange and black, but some are uniformly blond, brown, rust, or tan. Their stiff bristles are mostly uniform in length. Nearly mature caterpillars seek shelter in leaf litter or beneath objects on the ground to overwinter, and are often seen crossing roads and driveways in fall. They emerge the following spring to continue feeding, construct a cocoon, and pupate. Two generations are produced annually in most of the range, probably more in the Gulf Coast states. Adults are attracted to light.

Larva

Cicada Killer

Sphecius speciosus L 1.2–2 in (30–50 mm)

The female Cicada Killer hunts, captures, stings, and paralyzes cicadas as provisions in her subterranean brood chambers. The Cicada Killer seldom stings, but is a nuisance in yards and play areas.

KEY FACTS

This large and robust wasp has a reddish and black thorax, amber wings, black or brown abdomen with bright yellow markings, and orangish legs.

+ **habitat:** Gardens, parks, and forest edges

+ **range:** Widespread east of the Rockies

+ **food:** Adults sip nectar; larvae eat paralyzed cicadas.

In the heat of summer, the male Cicada Killer emerges first and quickly stakes out an area where females are likely to appear. He defends the territory, confronting rival males with vigorous bouts of head-butting, grappling, and biting. Newly emerged and receptive females are pursued and grasped from behind. The female hunts for cicadas and transports her paralyzed victims by alternately dragging them on the ground and gliding short distances to her burrow. Burrows in sandy soils in embankments, or along sidewalks and roads, may reach 4 feet (1.2 m) deep and consist of multiple branches that serve as brood chambers.

With cicada

Black-and-yellow Mud Dauber

Sceliphron caementarium L 1 in (25 mm)

The female Black-and-yellow Mud Dauber builds mud nests, especially under eaves and rafters in buildings and picnic shelters. She provisions nests with paralyzed spiders as food for larvae.

KEY FACTS

The body is black and yellow with a long, threadlike waist; dark wings folded flat over back; mostly black abdomen; and mostly yellow legs.

+ **habitat:** Gardens, parks, fields, and openings in wooded areas

+ **range:** Southern Canada and throughout the U.S.

+ **food:** Adults sip nectar; larvae eat spiders.

Hardly aggressive, the female Black-and-yellow Mud Dauber can defend herself with a mild sting. In late summer, she visits flowers and carries mud from puddles to build her nest. Larvae develop rapidly and pass the winter as grubs enclosed in a brown, parchment-like cocoon within a mud cell. Pupation takes place in spring, and adults emerge about three weeks later. Shiny metallic Blue Mud Dauber (*Chalybion californicum*) females renovate abandoned *Sceliphron* nests and provision them with paralyzed spiders for their brood. They visit puddles and carry water to soften and refashion the existing nest.

Nest cells with larvae, spiders

European Honey Bee

Apis mellifera L worker 0.4–0.7 in (9–18 mm), drone 0.6 in (15–16 mm), queen 0.7–0.8 in (18–20 mm)

The European Honey Bee, introduced into the New World by British colonists, is now nearly cosmopolitan. It is valued for its pollination services, especially by fruit and nut growers.

KEY FACTS

The workers of this species are golden brown and black; they have clear wings, an abdomen with pale yellow bands, and hind legs that are modified for carrying pollen.

+ **habitat:** Wherever flowers bloom

+ **range:** U.S. and southern Canada

+ **food:** Adults sip nectar and honey; larvae eat pollen and honey.

European Honey Bees live in colonies up to 80,000 individuals led by a queen. Defending the colony, workers use and then lose their barbed stinger, dying soon afterward. Besides pollination, the bee is highly prized for honey, wax, pollen, bee venom, and other products. Wild colonies establish hives in hollow trees, and under bridges, overhangs, and other protected places. Habitat destruction, pesticides, parasites, and changing weather patterns have negatively affected honey bees, seriously threatening the bees' ability to pollinate food crops for humans and domestic animals.

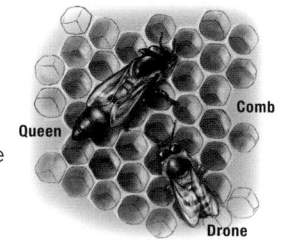

Queen　**Comb**　**Drone**

Yellow Bumble Bee

Bombus fervidus L worker 0.5–0.7 in (13–18 mm), drone 0.4–0.6 in (10–15 mm), queen 0.7–0.9 in (18–23 mm)

The Yellow Bumble Bee thrives in cold climates by vibrating its thoracic muscles to raise its body temperature. Unlike the European Honey Bee, workers retain their stinger and use it repeatedly.

KEY FACTS

Worker has a black head; yellow thorax, with a black band across rear half; smoky wings; and mostly yellow abdomen, with a black tip.

+ **habitat:** Grassy areas in gardens, parks, meadows, roadsides, forest openings

+ **range:** U.S. and southern Canada

+ **food:** Adults sip nectar and honey; larvae eat pollen and honey.

The Yellow Bumble Bee, also called the Golden Northern Bumble Bee, is social, living in colonies of 50 to 125 individuals led by a queen. Unlike those of the European Honey Bee, Yellow Bumble Bee colonies last only one year. Workers and drones die at the end of the season. Newly mated queens overwinter and emerge in spring to establish new colonies. They build nests of grass aboveground or in abandoned rodent burrows and other sheltered spaces underground. The nest floor contains potlike cells of wax containing honey or the brood. Like other bumble bees, this species is an important pollinator in decline and in need of conservation.

Nest

Eastern Carpenter Bee

Xylocopa virginica L 0.8–1 in (20–25 mm)

The female Eastern Carpenter Bee chews tunnels into dead limbs or unfinished wooden structures and partitions them into several brood cells, each provisioned with pollen and a single egg.

KEY FACTS

This large, solitary bee is shiny blue-black, with a head nearly as wide as the thorax, a black (female) or white (male) face, and short or absent abdominal setae.

+ **habitat:** Urban and rural areas, woodlands

+ **range:** New England to central Florida, west to Nebraska and Texas

+ **food:** Adults eat nectar; larvae eat pollen.

The adult Eastern Carpenter Bee flies in late spring and briefly in late summer. Instead of wood, the larvae consume pollen, and adults rely on nectar for nourishment. As they forage, carpenter bees mark each flower visited with a pheromone that lasts up to ten minutes. By skipping marked flowers, other bees can avoid those recently depleted of nectar. The dark-faced female is capable of delivering a defensive sting via a modified egg-laying tube, or ovipositor but seldom does so. The white-faced male cannot sting but can drive off rival males and intimidate humans who venture into his territory.

Male face

Cow Killer

Dasymutilla occidentalis L 0.6–1 in (15–25 mm)

Our most widely recognized velvet ant, the Cow Killer is a solitary wasp that does not kill cows. Fully winged males fly in search of wingless females scurrying on the ground.

KEY FACTS

Hairy and antlike, this species is bright red with black markings; males are winged with a black front half of abdomen, and females are wingless with a black band across the middle of abdomen.

+ **habitat:** Open areas and woodland edges

+ **range:** Eastern U.S.

+ **food:** Adults sip nectar; larvae eat bumblebee larvae and pupae.

The Cow Killer is an example of aposematic coloration, meaning its black-and-red body is a warning to potential predators. The female defends herself with an excruciating sting that is more painful than a European Honey Bee's. The stingless male benefits from a similarly bold pattern, which misleads his enemies into thinking that he, too, can sting. Larvae parasitize grubs and pupae of ground-nesting Bumble Bees and occasionally, Great Golden Digger Wasps. The female often lives more than 500 days in captivity, suggesting that she might overwinter in the wild as an adult and persist through almost an entire second season.

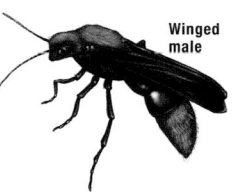

Winged male

Tarantula Hawks

Pepsis species L 1.7 in (43 mm)

Large, iridescent, and borne on bright reddish orange wings, Tarantula Hawks are a familiar sight in the West as they drink nectar from milkweeds and other flowers on hot days.

KEY FACTS

This robust wasp is black, sometimes with bluish or greenish reflections; it has straight (male) or curled (female) antennae and dark to bright reddish orange wings.

+ **habitat:** Open habitats in arid regions

+ **range:** Southern half of U.S.

+ **food:** Adults sip nectar and sap; larvae eat tarantulas and large spiders.

The hard, smooth body and quick reflexes of the female Tarantula Hawk give her an advantage over larger spider prey, which she paralyzes with stings and stuffs down burrows as food for developing wasp grubs. The incredible pain generated by the female's sting is most likely the evolutionary basis for mimicry by similarly colored yet unrelated insects such as the longhorn beetle (*Tragidion*) and a large fly (*Mydas luteipennis*). The genera *Pepsis* and *Hemipepsis* are similar and best distinguished by microscopic examination. They contain 19 species north of Mexico, all commonly referred to as tarantula hawks.

With tarantula

European Hornet

Vespa crabro L 0.7–1.5 in (18–38 mm)

The European Hornet prefers to build its enclosed paper nest in a tree hollow and is a nuisance when its nest is under porches and overhangs, or in outbuildings and basements.

KEY FACTS

This large, robust wasp has a mostly reddish brown head and thorax; narrow amber wings; and a reddish brown abdomen that is mostly black in front, yellow in rear.

+ habitat: Gardens, parks, woodlands

+ range: Established in southern Canada, eastern U.S.

+ food: Adults sip nectar and sap; larvae eat insects.

The European Hornet lives in colonies dominated by a fertile queen and sterile female workers. Though sometimes considered pests when they strip bark off young branches from apple trees and lilac bushes, these wasps are significant predators of insect pests in gardens, helping to control their populations. Although they can attack insects as large as cicadas and dragonflies, up to 80 percent of their diet consists of flies. Prey are chewed into a pulp and fed to larvae suspended upside down in honeycombed brood cells stacked in the nest. Active day and night, workers are sometimes attracted to light.

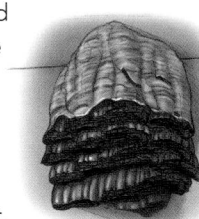

Brood cell layers inside nest

Paper Wasp

Polistes aurifer L 0.4–1 in (10–25 mm)

Multiple Paper Wasp queens establish a single nest, but one queen becomes dominant. She destroys the others' eggs and replaces them with her own, which the other queens will tend.

KEY FACTS

It has a mostly bright yellow face and abdomen; mostly black thorax with narrow yellow markings; smoky wings; and mostly yellow legs with brownish bases.

+ habitat: Gardens, parks, open and wooded areas

+ range: Western U.S. and Canada

+ food: Adults sip nectar, sap, and fruit juices; larvae eat chewed-up insects.

Overwintering queens emerge in spring and build nests under twigs and branches. They tend to select protected sites associated with buildings, especially under eaves and in attics. The nest consists of a layer of honeycombed brood cells, expanded as the colony grows to 50 to 200 workers at the height of summer. Insect hunting by foraging workers is considered beneficial. Northern populations are generally black, boldly marked with yellow; those in the south have a largely yellow abdomen or are mostly reddish brown with moderate to strong yellow patterns. There are 24 *Polistes* species in Canada and the U.S.

Nest

Potter Wasp

Eumenes fraternus L 0.4–0.75 in (9–19 mm)

The female Potter Wasp builds a marble-size mud nest with a spout. The nest consists of an internal brood chamber containing one egg provisioned with paralyzed moth caterpillars.

KEY FACTS

This wasp is black with white markings; the male has hooked antennae, clawlike at the tips; dark wings; and a slender first abdominal segment that is as long as the thorax.

+ **habitat:** Open areas and woodlands

+ **range:** Eastern U.S. and Canada

+ **food:** Adults sip nectar; larvae eat caterpillars.

The female Potter Wasp attaches her nest singly or in groups of two to five on rocks, stems, and tree trunks. She will also use the surfaces of picnic tables, railings, and other structures made of wood. Once the nest is completed, she inserts the tip of her abdomen into its narrow opening and lays a single egg inside, then flies off in search of spring cankerworms and parsnip webworms. After locating and stinging the caterpillars, she stuffs their paralyzed bodies through the funnel-shaped spout and seals the opening with mud. Eight species of *Eumenes* occur in Canada and the United States.

Pot nests

Red Harvester Ant

Pogonomyrmex barbatus L worker 0.2–0.5 in (6–13 mm)

The genus *Pogonomyrmex* is derived from two Greek words meaning "bearded ant." The beard-like structure under the head of many "pogos" helps the workers to carry seeds, eggs, and sand.

KEY FACTS

Uniformly light to dark red, this ant has a broad, blocky head; hind tibiae with finely comblike spurs; and an abdomen with threadlike first segments.

+ **habitat:** Arid deserts and scrublands

+ **range:** Southern Nevada and Arizona, east to Kansas, Arkansas, and Louisiana

+ **food:** Seeds and insects

Colonies of Red Harvester Ants have one queen that lives for 15 to 20 years and may include up to 12,000 workers. Workers clear large areas around craterlike mounds of pebbles surrounding the nest entrance. They defend the colony from intruders by inflicting venomous and painful stings. Foraging workers return from the field with seeds and dead insects to stock their food stores. Genomic research conducted on *Pogonomyrmex* has shed light on regulatory mechanisms that govern the division of labor among reproductive and worker castes. There are 25 species of *Pogonomyrmex* north of Mexico.

Close-up carrying sand particle

Little Black Ant

Monomorium minimum L worker 0.04–0.08 (1–2 mm); queen 0.16–0.2 in (4–5 mm)

The Little Black Ant invades houses in small numbers to search for sweets and greasy foods. Its foraging columns are conspicuous across kitchen walls, counters, cabinets, and floors.

KEY FACTS

Small, shiny reddish brown to jet-black, this ant is similar to *Solenopsis*, but it has 11-segmented antennae with a 3-segmented club and a slender "waist" with two nodes.

+ **habitat:** Disturbed areas

+ **range:** Throughout much of U.S. and Canada

+ **food:** Honeydew, nectar, pollen, insects

Highly adaptive, the Little Black Ant builds its nest in open areas in lawns with or without small craters of fine soil, or under rocks and in decayed wood. It will also establish nests in woodwork and masonry. Colonies consist of just one form of worker (no soldiers) and many queens. For most of the year, the colony's inhabitants consist primarily of workers and the brood. Winged males and females are produced in the summer. The males die soon after mating, and the queens shed their wings and seek a site to start a new colony. Workers hunt for insects, and they tend aphids to obtain honeydew.

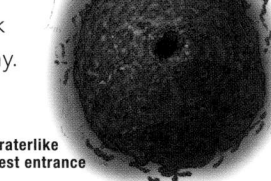

Craterlike nest entrance

Red Imported Fire Ant

Solenopsis invicta L worker 0.08–0.2 in (2–6 mm)

The Red Imported Fire Ant delivers venomous stings that produce a burning sensation and a red, fluid-filled pustule. Multiple stings can seriously injure or kill caged animals and nestlings.

KEY FACTS

This ant is smooth, shiny, reddish brown to black; it has 10-segmented antennae with a 2-segmented club, and a "waist" between thorax and abdomen with 2 nodes.

+ **habitat:** Developed areas

+ **range:** Virginia to Florida, west to New Mexico, southern California

+ **food:** Plant tissue, honeydew, insects, and small vertebrates

First reported from Alabama in the 1930s, this ant is a South American import that has spread throughout the southeastern United States. Infestations are reported from California, Australia, and Asia. They live in colonies with millions of individuals and one or more queens that produce as many as 200 eggs a day. Fire ants are serious economic pests that cause millions of dollars of damage to crops and livestock. Efforts to control Red Imported Fire Ants with pesticides have severely affected native ants and other wildlife. There are 40 mostly native species of *Solenopsis* in the United States.

Nest mound

Argentine Ant

Linepithema humile L 0.08–0.1 in (2–3 mm)

The Argentine Ant is native to South America, and through human activities is now established on six continents and many oceanic islands, where it often displaces native ant species.

KEY FACTS

A small ant, this species is light to dark brown and has 10-segmented antennae with 2-segmented club at tips, and a narrow waist with 1 distinct knob.

+ **habitat:** Urban and suburban areas, agricultural fields, and other disturbed areas

+ **range:** Across the U.S.

+ **food:** Honeydew and dead insects

The tiny Argentine Ant is a persistent pest adapted for living in small spaces, including cracks in walls and other gaps in buildings. It tends aphids and their sap-sucking relatives to obtain honeydew and will protect them from predators and parasitoids, hampering biological control efforts that target the ants. Unlike other species, Argentine Ants from different colonies seldom attack one another. Instead, colonies of closely related individuals may form a giant "supercolony." Supercolonies have been found around the world, including one stretching 560 miles (900 km) along the California coast.

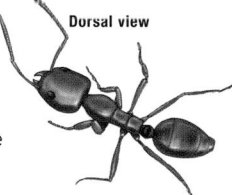

Dorsal view

Desert Blond Tarantula

Aphonopelma chalcodes L to 2.8 in (70 mm)

The Desert Blond Tarantula male searches on overcast days and evenings for females in vertical, silk-lined burrows. They deter predators with abdominal hairs that cause severe itching.

KEY FACTS

The male is mostly black with a brownish gray carapace and rust-colored hairs on its abdomen; the female is larger, heavier, and paler.

+ **habitat:** Deserts and mesquite woodlands

+ **range:** Arizona, New Mexico, and adjacent Mexico

+ **food:** Honeydew and dead insects

The Desert Blond Tarantula breeds in summer and generally avoids high daytime temperatures by remaining in its silk-lined burrow. The entrance to the burrow is often covered with a thin veil of silk or plugged with soil. The mature female seldom ventures far from the burrow. When threatened, the tarantula uses its legs to brush hairs off the abdomen that cause itchy rashes and irritate the eyes, nose, and mouth. In captivity, *Aphonopelma* females are known to live 30 years or more, but are unlikely to survive this long in the wild. More than 50 species of *Aphonopelma* occur mostly west of the Mississippi River.

Close-up of fangs

Brown Recluse Spider

Loxosceles reclusa L 0.3–0.4 in (7–10 mm)

The Brown Recluse Spider and other *Loxosceles* species are notorious for their potent venom. The bite from these spiders causes lesions that are slow to heal, but seldom results in death.

KEY FACTS

This small spider is brown with 3 pairs of eyes that form a strong arc across its face; a carapace with faint or distinct "fiddle" marking; and legs with 2 claws.

+ **habitat:** Natural and building crevices

+ **range:** Native to the Mississippi River Basin and adjacent areas

+ **food:** Insects and spiders

The Brown Recluse, or Violin Spider, lines its crevice with sticky silk that extends beyond the entrance to ensnare insect prey. One generation is produced annually, and adults may live three years in captivity. The bite is often painless, but other symptoms begin to appear in six to eight hours. The wound may eventually become necrotic and is prone to secondary infections. Most native species of *Loxosceles* live in the arid Southwest. The South American *L. laeta* is established in a few localities on the East and West Coasts, and the European *L. rufescens* occurs throughout most of the South and New York.

Fiddle pattern on cephalothorax

Long-bodied Cellar Spider

Pholcus phalangioides L 0.2–0.4 in (6–9 mm)

In urban legend, the Long-bodied Cellar Spider is highly venomous, but it is unable to bite humans. Its small, weak fangs cannot puncture human skin, and the venom is not very potent.

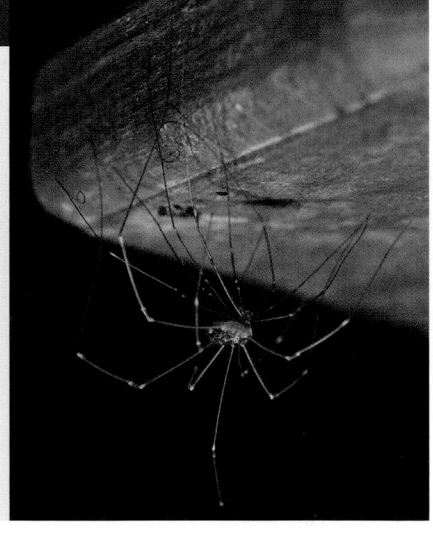

KEY FACTS

The distinct abdomen of this pale yellow spider is twice as long as wide; it also has 8 eyes; a carapace that is gray on the center; and long legs without rings.

+ **habitat:** Caves and rock overhangs, also dark basements and cellars

+ **range:** Widespread in the U.S. and Europe

+ **food:** Insects and other spiders

The Long-bodied Cellar Spider, also called the daddy longlegs, is the most common cellar spider in the United States. They prefer to live in dark, undisturbed places and hang upside down in tangled webs that lack any apparent design. When disturbed, they make themselves difficult to see by rapidly vibrating back and forth. Females encase their eggs in an incredibly thin silk sac and carry it in their mouthparts until they hatch. The Marbled Cellar Spider (*Holocnemus pluchei*), an Old World species now established throughout the United States, is similar in form, but its legs have distinct white rings.

Female tending egg sac

Common House Spider
Parasteatoda tepidariorum L 0.1–0.2 in (3–6 mm)

The Common House Spider is widespread and well known. The female prefers to spin irregular webs in corners. The webs contain one or more egg cases wrapped in coarse, parchment-like silk.

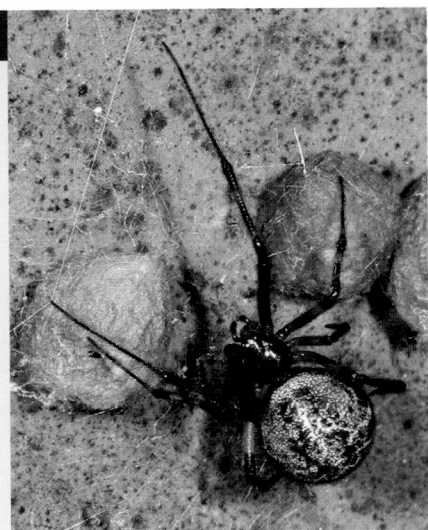

KEY FACTS

This spider is gray or tan, with a yellow-ish brown carapace; a whitish to brown abdomen with indistinct, gray chevrons; and banded yellowish (female) or orange (male) legs.

+ habitat: Homes, outbuildings, fences, bridges, rock faces, under stones

+ range: Southern Canada and the U.S.

+ food: Insects

The Common House Spider, also called the American House Spider, was previously included in the genus *Achaearanea*. Adults may live a year or more after reaching maturity. Males and slightly larger, heavier-bodied females occur year-round in association with human habitations. Each tangled "cobweb" is typically occupied by a female suspended upside down, with brown, pear-shaped egg sacs, flies, mosquitoes, ants, and other prey encased mummy-like in silk. Old prey items wrapped in silk eventually drop from the web. The spider will hide behind debris or drop from the web on a silk line when threatened.

Egg sac

Southern Black Widow
Latrodectus mactans L male 0.1–0.2 in (3–6 mm); female 0.3–0.4 in (8–10 mm)

A Southern Black Widow will usually inflict its neurotoxic bite only when accidentally trapped in clothing or when its web is disturbed. The bite requires immediate medical attention.

KEY FACTS

Mature female is shiny black with a distinctive and usually complete reddish hourglass marking underneath globose abdomen; male and immatures are smaller and variably marked.

+ habitat: Rock walls, woodpiles, meter boxes, and animal burrows

+ range: Eastern U.S.

+ food: Crawling insects and other arthropods

Notorious for her venomous bite, the female Southern Black Widow seldom leaves her web and is reluctant to attack humans unless threatened. Her messy web is constructed of strong and sticky silk, including runners to the ground that snag crawling insects and other arthropods. Entangled prey are hauled up into the web, killed, and eaten. The eggs are enveloped in a whitish to beige pear-shaped egg sac made from parchment-like silk, suspended in the web and sometimes guarded by the mother. The much smaller male has white lines along the sides of the abdomen. It is usually found in or near the web.

Male

Golden Silk Orb Weaver

Nephila clavipes L male 0.2–0.3 in (4–8 mm); female 0.9–1 in (22–25 mm)

The silk spokes radiating from the center of the Golden Silk Orb Weaver's sticky golden web appear notched, not straight. The silk has a tensile strength nearly equal to that of steel.

KEY FACTS

The abdomen of this colorful orb weaver species is long, yellowish, orangish, or greenish with pairs of white spots; the female has tufts on all except her third pair of legs.

+ **habitat:** Swamps, shaded hardwood, and palm forests

+ **range:** Southeastern U.S.

+ **food:** Flying insects

The Golden Silk Orb Weaver, also called calico spider or banana spider, lives in warm, humid regions, especially along the coasts. Females construct huge and slightly tilted webs that stretch up to three 3 feet (1 m) across. Freshly produced silk is sticky and yellow but soon fades to white as it dries out. The tensile strength of the silk is much greater than steel's per unit of weight. The female spider hangs head down on the lower side of the web and waits for flying insects to become entangled. The relatively tiny male lives at the margins of the web and sometimes clings to the underside of the female's abdomen.

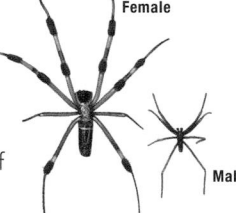

Female

Male

Yellow Garden Spider

Argiope aurantia L male 0.2–0.3 in (5–8 mm); female 0.7–1.1 in (19–28 mm)

The adult female Yellow Garden Spider, large and conspicuous, typically suspends herself head down at the center of her ornate orb web. The smaller, paler male lurks at the web's fringes.

KEY FACTS

The large female has a silvery carapace; her abdomen is boldly marked in black and yellow; her legs are black with variable orange, red, and yellow markings.

+ **habitat:** Gardens, parks, and riparian woodlands

+ **range:** Throughout the U.S. and in southern Canada

+ **food:** Flying insects

The intricate orb web of the Yellow Garden Spider consists of a spiral of sticky silk laid on a platform of relatively dry and rigid spokes radiating out from the center. Incorporated into the web is a zigzag ribbon of silk called the stabilimentum, the purpose of which is still under investigation. Among suggestions are that it makes the web easier to see and keeps other animals from inadvertently destroying it, helps to conceal the spider in the web, or has reflective qualities that attract insects. The marble-size egg sacs are constructed from tough, papery brown silk and suspended in the web.

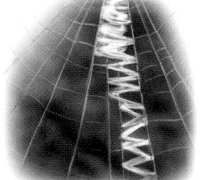

Stabilimentum

Spiny-backed Orb Weaver

Gasteracantha cancriformis L male 0.08–0.1 in (2–3 mm); female 0.3–0.4 in (8–10 mm)

The female Spiny-backed Orb Weaver has a brightly colored abdomen conspicuously ringed with six stout reddish or black spines and is more commonly seen than the smaller spineless male.

KEY FACTS

This orb weaver has a dark brown to brownish black carapace and legs; a red, white, orange, or yellow hard abdomen; and a black underside speckled with yellow.

+ **habitat:** Scrubby and wooded areas

+ **range:** Southeastern Virginia to Florida, west to southern California

+ **food:** Flying insects

Also known as the star spider or jewel box spider, the Spiny-backed Orb Weaver is generally a summer resident in most of its range but is commonly seen year-round in southern Florida and Texas. It builds its orb web between 3 and 6 feet (1 to 1.8 m) off the ground in open woods and woodland edges. The female generally rests head down in the center of the web, which may span 16 inches (41 cm) or more across. The spine and bright colors are thought to discourage birds from eating it. It attaches its yellow silken egg sac, which has a green stripe running along its entire length, to the surface of a nearby leaf.

Different color form

Giant Wolf Spider

Hogna carolinensis L male 0.7–0.8 in (18–20 mm); female 0.9–1.4 in (22–35 mm)

The Giant Wolf Spider, the largest of its kind in the U.S., builds chimneylike entrances to its burrow lined with a thin sheet of silk and festooned with bits of vegetation.

KEY FACTS

This spider has a dark brown carapace (lighter in male) with gray hairs and a pale stripe; a similar abdomen, with a darker stripe; and a black underside and leg bases.

+ **habitat:** Sandy soils in sand hills, grassy fields, prairies, hillsides, and along coastlines

+ **range:** Throughout the U.S.

+ **food:** Flying insects

The Giant Wolf Spider is the largest species of wolf spider in the United States. It is a nocturnal hunter that preys on crickets, grasshoppers, and other insects. Its days are spent hiding in a burrow that it may or may not construct itself. Its own burrow is usually inclined past the entrance and then becomes more winding. Mating takes place in late summer, and the female is soon seen with a round egg sac carried under her abdomen. Up to 100 spiders may emerge from a single egg sac and crawl up onto the mother's back, where they remain for about two weeks before venturing out on their own.

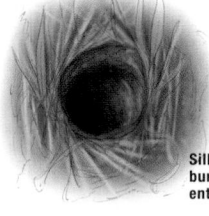

Silk-lined burrow entrance

Six-spotted Fishing Spider

Dolomedes triton L male 0.4–0.5 in (9–13 mm); female 0.7–0.8 in (17–20 mm)

The Six-spotted Fishing Spider rests head down on emergent aquatic vegetation, ready to pounce on an insect on the surface, and will dive underwater to capture prey or avoid predators.

KEY FACTS

This species is greenish to tan, with short bristles; it has lighter margins on the carapace, an abdomen with pairs of white spots, a pale underside, and water-repellent hairs.

+ habitat: Edges of ponds, lakes, and stream pools

+ range: U.S. and Canada, more common in the East

+ food: Insects

The Six-spotted Fishing Spider prefers standing-water habitats with herbaceous vegetation near the surface. It sometimes rests with its front legs on the water and hind legs on a stem. Adults will dive underwater to capture small fish, and young spiders behave similarly to capture mosquito larvae. Fishing spiders produce their egg sacs in spring and summer, and like other species in the family Pisauridae place them in sheets of silk called nursery webs. The female constructs a nursery web above the water surface and remains nearby to protect her young. There are seven species of *Dolomedes* in the United States and Canada.

Female with egg sac

Green Lynx Spider

Peucetia viridans L male 0.47–0.51 in (12–13 mm); female 0.55–0.6 in (14–16 mm)

The Green Lynx Spider is adapted for hunting insect prey on flowers and among bright green vegetation of low shrubs and other herbaceous growth. It is fleet of foot and an able jumper.

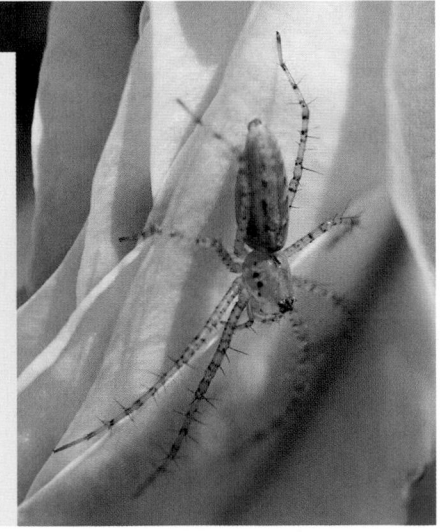

KEY FACTS

This bright green spider has a tapered abdomen with cream-colored chevrons edged reddish brown; pale green or yellow legs with black spines and spots.

+ habitat: Parks, agricultural fields, thorn scrub

+ range: Virginia to Florida, west to California

+ food: Insects, especially bees, wasps, caterpillars

In addition to its color, the Green Lynx Spider is usually distinguished by having reddish spots near the eyes and elsewhere on the body. The spiderlings hatch in late spring and reach maturity in summer. This spider pounces on prey and envelops it with long, spiny legs. In fall, the smaller, lighter-bodied male seeks a larger, heavier-bodied female. The female lays up to 600 bright orange eggs in a yellowish or brownish sac and attaches the sac to an upper branch of vegetation and shrubs. She tenaciously guards the egg sac until the spiderlings hatch, and will spit venom from her fangs to defend their brood.

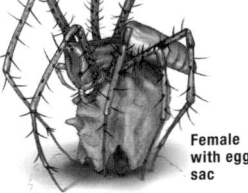

Female with egg sac

White-banded Crab Spider

Misumenoides formosipes L male 0.08–0.2 in (2–4 mm); female 0.2–0.4 in (5–11 mm)

The White-banded Crab Spider quietly lies in wait among flowers to ambush bees and flies. Over time, the female has a limited ability to change her color to match the background better.

KEY FACTS

This spider can be yellow, yellowish brown, creamy white, or pinkish, with darker carapace sides; prominent white ridges above eyes; and plain or variously marked abdomen.

+ **habitat:** Gardens, parks, woodland edges, other open areas

+ **range:** Throughout Canada and U.S.

+ **food:** Flower-visiting insects

The crablike appearance of the White-banded Crab Spider is enhanced by its enlarged first two pairs of legs and ability to run forward, backward, and sideways with equal rapidity. "White-banded" refers to the raised ridge over its eyes. The White-banded Crab Spider does not build webs or retreats. It lies in ambush among flowers, especially composites such as goldenrods, asters, and sunflowers. Young spiders mature in midsummer. The male is smaller, more slender, does not change color, and actively searches for the sedentary female. Up to 100 eggs are deposited in white silken sacs attached to vegetation.

Yellow form

Bold Jumping Spider

Phidippus audax L male 0.2–0.5 in (6–13 mm); female 0.3–0.6 in (8–15 mm)

Robust and brash, the Bold Jumping Spider attracts attention when encountered in the home. This common spider is typically found outdoors under boards, on tree trunks, and on buildings.

KEY FACTS

This black and hairy species has an abdomen with a white band at the base, followed by a variable central white spot and a pair of white spots at the rear; immatures have yellow or orange spots.

+ **habitat:** Developed and natural locations

+ **range:** Southern Canada and the U.S.

+ **food:** Insects

The Bold Jumping Spider is especially common in the eastern and midwestern U.S. Southern populations may have two white bands on the carapace that extend back from the eyes. The male is smaller than the female and has white-and-black bands on the front legs and metallic green mouthparts. This species is sometimes common in agricultural fields, preying on insect pests. The young build silk-lined retreats in curled leaves, and up to 30 may gather to overwinter in protected places, such as an old log. They emerge and reach maturity in spring, mate, and lay orange eggs in a lens-shaped sac hidden under bark in summer.

Metallic fang bases of male

Brown Harvestman

Phalangium opilio L 0.16–0.24 in (4–6 mm)

The Brown Harvestman is distinguished from other spiders by the broad fusion of its frontal body and segmented abdomen, giving it the appearance of having a single body region.

KEY FACTS

Reddish brown, this species has a segmented body with a broad, dark stripe down the middle and eight dark legs with pale bases; males have large, pincherlike mouthparts.

+ **habitat:** Open and wooded areas

+ **range:** Throughout U.S. and Canada

+ **food:** Small arthropods and decaying vegetation

The Brown Harvestman, also known as daddy longlegs, is found resting on leaves and tree trunks in spring and summer. On hot days, harvestmen sometimes congregate on the shady sides of homes and outbuildings. Unlike spiders, harvestmen engage in direct copulation. With an extendable egg-laying tube or ovipositor, the female will lay several hundred eggs at a time in moist soil, mosses, and rotten wood. The Brown Harvestman uses its long and slender legs for walking, breathing, smelling, and capturing prey. Its second pair of legs is used primarily as a sense organ. It is not venomous and is harmless to people and pets.

Close-up of body

Arizona Desert Hairy Scorpion

Hadrurus arizonensis L to 5.5 in (14 cm)

The Arizona Desert Hairy Scorpion is the largest scorpion species in the U.S. It is adapted to hot, dry climates. Despite its size, its sting is not life-threatening.

KEY FACTS

This scorpion is large and hairy, with crablike claws; a long, bristly, stinger-tipped tail; and a yellowish body, with dark back and pale between the eyes.

+ **habitat:** Rocky deserts and arid foothills

+ **range:** Southwestern Utah and Arizona to southern Nevada and California

+ **food:** Insects, arachnids, other invertebrates

The Arizona Desert Hairy Scorpion is a nocturnal predator that spends its days under boards, rocks, and other debris on the ground, or in a deep burrow. Although venomous, the sting causes only local pain similar to that of the European Honey Bee. Potential mates grasp each other with their claws and pull back and forth in a mating "dance" called a *promenade à deux*. The male deposits a sperm packet on the ground, which the female picks up with her reproductive organs. Young scorpions are born alive and crawl onto their mother's back for about two weeks. There are four species of *Hadrurus* in their region.

Fluorescent under black light

Windscorpions

Eremobates species L 0.6–1.3 in (15–32 mm)

Despite their lightning speed and fearsome appearance, Pale Windscorpions are not venomous, and bites only if handled. They run rapidly in a zigzag pattern searching for prey at night.

KEY FACTS

This fast runner is brownish yellow, clothed in stiff setae, with pincherlike mouthpart, segmented abdomen, and eight legs, including antenna-like forelegs; male smaller than female.

+ habitat: Arid scrub and deserts

+ range: California and Idaho to North Dakota and Texas

+ food: Insects, spiders, other arthropods

Pale Windscorpions have enlarged mouthparts, or chelicerae, that project forward and are used to capture and cut up prey. A pair of eyes tops the small carapace that covers the head and frontal part of the body. Its long forelegs are not used for locomotion but as sensory organs. Windscorpions mate face-to-face as the male passes a sperm packet to the female. Up to 50 eggs are laid in a burrow, but the female will not guard or care for them. Young windscorpions, which are not true scorpions, resemble adults and begin hunting soon after they hatch. Windscorpions are sometimes called camel spiders in the Middle East.

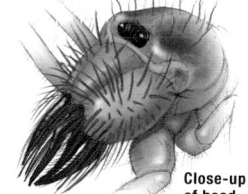

Close-up of head

Giant Whipscorpion

Mastigoproctus giganteus L to 3 in (75 mm)

True to its name, the Giant Whipscorpion is the largest of its kind in the world. Also known as a vinegaroon, it can accurately spray a defensive, vinegar-like solution from its anus.

KEY FACTS

This brown and black species has clawlike mouthparts; front of body covered by carapace; long, whiplike flagellum on abdomen; and eight legs, including antenna-like forelegs.

+ habitat: Thorn scrub, oak and pine forest

+ range: Arizona to Florida

+ food: Termites, cockroaches, and crickets

The Giant Whipscorpion comes out from under leaf litter, logs, and rocks on warm summer nights to hunt for prey. After a long and complex courtship dance, the female buries herself in a subterranean chamber and lays up to 30 large, yellowish eggs in a membranous sac carried beneath the body. The young hatch, ride on their mother's back for several molts, and take two to three years to mature. The Whipscorpion, which is not a true scorpion, is not venomous. It defends itself by raising its abdomen and whipping its flagellum back and forth while releasing an 84 percent solution of acetic acid from its turretlike anus.

Close-up of head

Chigger

Eutrombicula alfreddugesi L 0.04 in (1 mm)

The Chigger does not burrow into human skin, but instead feeds by piercing the skin near the openings of hair follicles, especially in constricted areas around the ankles, waist, and crotch.

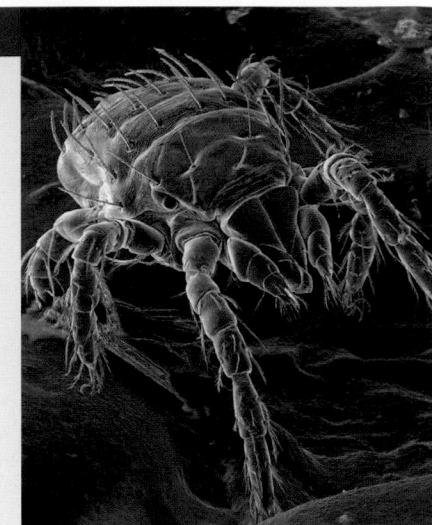

KEY FACTS

Larvae are red to reddish orange with scattered setae; nymphs and adults are bright red and hairy, and body is constricted behind a second pair of legs.

+ **habitat:** Fields and grasslands, especially in humid regions

+ **range:** Canada to Argentina

+ **food:** Adults and larvae eat vertebrate lymph and cell tissue.

Only the larval stage of the Chigger is parasitic, latching onto amphibians, reptiles, birds, and mammals to feed on partially digested skin cells and lymph. Humans are sometimes infested with large numbers of Chiggers. The larvae soon leave, but not before their bites develop into itchy lesions that may take several days to heal, or longer if they become infected by persistent scratching. Although it causes dermatitis in humans, the Chigger is not a vector of disease. The larvae are common in late summer and early fall in temperate climates, but they are present year-round in southern Florida.

Chigger bites

Giant Red Velvet Mites

Dinothrombium species L to 0.4 in (10 mm)

Pea-size Giant Red Velvet Mites are huge in the mite world. Massive emergences triggered by summer thundershowers carpet the desert floor with crimson and are visible from airplanes.

KEY FACTS

This mite is large, fuzzy, bright red or orangish, sometimes with whitish or creamy markings. Its wrinkled body is widest just behind the head, with a waistlike constriction at the middle and eight short legs.

+ **habitat:** Deserts

+ **range:** Southern California to Texas

+ **food:** Termites

Giant Red Velvet Mites spend most of their life in silk-lined subterranean chambers in a diapause-like state waiting for rain. Their annual emergence is timed to coincide with that of its primary prey, the termite. The opportunity to gorge lasts only hours because termites quickly shed their wings and bury themselves after mating. The thick, bright coat of red hairlike setae apparently serves as a warning to would-be predators. When offered these mites, insect-eating animals either ignore them or quickly spit them out. At least two species of *Dinothrombium* occur in the desert regions of the Southwest.

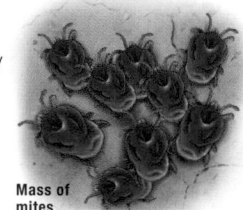

Mass of mites

Lone Star Tick

Amblyomma americanum L to 0.2 in (5 mm) (unengorged)

The Lone Star Tick is an important vector of ehrlichiosis and anaplasmosis, but it is unlikely to spread Lyme disease. The tick's name refers to the female's distinct spot on the back.

KEY FACTS

This tick has eyes present; long mouthparts; ornately patterned shield (scutum); turquoise spot (females); and posterior margined with 11 rectangles (festoons).

+ **habitat:** Wooded areas with thick underbrush

+ **range:** New England to Florida, west to Iowa and Texas

+ **food:** Blood of mammals and birds

Overwintering nymphs and adults of the Lone Star Tick emerge in spring to feed; the larvae generally appear in late summer. Unlike species of *Dermacentor*, this "three-host tick" bites humans, pets, and wildlife in the larval, nymphal, and adult stages. In fact, any mammal and ground-feeding bird is a potential host for this infamous pest. Lone Star Ticks are vectors of Rocky Mountain spotted fever and tularemia. Recent research has demonstrated that the bite of this tick in southeastern United States can lead to the development of an allergy to beef, pork, and lamb in humans that is known as alpha-gal.

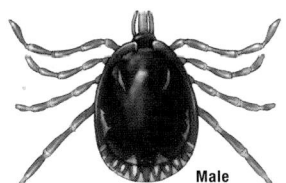

Male

Rocky Mountain Wood Tick

Dermacentor andersoni L to 0.2 in (5 mm) (unengorged)

The Rocky Mountain Wood Tick and its eastern counterpart, the American Dog Tick, are the primary vectors of Rocky Mountain spotted fever. Toxins in their saliva can cause tick paralysis.

KEY FACTS

Eyes are present; an ornately patterned shield (scutum); margined posterior with 11 rectangles (festoons); and a weak (male) or strong (female) body pattern.

+ **habitat:** Open habitats with low, brushy vegetation

+ **range:** Western U.S. and Canada

+ **food:** Blood of wild and domestic animals, and humans

After mating on the host, the blood-engorged female Rocky Mountain Wood Tick drops to the ground and produces a mass of more than 6,000 eggs. The six-legged larvae hatch in just over a month and begin searching for a suitable host. Larval ticks attack rodents, and eight-legged nymphs and adults parasitize almost any medium- to large-size mammals. Because three different hosts are required to complete the life cycle, this is a "three-host tick." The life cycle may take three months, but may extend two or more years in harsh conditions. This species also transmits anaplasmosis in cattle, goats, and sheep.

Gravid female

Black-legged Tick

Ixodes scapularis L to 0.1 in (3 mm) (unengorged)

Black-legged Ticks, or Deer Ticks, are active in spring and fall. This species and their western counterparts, the Western Black-legged Ticks, are the primary vectors of Lyme disease.

KEY FACTS

Eyes are absent; shield (scutum) with no ornate pattern; posterior margin without rectangles (festoons); and body with no pattern. The male is smaller than the female.

+ **habitat:** Hardwood and mixed forests

+ **range:** Eastern U.S. and Canada

+ **food:** Blood of rodents, deer, and humans

The Black-legged Tick has a two-year life cycle in three stages of development: larva, nymph, and adult. A blood meal is required before development into the next stage. After finding a host, the tick will feed for four or five days until its blue-gray abdomen is engorged with blood. Immature stages feed on small mammals, birds, and lizards. Adults are most common on white-tailed deer. Although all stages will bite humans, the nymphal stage is most likely to transmit Lyme disease in late spring and early summer. This tick is also the primary vector for human babesiosis and human granulocytic ehrlichiosis.

Bull's-eye rash

House Centipede

Scutigera coleoptrata L to 1.4 in (35 mm)

House Centipedes are often trapped in basins and tubs. They seldom bite and cause only temporary local pain. They hide outdoors under rocks, logs, and other moist, protected areas.

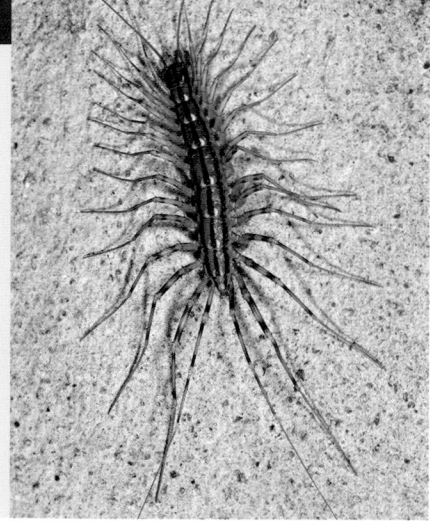

KEY FACTS

This fast, flattened, leggy creature is pale with dark stripes, a large capsule-shaped head, bulging compound eyes, and 15 pairs of long, banded legs.

+ **habitat:** Various outdoor areas; also in homes, schools, and office buildings

+ **range:** Mediterranean region; established throughout North America

+ **food:** Insects

The House Centipede's speed is partly due to a modified respiratory system that is intermediate between the tracheal system of other centipedes and the lungs of spiders. It often darts across floors and on walls in homes, and should be considered beneficial because it eats silverfish, cockroaches, clothes moths, and other household insect pests. High numbers in buildings are an indication of insect infestations. The House Centipede is often seen outdoors on humid summer nights hunting for insects attracted to lights. The female lives several years and can produce more than 100 offspring in her lifetime.

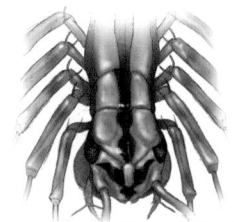

Close-up of head

Giant Centipede

Scolopendra heros L to 6.5 in (16.5 cm)

The largest centipede in North America, the Giant Centipede is variable in color and is also known as the Texas Red-headed Centipede and Black-tailed Centipede.

KEY FACTS

This very large centipede has four simple eyes at the base of each antenna, breathing holes with three-part covers, and 21 pairs of legs.

+ habitat: Deserts, thorn scrub, oak woodlands

+ range: Kansas, Missouri, Arkansas, and Louisiana, to southern Arizona

+ food: Arthropods and occasionally small vertebrates

The Giant Centipede is primarily nocturnal, but is sometimes seen during the day, especially in the morning after heavy rains. The female lays clutches of 15 to 35 eggs and wraps her body around them until they hatch. Individuals in Arizona and Texas are yellowish, or yellowish with bluish black bands. Dark posterior segments in some populations are thought to be a "pseudo-head" that directs predators to attack a less vital part of the body, leaving the head free to bite the attacker with modified front legs called gnathopods. The bite of this species is very painful, but it is not considered life-threatening.

Different color form

Xystodesmid Millipede

Apheloria virginiensis L to 2.2 in (57 mm)

Apheloria virginiensis has five subspecies, and like other brightly colored millipedes it produces a fruity-smelling defensive secretion containing benzaldehyde and hydrogen cyanide.

KEY FACTS

The body trunk of this millipede has 18 to 22 somewhat flattened segments, each of which are colored black, with pink, red, and yellow markings and bands.

+ habitat: Moist woodlands under rocks, logs, and leaf litter

+ range: Southern Quebec to Virginia, Indiana and Missouri

+ food: Arthropods and occasionally small vertebrates

Although not lethal to humans, the toxic secretions of adult *Apheloria* millipedes are highly repellent to small predators, especially ants. The Appalachian Mountains are home to complexes of about 100 similarly colored species. Their bright, contrasting patterns are aposematic, meaning that they boldly announce their unpalatability to predators. The complexes, called mimicry rings, are similar to those of some butterflies. Identification requires examination of the male's reproductive organs. Millipedes help to speed up the decomposition process, but are poorly studied because they are seldom pests.

Close-up of body

Narceus Millipede
Narceus species L to 5 in (12.6 cm)

The largest millipedes in the eastern United States, Narceus Millipedes are often seen on summer nights crawling over litter or climbing up tree trunks, boulders, and cliff faces.

KEY FACTS

It is thick and cylindrical, with 53-plus dark and pinkish or reddish edged segments, a segment behind the head with legs, and reddish or pinkish legs.

+ **habitat:** Moist woodlands, occasionally buildings

+ **range:** New England and southern Ontario to Florida, west to Minnesota and Texas

+ **food:** Decaying and living plant materials

This species consists of multiple species called the "*Narceus americanus/annularis* complex." Their days and winters are spent in soil or under rotting logs. Each egg is laid in a cup made from regurgitated food. Despite noxious defensive chemicals secreted along the sides of the body, the Narceus Millipede falls prey to beetle larvae known as glowworms. The larvae coil their body around the front of the millipede, deliver a paralyzing bite that liquefies the internal tissues, and then imbibe the millipede's body contents. The other U.S. species, *N. gordanus*, occurs in northern Florida.

Close-up of head

Desert Millipede
Orthoporus ornatus L to 5 in (12.6 cm)

Sometimes called rainworm, the Desert Millipede emerges in huge numbers after heavy summer downpours and feeds on the bark of joint firs, mesquites, Cholla cacti, and other desert plants.

KEY FACTS

Long and cylindrical, this millipede is uniformly shiny reddish brown or yellowish, or banded in one or both colors.

+ **habitat:** Sonoran and Chihuahuan Deserts, southern Great Plains

+ **range:** Arizona, New Mexico, western Texas, and Oklahoma

+ **food:** Bark, dead plant tissues, live shoots, animal feces

Seldom seen until the arrival of the summer monsoons, the Desert Millipede spends much of its time under rocks and debris or down in rodent burrows. This large and sometimes numerous detritivore spends its brief time on the surface feeding and looking for a mate. The female buries batches of up to 100 or more eggs, each coated with feces, in the soil. Although harmless, this millipede can defend itself. When threatened, it coils its body around its head and legs and secretes a dark, foul-smelling fluid from pores on the sides of its body. Bleached remains of dead millipedes are a common sight.

Defensive position

Common Pillbug

Armadillidium vulgare L to 0.7 in (18 mm)

To defend itself, a Common Pillbug rolls itself into a tight ball, its head covered by the tip of its abdomen. It is sometimes considered a pest in greenhouses and mushroom farms.

KEY FACTS

The elongate-oval Pill-bug is straight, gray or black, and sometimes nearly yellow, with carapace of arched, smooth, overlapping plates. The Pillbug has seven pairs of legs.

+ habitat: Moist areas near human dwellings, greenhouses, agricultural fields

+ range: Southern Canada and the U.S.

+ food: Plant and animal tissues

Also called roly-poly and woodlouse, the Pillbug and its relatives are related to marine crustaceans. Although adapted to life on land, its life is controlled by the need to maintain proper water balance for respiration and waste elimination. Waste products stored in its body during dry periods appear as yellow patterns on the carapace. After mating, the female releases up to 100 eggs in a fluid-filled sac, or marsupium, on the underside of the abdomen. Pale juveniles hatch in two or three months and become darker and larger with each molt. The total life span is one or two years.

Female with young

Rough Sowbug

Porcellio scaber L 0.5–0.7 in (14–17 mm)

Known as roly-poly, potato bug, and wood louse, the Rough Sowbug is an omnivorous scavenger that sometimes congregates in large numbers under loose bark. It cannot roll up into a ball.

KEY FACTS

This species is oblong, somewhat flattened, slate blue, and sometimes with lighter patterns; it is covered with arched, overlapping, roughly sculptured plates and has seven pairs of legs.

+ habitat: Moderately damp areas in gardens and woodlands

+ range: Southern Canada and the U.S.

+ food: Plant and animal materials

The Rough Sowbug is typically found under debris on the ground, or under leaf litter and loose bark. It is mainly nocturnal on warmer, hotter days but will venture out in the day in cooler, wetter weather. Its roughened carapace serves as an "anti-adhesive structure" that minimizes the effects of surface tension when it comes into contact with wet surfaces. The Sowbug is less likely to stick to wet surfaces, and wet particles are less likely to stick to the Sowbug. Bluish or purplish individuals carry iridoviruses that infect many kinds of invertebrates but are harmless to people and pets.

Infected with iridovirus

Further Resources

■ Mammals

BOOKS

Elbroch, Mark, and Kurt Rinehart. *Behavior of North American Mammals.* Peterson Reference Guides. Houghton Mifflin, 2011.

Mammal Species of the World: A Taxonomic and Geographic Reference, 3rd ed. Don E. Wilson & DeeAnn M. Reeder (editors). Johns Hopkins University Press, 2005.

National Geographic Book of Mammals. National Geographic Society, 1998.

Reid, Fiona A. *A Field Guide to the Mammals of North America,* 4th ed. Peterson Field Guides. Houghton Mifflin, 2006.

Rezendes, Paul. *Tracking and the Art of Seeing: How to Read Animal Tracks and Sign,* 2nd ed. HarperCollins, 1999.

Whitaker, John O., Jr. *National Audubon Society Field Guide to North American Mammals.* Alfred A. Knopf, 1996.

WEBSITES

Bat Conservation International. Available online at batcon.org.

eNature.com. Available online at enature.com.

National Geographic Society. Available online at animals .nationalgeographic.com/animals/mammals.

Smithsonian Institution. North American Mammals. Available online at www.mnh.si.edu/mna.

APPS

Audubon Mammals: A Field Guide to North American Mammals. Green Mountain Digital.

■ Birds

BOOKS

Alderfer, Jonathan, ed. *Complete Birds of North America.* National Geographic Society, 2005.

Alderfer, Jonathan, and Jon L. Dunn. *Birding Essentials.* National Geographic, 2007.

Alderfer, Jonathan, and Paul Hess. *Backyard Guide to the Birds of North America.* National Geographic, 2011.

Dunn, Jon L., and Jonathan Alderfer, eds. *Field Guide to the Birds of North America,* 6th edition. National Geographic Society, 2011.

Dunne, Pete. *Pete Dunne's Essential Field Guide Companion.* Houghton Mifflin, 2006.

Elbroch, Mark, and Eleanor Marks. *Bird Tracks & Sign.* Stackpole Books, 2001.

Erickson, Laura. *The Bird Watching Answer Book.* Workman Publishing Company, 2009.

Erickson, Laura. *101 Ways to Help Birds.* Stackpole Books, 2006.

Kaufman, Kenn. *Lives of North American Birds.* Houghton Mifflin, 1996.

Kroodsma, Donald. *The Backyard Birdsong Guide.* Chronicle Books, 2008.

Poole, Alan, and Frank Gill, eds. *The Birds of North America.* The Academy of Natural Sciences and the American Ornithologists' Union, 1992–2002. Available online at bna.birds.cornell.edu/BNA (fee).

Roth, Sally. *Bird-by-Bird Gardening.* Rodale, 2006.

Sibley, David A. The Sibley Guide to Birds. Alfred A. Knopf, 2000.

Stokes, Donald and Lillian. *The Stokes Field Guide to the Birds of North America.* Little, Brown and Company, 2010.

WEBSITES

American Birding Association. Available online at aba.org.

Birding News. Available online at birding.aba.org.

Christmas Bird Count. Available online at birds.audubon .org/christmas-bird-count.

Cornell Lab of Ornithology. Available online at birds.cornell.edu.

eBird. Available online at ebird.org.

Jonathan Alderfer. Available online at jonathanalderfer .com.

Laura Erickson's For the Birds. Available online at laura erickson.com.

APPS

National Geographic Birds for iPhone and iPad.

■ Reptiles & Amphibians

BOOKS

Behler, John L., and F. Wayne King. *National Audubon Society Field Guide to the Reptiles and Amphibians of North America.* Alfred A. Knopf, 1997.

Conant, Roger, and Joseph T. Collins. *A Field Guide to Reptiles and Amphibians: Eastern/Central North America,* 3rd ed. Peterson Field Guides. Houghton Mifflin, 1998.

Powell, Robert, Joseph T. Collins, and Errol D. Hooper, Jr. *Key to the Herpetofauna of the Continental United States and Canada,* 2nd ed. University Press of Kansas, 2012.

Stebbins, Robert C. *A Field Guide to Western Reptiles and Amphibians,* 3rd ed. Peterson Field Guides. Houghton Mifflin, 2003.

WEBSITES

eNature.com. Available online at enature.com.

National Geographic Society. Available online at animals.nationalgeographic.com/animals/amphibians.

National Geographic Society. Available online at animals.nationalgeographic.com/animals/reptiles.

Smithsonian Institution. Department of Vertebrate Zoology, Division of Reptiles and Amphibians. Available online at vertebrates.si.edu/herps.

Society for the Study of Amphibians and Reptiles. Available online at ssarherps.org.

APPS

Audubon Reptiles and Amphibians: A Field Guide to North American Reptiles and Amphibians. Green Mountain Digital.

■ Aquatic Life

BOOKS

Leatherman, Stephen, and Jack Williams. *Field Guide to the Water's Edge.* National Geographic Society, 2012.

Nelson, Joseph S. *Fishes of the World,* 4th ed. John Wiley and Sons, 2006.

Schullery, Paul. *American Flyfishing: A History.* Lyons Press, 1987.

WEBSITES

National Marine Fisheries Service. Available online at nmfs.noaa.gov.

National Geographic Society: Fish. Available online at animals.nationalgeographic.com/animals/fish.

National Geographic Society: The Ocean. Available online at ocean.nationalgeographic.com/ocean.

American Cetacean Society. Available online at acsonline.org.

Washington State University Beach Watchers. Available online at beachwatchers.wsu.edu/regional/about.

Texas Parks and Wildlife. Available online at tpwd.state.tx.us/landwater/water/aquaticspecies.

Minnesota Department of Natural Resources. Available online at dnr.state.mn.us/fish/index.html.

Commercial Fisheries. Available online at fishery.about.com.

■ Insects & Spiders

BOOKS

Cranshaw, Whitney. *Garden Insects of North America: The Ultimate Guide to Backyard Bugs.* Princeton University Press, 2004.

Eaton, Eric R., and Kenn Kaufman. *Kaufman Field Guide to Insects of North America.* Houghton Mifflin Harcourt, 2007.

Evans, Arthur V. *National Wildlife Federation Field Guide to Insects and Spiders of North America.* Sterling Publishing, 2007.

Howell, W. Mike, and Ronald L. Jenkins. *Spiders of the Eastern United States: A Photographic Guide.* Pearson Education, 2004.

Jackman, John A. *A Field Guide to Spiders and Scorpions of Texas.* Gulf Publishing,1999.

Marshall, Stephen. *Insects: Their Natural History and Diversity, with a Photographic Guide to Insects of Eastern North America.* Firefly Books, 2006.

Murray, Tom. *Insects of New England and New York.* Kollath-Stensaas Publishing, 2012.

WEBSITES

BugGuide.net. An outstanding resource for identified insect images, taxonomic information, and other online resources. Available online at bugguide.net.

National Geographic Society. Bugs. Available online at animals.nationalgeographic.com/animals/bugs.

Insect Identification for the Casual Observer. Available online at insectidentification.org.

ORGANIZATIONS

Entomological Society of America. The largest organization in the world serving the needs of professional entomologists, educators, students, and amateur naturalists. Available online at entsoc.org.

The Xerces Society. An international organization that promotes the conservation of invertebrates. Available online at xerces.org.

About the Contributors

■ About the Authors

Jonathan Alderfer, artist, author, and editor, has contributed paintings to the *National Geographic Field Guide to the Birds of North America*, and co-wrote the sixth edition in 2011. He has written or edited numerous bird books for National Geographic, most recently *Backyard Guide to the Birds of North America* and *Bird-watcher's Bible*. A selection of his original art can be viewed at jonathanalderfer.com. He and his wife live in Washington, D.C.

Laura Erickson has written six books on birds—including *Sharing the Wonder of Birds with Kids,* winner of the National Outdoor Book Award—writes for birding magazines, and blogs for the American Birding Association. She has been a licensed wildlife rehabilitator and previously served as science editor at the Cornell Lab of Ornithology. Her radio program, *For the Birds,* airs on public and community radio stations, and is podcast via iTunes. She lives in Duluth, Minnesota, with her husband.

Entomologist **Arthur V. Evans** is an author, photographer, and radio broadcaster, and teaches entomology at several universities in Virginia. Evans has published numerous scientific papers, popular articles, and books on insects and spiders, including the *National Wildlife Federation Field Guide to Insects and Spiders of North America* (2007). His next book, *Field Guide to Beetles of Eastern North America,* will be published by Princeton University Press. He lives in Richmond, Virginia.

Noel Grove is a former National Geographic magazine writer who has written more than two dozen articles for the magazine and several books and chapters for the Book Division. He was the first senior assistant editor for the environment for the magazine, and many of his publications are on that subject. Among his books are one on The Nature Conservancy and another on the National Wildlife Refuge System. He lives with his wife in the Blue Ridge Mountains of Virginia.

A former National Geographic staff member, **Catherine Herbert Howell** writes about natural history for adults and children. She explored the relationships between people and plants in *Flora Mirabilis: How Plants Have Shaped World Knowledge, Health, Wealth, and Beauty* (2009) and the significance of birds in world cultures in *National Geographic Bird-watcher's Bible* (2012). Howell shares her fascination with urban wildlife as a master naturalist volunteer in Arlington, Virginia.

■ About the Consultants

Gene Helfman taught ichthyology (fish biology), animal behavior, and conservation ecology for 30 years at the University of Georgia in Athens, Georgia. His research included diving observations of predator–prey interactions, behavior of eels, and effects of land use practices on stream fishes. He has co-authored textbooks on ichthyology, fish conservation, and sharks.

Bernadette Holthuis is a research associate at the Florida Museum of Natural History, University of Florida, studying evolution in marine and freshwater invertebrates. She received her doctorate from the University of Washington, and has worked on invertebrates in the tropical and temperate Pacific Ocean. She has taught classes in natural history at the Florida Museum and in marine invertebrate zoology at Friday Harbor Laboratories, Washington.

Robert Powell is professor of biology at Avila University in Kansas City, Missouri. Believing that biologists must "do" biology by getting their hands dirty, he has been chasing amphibians and reptiles since a university herpetology course in 1968 turned him on to field trips. He is co-author or co-editor of eight books and hundreds of scientific publications, many co-authored with students.

Samuel Zeveloff is a professor in the Department of Zoology at Weber State University in Ogden, Utah. He received his Ph.D. in zoology from the University of Wyoming. Zeveloff teaches mammalogy and conservation biology and has studied raccoons and mountain goats. He is the author of two books: *Mammals of the Intermountain West* and *Raccoons: A Natural History.*

■ About the Artists

Fernando G. Baptista graduated in fine arts from País Vasco in Spain. He worked as a graphic artist for a newspaper in Bilbao, as a freelancer creating reconstructive pieces for museums, and spent six years as a professor of infographics at the University of Navarra. Baptista has won more than 125 international awards. In 2012 he was named one of the five most influential infographic artists of the past 20 years.

Mesa Schumacher is an artist and illustrator specializing in scientific and information art. Her art has taken her to archaeological sites in Turkey, Belize, Portugal, and Peru, laboratories, and the archives of the Smithsonian. She has exhibited work in Europe, Asia, and the Americas, and is currently a freelance artist living in Washington, D.C.

Jared Travnicek is a scientific and medical illustrator. He received his M.A. in biological and medical illustration from the Johns Hopkins University School of Medicine in Baltimore, Maryland. Travnicek is a certified medical illustrator and a professional member of the Association of Medical Illustrators. He contributed artwork to *National Geographic Illustrated Guide to Nature.*

Illustrations Credits

New artwork appearing in this book was created by Fernando G. Baptista, Mesa Schumacher, and Jared Travnicek.

Front Cover (Top Row, Left to Right), Patti Sullivan Schmidt/Getty Images; Paul Nicklen/Getty Images; Matthias Breiter/Minden Pictures/National Geographic Creative; Scott Linstead; (Middle Row, Left to Right), Medford Taylor/Getty Images; Matthew Crowley Photography/Getty Images; tomh1000/iStockphoto; Jeanne White/Getty Images; (Bottom Row, Left to Right), Darrell Gulin/Getty Images; Paul Nicklen/Getty Images; Alex Saberi/Getty Images; George Grall/Getty Images.

Back Cover (Top Row, Left to Right), H.H. Fox Photography/Getty Images; Michael Durham/Minden Pictures/National Geographic Creative; Phil Lanoue/National Geographic Your Shot; George Grall/National Geographic Creative; (Middle Row, Left to Right), George Grall/Getty Images; Tim Fitzharris/Minden Pictures/National Geographic Creative; Raul Touzon/National Geographic Creative; Mirenchu A Fernandez/Getty Images; (Bottom Row, Left to Right), Michael Parrish Photography/Getty Images; Brian J. Skerry/National Geographic Creative; George Grall/National Geographic Creative; Magdalena Garneau Photography/Getty Images.

Interior 2-3, Steve Courson; 4, Danita Delimont/Alamy; 6, Carr Clifton/Minden Pictures/National Geographic Creative; 8, Paul Souders/Getty Images; 11, William Church/Getty Images, 12, Sean Russell/Getty Images; 14 (UP), David Courtenay/Getty Images; 14 (LO), Stan Tekiela/NatureSmart Wildlife; 15 (UP), Wayne Lynch/Getty Images; 15 (LO), Visuals Unlimited, Inc./Don Grall/Getty Images; 16 (UP), Michael G. Mill/Shutterstock; 16 (LO), Thomas & Pat Leeson/Science Source; 17 (UP), mandj98/iStockphoto; 17 (LO), PHOTO 24/Getty Images; 18 (UP), Stan Tekiela/NatureSmart Wildlife; 18 (LO), Ed Cesar/Science Source; 19 (UP), Nicholas Bergkessel, Jr./Science Source; 19 (LO), Stubblefield Photography/Shutterstock; 20 (UP), Martha Marks/Shutterstock; 20 (LO), Robert Franz/Getty Images; 21 (UP), Henk Bentlage/Shutterstock; 21 (LO), Fremme/Shutterstock; 22 (UP), Craig K Lorenz/Getty Images; 22 (LO), Andreas Resch/Shutterstock; 23 (UP), Tom Reichner/Shutterstock; 23 (LO), Stan Tekiela/NatureSmart Wildlife; 24 (UP), Ed Reschke/Getty Images; 24 (LO), worldswildlifewonders/Shutterstock; 25 (UP), KENNETH W. FINK/Getty Images; 25 (LO), Sam Fried/Science Source; 26 (UP), Stan Tekiela/NatureSmart Wildlife; 26 (LO), Tim Stirling/Shutterstock; 27 (UP), brokentone/iStockphoto; 27 (LO), Stan Tekiela/NatureSmart Wildlife; 28 (UP), Howard Stapleton/Alamy; 28 (LO), Tom Reichner/Shutterstock; 29 (UP), Francis Bossé/Shutterstock; 29 (LO), Danita Delimont/Getty Images; 30 (UP), Charles Schug/iStockphoto; 30 (LO), Purestock/Getty Images; 31 (UP), Yva Momatiuk & John Eastcott/Minden Pictures/National Geographic Creative; 31 (LO), C. Allan Morgan/Getty Images; 32 (UP), Stan Tekiela/NatureSmart Wildlife; 32 (LO), Shattil & Rozinski/NPL/Minden Pictures; 33 (UP), John Cancalosi/Getty Images; 33 (LO), E.R. Degginger/Alamy; 34 (UP), Gary Meszaros/Science Source; 34 (LO), Wayne Lynch/Getty Images; 35 (UP), TOM MCHUGH/Getty Images; 35 (LO), Daniel Cox/Getty Images; 36 (UP), Kenneth L. Crowell/Mammal Images Library; 36 (LO), Stan Tekiela/NatureSmart Wildlife; 37 (UP), Stan Tekiela/NatureSmart Wildlife; 37 (LO), Rob & Ann Simpson/Visuals Unlimited, Inc.; 38 (UP), Robert J Erwin/Getty Images; 38 (LO), Barry Mansell/Minden Pictures; 39 (UP), Brian Lasenby/Shutterstock; 39 (LO), David Moskowitz, davidmoskowitz.net; 40 (LO), Stan Tekiela/NatureSmart Wildlife; 41 (UP), Rolf Kopfle/Getty Images; 41 (LO), Rob & Ann Simpson/Visuals Unlimited, Inc.; 42 (UP), Barry Mansell/naturepl.com; 42 (LO), Stan Tekiela/NatureSmart Wildlife; 43 (UP), Stan Tekiela/NatureSmart Wildlife; 43 (LO), Stan Tekiela/NatureSmart Wildlife; 44 (UP), Gary Meszaros/Visuals Unlimited/Getty Images; 44 (LO), Michael Durham/Minden Pictures; 45 (UP), Rick & Nora Bowers/Alamy; 45 (LO), Louise Murray/age fotostock; 46 (UP & LO), Stan Tekiela/NatureSmart Wildlife; 47 (UP), Rodger Jackman/Getty Images; 47 (LO), Erni/Shutterstock; 48 (UP), Stan Tekiela/NatureSmart Wildlife; 48 (LO), Tom Reichner/Shutterstock; 49 (UP), Steven Kazlowski/Getty Images; 49 (LO), Stan Tekiela/NatureSmart Wildlife; 50 (UP), Howard Sandler/Shutterstock; 50 (LO), Jim Brandenburg/Minden Pictures/National Geographic Creative; 51 (UP), twildlife/iStockphoto; 51 (LO), David Tipling/Getty Images; 52 (UP), Pyshnyy Maxim Vjacheslavovich/Shutterstock; 52 (LO), A & J Visage/Alamy; 53 (UP), Stan Tekiela/NatureSmart Wildlife; 53 (LO), Hal Beral/Visuals Unlimited, Inc.; 54 (UP), Tom Reichner/Shutterstock; 54 (LO), altrendo nature/Getty Images; 55 (UP), visceralimage/Shutterstock; 55 (LO), John Macgregor/Getty Images; 56 (UP), Rob & Ann Simpson/Visuals Unlimited, Inc.; 56 (LO), Phil Myers; 57 (UP), Tom & Pat Leeson; 57 (LO), Stan Tekiela/NatureSmart Wildlife; 58 (LO), Gary Meszaros/Visuals Unlimited, Inc.; 59 (LO), Ronald G. Altig; 60 (UP), Dwight R. Kuhn; 60 (LO), Ken Catania/Visuals Unlimited, Inc.; 61 (UP), Dwight R. Kuhn; 61 (LO), Gerry Ellis/Minden Pictures/National Geographic Creative; 62 (UP), Michael Durham/www.DurmPhoto.com; 62 (LO), Merlin Tuttle/BCI/Getty Images; 63 (UP), Merlin Tuttle/Getty Images; 63 (LO), Pete Oxford/Getty Images; 64 (UP), Joel Sartore/Getty Images; 64 (LO), Jared Hobbs/Getty Images; 65 (UP), Merlin D. Tuttle/Science Source; 65 (LO), Michael Durham/Getty Images; 66 (UP), Visuals Unlimited, Inc./John Abbott/Getty Images; 66 (LO), Merlin D. Tuttle/Science Source; 67 (UP), Michael Durham/Getty Images; 67 (LO), Merlin D. Tuttle/Science Source; 68 (UP), Michael Durham/Getty Images; 68 (LO), Jack Milchanowski/

Visuals Unlimited, Inc.; 69 (UP), Rick & Nora Bowers/Visuals Unlimited, Inc.; 69 (LO), Michael Durham/Getty Images; 70 (UP), Michael Durham/Getty Images; 70 (LO), Helen E. Grose/Shutterstock; 71 (UP), Stan Tekiela/NatureSmart Wildlife; 71 (LO), Scott E Read/Shutterstock; 72 (UP), Dennis Donohue/Shutterstock; 72 (LO), Geoffrey Kuchera/Shutterstock; 73 (UP), Holly Kuchera/Shutterstock; 73 (LO), jeanro/iStockphoto; 74 (UP), visceralimage/iStockphoto; 74 (LO), visceralimage/Shutterstock; 75 (UP), Stan Tekiela/NatureSmart Wildlife; 75 (LO), Darrell Gulin/Getty Images; 76 (UP), driftlessstudio/iStockphoto; 76 (LO), Yva Momatiuk & John Eastcott/Minden Pictures; 77 (UP), Gleb Tarro/Shutterstock; 77 (LO), outdoorsman/Shutterstock; 78 (UP), Michael Quinton/Minden Pictures/National Geographic Creative; 78 (LO), visceralimage/Shutterstock; 79 (UP), Steve Brigman/Shutterstock; 79 (LO), Stan Tekiela/NatureSmart Wildlife; 80 (UP), Stan Tekiela/NatureSmart Wildlife; 80 (LO), VanH/iStockphoto; 81 (UP), John E Marriott/Getty Images; 81 (LO), Alucard2100/Shutterstock; 82 (UP), Christopher Mills/Shutterstock; 82 (LO), Stan Tekiela/NatureSmart Wildlife; 83 (UP), Claus Meyer/Getty Images; 83 (LO), Anthony Mercieca/Getty Images; 84 (UP), Tim Fitzharris/Minden Pictures/National Geographic Creative; 84 (LO), Jared Hobbs/Getty Images; 85 (UP), David Welling/naturepl.com; 85 (LO), l i g h t p o e t/Shutterstock; 86 (UP), Heiko Kiera/Shutterstock; 86 (LO), Gerald & Buff Corsi/Visuals Unlimited, Inc.; 87 (UP), mariait/Shutterstock; 87 (LO), IbajaUsap/Shutterstock; 88 (UP), twildlife/iStockphoto; 88 (LO), Arnold John Labrentz/Shutterstock; 89 (UP), Brian Lasenby/Shutterstock; 89 (LO), Tom Reichner/Shutterstock; 90 (UP), John E Marriott/Getty Images; 90 (LO), Alfie Photography/Shutterstock; 91 (UP), Alan Scheer/Shutterstock; 91 (LO), clark42/iStockphoto; 92 (UP), Josh Schutz/Shutterstock; 92 (LO), l i g h t p o e t/Shutterstock; 93 (UP), Tom Reichner/Shutterstock; 93 (LO), Richard Seeley/Shutterstock; 94, Jim Zipp; 95, John C. Pitcher; 96, Jonathan Alderfer; 97, Alan Murphy; 100 (UPLE), Michel Lamarche/National Geographic Your Shot; 100 (LOLE), Oleg Slyusarchuk/National Geographic Your Shot; 100 (UPRT & LORT), Cynthia J. House; 101 (UPLE), Robert Royse; 101 (LOLE), Tim Zurowski; 101 (UPRT & LORT), Cynthia J. House; 102 (UPLE), Bert De Tilly/National Geographic Your Shot; 102 (LOLE), Homer Caliwag; 102 (UPRT & LORT), Cynthia J. House; 103 (UPLE), Marie Read; 103 (LOLE), E.J. Peiker; 103 (UPRT & LORT), Cynthia J. House; 104 (UPLE), AvianArt Images by David Hemmings; 104 (LOLE), E.J. Peiker; 104 (UPRT & LORT), Cynthia J. House; 105 (UPLE), Michael Ohaion/National Geographic Your Shot; 105 (LOLE), Bob Gress; 105 (UPRT & LORT), Kent Pendleton; 106 (UPLE), Rob Kemp/Shutterstock; 106 (LOLE), Phillip Holland/Shutterstock; 106 (UPRT & LORT), Kent Pendleton; 107 (UPLE), Richard Walsh/National Geographic Your Shot; 107 (UPRT), Kent Pendleton; 107 (LOLE), Sylvie Goguen/National Geographic Your Shot; 107 (LORT), David Quinn; 108 (UPLE), James Cumming/National Geographic Your Shot; 108 (UPRT), Jonathan Alderfer; 108 (LOLE), Marie Read; 108 (LORT), H. Jon Janosik; 109 (UPLE), Homer Caliwag; 109 (UPRT), Jonathan Alderfer; 109 (LOLE), Jim Neiger/www.flightschoolphotography.com; 109 (LORT), H. Jon Janosik; 110 (UPRT), Danny Brown/National Geographic Your Shot; 110 (UPRT), H. Jon Janosik; 110 (LOLE), E.J. Peiker; 110 (LORT—Immature & Non-Breeding Adult), H. Jon Janosik; 110 (LORT—Breeding Adult), Jonathan Alderfer; 111 (UPLE), Kevin T. Karlson; 111 (LOLE), E.J. Peiker; 111 (UPRT & LORT), Diane Pierce; 112 (UPLE), Steve Ellwood/National Geographic Your Shot; 112 (LOLE), Mia McPherson/onthewingphotography.com; 112 (UPRT & LORT), Diane Pierce; 113 (UPLE), Jim Zipp; 113 (UPRT), Peter Burke; 113 (LOLE), Larsek/Shutterstock; 113 (LORT), Diane Pierce; 114 (UPLE), Harry Moulis/National Geographic Your Shot; 114 (LOLE), Robert Royse; 114 (UPRT & LORT), Diane Pierce; 115 (UPLE), Bob Steele; 115 (LOLE), Bret Douglas/National Geographic Your Shot; 115 (UPRT & LORT), Donald L. Malick; 116 (UPLE), Clay Billman; 116 (UPRT), Donald L. Malick; 116 (LOLE), Kenneth Rush/Shutterstock; 116 (LORT), N. John Schmitt; 117 (UPLE), Alan Murphy; 117 (LOLE), Kevin T. Karlson; 117 (UPRT & LORT), Donald L. Malick; 118 (UPLE), Jim Neiger/www.flightschoolphotography.com; 118 (LOLE), Lloyd Spitalnik; 118 (UPRT & LORT), N. John Schmitt; 119 (UPLE), AvianArt Images by David Hemmings; 119 (LOLE), Jim Zipp; 119 (UPRT & LORT), Donald L. Malick; 120 (UPLE), RobertMcCaw.com; 120 (LOLE), David March/National Geographic Your Shot; 120 (UPRT & LORT), Marc R. Hanson; 121 (UPLE), Maxis Gamez/www.gvisions.org; 121 (UPRT), Diane Pierce; 121 (LOLE), Lloyd Spitalnik; 121 (LORT), Jonathan Alderfer; 122 (UPLE), Jim Zipp; 122 (UPRT), John C. Pitcher; 122 (LOLE), RobertMcCaw.com; 122 (LORT), Killian Mullarney; 123 (UPLE), Richard Fitzer/Shutterstock; 123 (UPRT), H. Jon Janosik; 123 (LOLE), Homer Caliwag; 123 (LORT), John C. Pitcher; 124 (UPLE), Kevin T. Karlson; 124 (UPRT), John C. Pitcher; 124 (LOLE), Bob Steele; 124 (LORT), Michael O'Brien; 125 (UPLE), E.J. Peiker; 125 (UPRT), Daniel S. Smith; 125 (LOLE), Alan Murphy; 125 (LORT), John C. Pitcher; 126 (UPLE), Bob Steele; 126 (UPRT), Thomas R. Schultz; 126 (LOLE), Lloyd Spitalnik; 126 (LORT), John C. Pitcher; 127 (UPLE), RobertMcCaw.com; 127 (UPRT), Jonathan Alderfer; 127 (LOLE), Alan Murphy; 127 (LORT), Thomas R. Schultz; 128 (UPLE), Homer Caliwag; 128 (LOLE), Glenn Bartley; 128 (UPRT & LORT), Thomas R. Schultz; 129 (UPLE), Homer Caliwag; 129 (LOLE), Bob Steele; 129 (UPRT & LORT), Thomas R. Schultz; 130 (UPLE), Lloyd Spitalnik; 130 (UPRT), Thomas R. Schultz; 130 (LOLE), E.J. Peiker; 130 (LORT), Chuck Ripper; 131 (UPLE), Pooja C. Raghavendra/National Geographic Your Shot; 131 (UPRT), H. Douglas Pratt; 131 (LOLE), Bob Steele; 131 (LORT), N. John Schmitt; 132 (UPLE), Jim Zipp; 132 (LOLE), Alan Murphy; 132 (UPRT & LORT), N. John Schmitt; 133 (UPLE), Lloyd Spitalnik; 133 (UPRT),

N. John Schmitt; 133 (LOLE), Alan Murphy; 133 (LORT), H. Douglas Pratt; 134 (UPLE), Alan Murphy; 134 (UPRT), H. Douglas Pratt; 134 (LOLE), Deborah Smith/National Geographic Your Shot; 134 (LORT), Donald L. Malick; 135 (UPLE), Alan Murphy; 135 (LOLE), Robert Postma/National Geographic Your Shot; 135 (UPRT & LORT), Donald L. Malick; 136 (UPLE), AvianArt Images by David Hemmings; 136 (LOLE), Paul Ayick/National Geographic Your Shot; 136 (UPRT & LORT), Donald L. Malick; 137 (UPLE), Bob Steele; 137 (UPRT), Thomas R. Schultz; 137 (LOLE), Judd Patterson; 137 (LORT), N. John Schmitt; 138 (UPLE), Terence P. Brashear; 138 (UPRT), N. John Schmitt; 138 (LOLE), Alan Murphy; 138 (LORT), H. Douglas Pratt; 139 (UPLE), Brian E. Small; 139 (LOLE), Bob Steele; 139 (UPRT & LORT), H. Douglas Pratt; 140 (UPLE), Glenn Bartley; 140 (UPRT), H. Douglas Pratt; 140 (LOLE), Alan Murphy; 140 (LORT), Donald L. Malick; 141 (UPLE), FloridaStock/Shutterstock; 141 (LOLE), Marie Read; 141 (UPRT & LORT), Donald L. Malick; 142 (UPLE), Judd Patterson; 142 (LOLE), Steve Byland/Shutterstock; 142 (UPRT & LORT), Donald L. Malick; 143 (UPLE), Brian E. Small; 143 (UPRT), David Beadle; 143 (LOLE), Jack Sutton/National Geographic Your Shot; 143 (LORT), H. Douglas Pratt; 144 (UPLE), Brook Burling/National Geographic Your Shot; 144 (LOLE), Eileen Worman/National Geographic Your Shot; 144 (UPRT & LORT), H. Douglas Pratt; 145 (UPLE), John Stankewitz/National Geographic Your Shot; 145 (UPRT), Jonathan Alderfer; 145 (LOLE), Judd Patterson; 145 (LORT), H. Douglas Pratt; 146 (UPLE), Mia McPherson/onthewingphotography.com; 146 (LOLE), Terence P. Brashear; 146 (UPRT & LORT), H. Douglas Pratt; 147 (UPLE), Steven Smith/National Geographic Your Shot; 147 (LOLE), Tony Campbell/Shutterstock; 147 (UPRT & LORT), H. Douglas Pratt; 148 (UPLE), Bob Steele; 148 (UPRT), N. John Schmitt; 148 (LOLE), RobertMcCaw.com; 148 (LORT), H. Douglas Pratt; 149 (UPLE), Julie Bishop/National Geographic Your Shot; 149 (LOLE), Patricia Fiedler/National Geographic Your Shot; 149 (UPRT & LORT), N. John Schmitt; 150 (UPLE), Alan Murphy; 150 (UPRT), David Beadle; 150 (LOLE), Mark McMaster/National Geographic Your Shot; 150 (LORT—Eastern & Adult), H. Douglas Pratt; 150 (LORT—Western), Thomas R. Schultz; 151 (UPLE), Steve Hamilton/National Geographic Your Shot; 151 (LOLE), Bob Gress; 151 (UPRT & LORT), H. Douglas Pratt; 152 (UPLE), AvianArt Images by David Hemmings; 152 (UPRT), H. Douglas Pratt; 152 (LOLE), Marie Read; 152 (LORT), Michael O'Brien; 153 (UPLE), Steven Smith/National Geographic Your Shot; 153 (UPRT), Michael O'Brien; 153 (LOLE), Roger van Gelder; 153 (LORT), John P. O'Neill; 154 (UPLE), Peter Friedlieb/National Geographic Your Shot; 154 (LOLE), Jaromir Penicka/National Geographic Your Shot; 154 (UPRT & LORT), H. Douglas Pratt; 155 (UPLE), Roger van Gelder; 155 (UPRT), Thomas R. Schultz; 155 (LOLE), Alexander Viduetsky/National Geographic Your Shot; 155 (LORT), H. Douglas Pratt; 156 (UPLE), RobertMcCaw.com; 156 (LOLE), Bill Dalton/National Geographic Your Shot; 156 (UPRT & LORT), H. Douglas Pratt; 157 (UPLE), Jim Zipp; 157 (LOLE), Alan Murphy; 157 (UPRT & LORT), N. John Schmitt; 158 (UPLE), RobertMcCaw.com; 158 (LOLE), Jeff Brubaker/National Geographic Your Shot; 158 (UPRT & LORT), H. Douglas Pratt; 159 (UPLE), Jeff Pinkerton/National Geographic Your Shot; 159 (LOLE), Gary Hamilton/National Geographic Your Shot; 159 (UPRT & LORT), H. Douglas Pratt; 160 (UPLE), Rowland Willis/National Geographic Your Shot; 160 (UPRT), Thomas R. Schultz; 160 (LOLE), Beth Pirie/National Geographic Your Shot; 160 (LORT), H. Douglas Pratt; 161 (UPLE), Martha Marks/Shutterstock; 161 (LOLE), Alan Murphy; 161 (UPRT & LORT), H. Douglas Pratt; 162 (UPLE), Lloyd Spitalnik; 162 (UPRT), H. Douglas Pratt; 162 (LOLE), Eirini Pajak/National Geographic Your Shot; 162 (LORT), N. John Schmitt; 163 (UPLE), Erik Mandre/National Geographic Your Shot; 163 (LOLE), George Chiu/National Geographic Your Shot; 163 (UPRT & LORT), H. Douglas Pratt; 164 (UPLE), Marie Read; 164 (UPRT), Diane Pierce; 164 (LOLE), Lloyd Spitalnik; 164 (LORT), Thomas R. Schultz; 165 (UPLE), Michael G. Mill/Shutterstock; 165 (UPRT), Thomas R. Schultz; 165 (LOLE), Alan Murphy; 165 (LORT), H. Douglas Pratt; 166 (UPLE), RobertMcCaw.com; 166 (UPRT), Thomas R. Schultz; 166 (LOLE), Laura L. Erickson; 166 (LORT—Breeding), H. Douglas Pratt; 166 (LORT—Fall), Thomas R. Schultz; 167 (UPLE), RobertMcCaw.com; 167 (LOLE), Brian E. Small; 167 (UPRT & LORT), Peter Burke; 168 (UPLE), Marie Read; 168 (UPRT), Thomas R. Schultz; 168 (LOLE), Robert Royse; 168 (LORT), Diane Pierce; 169 (UPLE), Richard Cronberg/National Geographic Your Shot; 169 (UPRT), Diane Pierce; 169 (LOLE), David Seibel; 169 (LORT), Thomas R. Schultz; 170 (UPLE), Garth McElroy; 170 (LOLE), Alan Murphy; 170 (UPRT & LORT), Diane Pierce; 171 (UPLE), Lloyd Spitalnik; 171 (LOLE), Bob Steele; 171 (UPRT & LORT), Peter Burke; 172 (UPLE), Roy Orr/National Geographic Your Shot; 172 (LOLE), Brian Lasenby/Shutterstock; 172 (UPRT & LORT), Diane Pierce; 173 (UPLE), Danny Brown/National Geographic Your Shot; 173 (LOLE), Alan Murphy; 173 (UPRT & LORT), Thomas R. Schultz; 174 (UPLE), Stacey Huston/National Geographic Your Shot; 174 (UPRT), Thomas R. Schultz; 174 (LOLE), Steve Creek/National Geographic Your Shot; 174 (LORT), H. Douglas Pratt; 175 (UPLE), Judd Patterson; 175 (LOLE), Dennis Donohue/Shutterstock; 175 (UPRT & LORT), H. Douglas Pratt; 176 (UPLE), Roger van Gelder; 176 (LOLE), Beverly Cochran/National Geographic Your Shot; 176 (UPRT & LORT), H. Douglas Pratt; 177 (UPLE), Jose Hernandez/National Geographic Your Shot; 177 (UPRT), N. John Schmitt; 177 (LOLE), RobertMcCaw.com; 177 (LORT), Peter Burke; 178 (UPLE), Mark Lewer/National Geographic Your Shot; 178 (UPRT), Peter Burke; 178 (LOLE), Matthew Studebaker; 178 (LORT), Diane Pierce; 179 (UPLE), Stubblefield Photography/Shutterstock; 179 (UPRT), Thomas R. Schultz; 179 (LOLE), turtix/Shutterstock; 179 (LORT), N. John Schmitt; 180, Rolf Nussbaumer/NPL/Minden Pictures; 183, Michael Melford/National Geographic Creative; 186 (UP), Roy Toft/National Geographic Creative; 186 (LO), Wayne Lynch/age fotostock; 187 (UP), Raul Touzon/National Geographic Creative; 187 (LO), Paul Sutherland/National Geographic Creative; 188 (UP), George Grall/National Geographic Creative; 188 (LO), Bill Curtsinger/National Geographic Creative; 189 (UP), Bianca Lavies/National Geographic Creative; 189 (LO), Danita Delimont/Alamy; 190 (UP), Suzanne L. Collins/Science Source; 190 (LO), Jasper Doest/Foto Natura/Minden Pictures; 191 (UP), David A. Northcott/Corbis; 191 (LO), L Lee Rue/FLPA.co.uk; 192 (UP), James Harding; 192 (LO), Roderick Perkinson/National Geographic Your Shot; 193 (UP), Joel Sartore/National Geographic Creative; 193 (LO), Justin G. Coleman; 194 (UP), Visuals Unlimited, Inc./Michael Redmer/Getty Images; 194 (LO), Joe McDonald/Corbis; 195 (UP), Paul Sutherland/National Geographic Creative; 195 (LO), Joseph T Collins/Photo Researchers/Getty Images; 196 (UP), Tierfotoagentur/Alamy; 196 (LO), Brian Lasenby/Shutterstock; 197 (UP), Todd Pusser/naturepl.com; 197 (LO), All Canada Photos/Alamy; 198 (UP), Tom Ulrich/Visuals Unlimited, Inc./Getty Images; 198 (LO), Judd Patterson; 199 (UP), Jim Abernethy/National Geographic Creative; 199 (LO), Norbert Wu/Minden Pictures/National Geographic Creative; 200 (UP), Frans Lanting/National Geographic Creative; 200 (LO), George Grall/National Geographic Creative; 201 (UP), Michael Patrick O'Neill/Alamy; 201 (LO), Suzanne L. & Joseph T. Collins/Science Source; 202 (UP), Ryan M. Bolton/Shutterstock; 202 (LO), Matt Jeppson/Shutterstock; 203 (UP), Fabio Pupin/Visuals Unlimited/Corbis; 203 (LO), Robert Buquoi; 204 (UP), Dave Knowles; 204 (LO), Malgorzata Litkowska/Shutterstock; 205 (UP), Chris Mattison/FLPA/age fotostock; 205 (LO), Danilo Donadoni/age fotostock; 206 (UP), Roy Toft/National Geographic Creative; 206 (LO), John Cancalosi/National Geographic Creative; 207 (UP), Michael and Patricia Fogden/Minden Pictures/National Geographic Creative; 207 (LO), Joe McDonald/Visuals Unlimited, Inc.; 208 (UP), D. Robert & Lorri Franz/Corbis; 208 (LO), Lee Rentz; 209 (UP), Thomas Brennan; 209 (LO), Jack Goldfarb/age fotostock; 210 (UP), Hal Beral/Corbis; 210 (LO), FOTOSEARCH RM/age fotostock; 211 (UP), Robert Shantz/Alamy; 211 (LO), Thomas Brennan; 212 (UP), John Cancalosi/Getty Images; 212 (LO), James Gerholdt/Getty Images; 213 (UP), Matt Jeppson/Shutterstock; 213 (LO), ZSSD/Minden Pictures/National Geographic Creative; 214 (UP), StuPorts/iStockphoto; 214 (LO), Sari ONeal/Shutterstock; 215 (UP), Robert Shantz/Alamy; 215 (LO), Aurora Photos/Alamy; 216 (UP), Ken Wray; 216 (LO), Brent Landreth/Alamy; 217 (UP), Matt Jeppson/Shutterstock; 217 (LO), Ron Sanford/Getty Images; 218 (UP), Todd Pierson; 218 (LO), Bryce Garling; 219 (UP), Pat Briggs; 219 (LO), John Cancalosi/National Geographic Creative; 220 (UP), Roy Palmer/Shutterstock; 220 (LO), lunatic67/Shutterstock; 221 (UP), Larry Miller/Getty Images; 221 (LO), JasonOndreicka/iStockphoto; 222 (UP), MattiaATH/Shutterstock; 222 (LO), Matt Jeppson/Shutterstock; 223 (UP), Patrick K. Campbell/Shutterstock; 223 (LO), Don Johnston/All Canada Photos/Corbis; 224 (UP), Ian Shive/TandemStock.com; 224 (LO), Prisma Bildagentur AG/Alamy; 225 (UP), Suzanne L. Collins/Science Source; 225 (LO), Jay Ondreicka/Shutterstock; 226 (UP), Ron Grundwald; 226 (LO), Eitan Grundwald; 227 (UP), Gary Nafis; 227 (LO), Ryan M. Bolton/Shutterstock; 228 (UP), Mike Pingleton; 228 (LO), Robert Hamilton/Alamy; 229 (UP), George Grall/National Geographic Creative; 229 (LO), SteveByland/iStockphoto; 230 (UP), Rusty Dodson/Shutterstock; 230 (LO), Jason Patrick Ross/Shutterstock; 231 (UP), Ken Wray; 231 (LO), Andrew McKinney/Shutterstock; 232 (UP), Mike Pingleton; 232 (LO), Shoemcfly/iStockphoto; 233 (UP), Sam Martin; 233 (LO), Matt Jeppson/Shutterstock; 234 (UP), Jay Ondreicka/Shutterstock; 234 (LO), the4js/iStockphoto; 235 (UP), Joel Sartore/National Geographic Creative; 235 (LO), Jack Milchanowski/age fotostock/Getty Images; 236 (UP), Joel Sartore/National Geographic Creative; 236 (LO), Paul Chesley/National Geographic Creative; 237 (UP), Amy White & Al Petteway/National Geographic Creative; 237 (LO), Wayne Lynch/Getty Images; 238 (UP), Nature's Images/Getty Images; 238 (LO), Robert Noonan/Science Source; 239 (UP), Joel Sartore/National Geographic Creative; 239 (LO), Lynda Richardson/Corbis; 240 (UP), Nature's Images/Science Source; 240 (LO), John Cancalosi/naturepl.com; 241 (UP), Gerald & Buff Corsi/Visuals Unlimited, Inc.; 241 (LO), Joel Sartore/National Geographic Creative; 242 (UP), Roy Toft/National Geographic Creative; 242 (LO), George Grall/National Geographic Creative; 243 (UP), WILDLIFE GmbH/Alamy; 243 (LO), Rob Schell; 244 (UP), Byron Jorjorian/Alamy; 244 (LO), Barry Mansell/naturepl.com; 245 (UP), Gerold & Cynthia Merker/Visuals Unlimited, Inc.; 245 (LO), Joe McDonald/Visuals Unlimited, Inc.; 246 (UP), Ken Wray; 246 (LO), Paul Zahl/National Geographic Creative; 247 (UP), Rob & Ann Simpson/Visuals Unlimited, Inc.; 247 (LO), DENNIS, DAVID M./Animals Animals/Earth Scenes; 248 (UP), Jason Patrick Ross/Shutterstock; 248 (LO), Ryan M. Bolton/Shutterstock; 249 (UP), Fero Bednar/Alamy; 249 (LO), Suzanne L. Collins/Science Source; 250 (UP), Derek Dafoe/National Geographic Your Shot; 250 (LO), Gary Meszaros/Science Source; 251 (UP), Thibaut Claeys/Alamy; 251 (LO), Michael Durham/Minden Pictures/National Geographic Creative; 252 (UP), Michael & Patricia Fogden/Minden Pictures/National Geographic Creative; 252 (LO), Pierson Hill; 253 (UP), Scott Leslie/Minden Pictures; 253 (LO), Jack Goldfarb Photography; 254 (UP), John & Barbara Gerlach/Getty Images; 254 (LO), Michelle Gilders/Alamy; 255 (UP), Ryan M. Bolton/Shutterstock; 255 (LO), REUTERS/Tim Wimborne; 256 (UP), Bill Hilton/Hilton Pond Center for Piedmont Natural History; 256 (LO), Rolf Nussbaumer Photography/Alamy; 257 (UP), Larry Michael/naturepl.com; 257 (LO), Steve Bower/Shutterstock; 258 (UP), Custom Life Science Images/Alamy; 258 (LO), Ryan M. Bolton/Shutterstock; 259 (UP), George Grall/National Geographic Creative; 259 (LO), ER Degginger/Science Source; 260, Masa Ushioda/www.coolwaterphoto.com; 263, Sue Daly/NPL/Minden Pictures; 264, Yva Momatiuk & John Eastcott/Minden Pictures; 266 (UP), Eric Engbretson/Engbretson Underwater Photo; 266 (LO), Gary Retherford/Science Source; 267 (UP), Alvin E. Staffan/Science Source; 267 (LO), Franco Banfi/Science Source; 268 (UP), Gary Meszaros/Science Source; 268 (LO), Alvin E. Staffan/Science Source; 269 (UP), Ted Kinsman/Science Source; 269 (LO), Mark Conlin/Getty Images; 270 (UP),

Scott-Berthoule/Science Source; 270 (LO), Picavet/Getty Images; 271 (UP), William H. Mullins/Science Source; 271 (LO), H. Berthoule/Science Source; 272 (UP), Paul Vecsei/Engbretson Underwater Photography; 272 (LO), Tim Fitzharris/Minden Pictures/Corbis; 273 (UP), George Grall/National Geographic/Getty Images; 273 (LO), Eric Engbretson/Engbretson Underwater Photo; 274 (UP), Tom McHugh/Science Source; 274 (LO), Gary Meszaros/Visuals Unlimited, Inc.; 275 (UP), blickwinkel/Alamy; 275 (LO), Mark Conlin/Alamy; 276 (UP), Gary Meszaros/Science Source; 276 (LO), Steve Maslowski/Science Source; 277 (UP), Ted Kinsman/Science Source; 277 (LO), Steve & Dave Maslowski/Science Source; 278 (UP), Alvin E. Staffan/Science Source; 278 (LO), William Thomas Cain/Getty Images; 279 (UP), Hellio & Van Ingen/Science Source; 279 (LO), Lynda Richardson/Corbis; 280 (UP), John Maraventano/Science Source; 280 (LO), Robert S. Michelson/Tom Stack & Associates; 281 (UP), Paul Nicklen/National Geographic Creative; 281 (LO), Brandon D. Cole/Corbis; 282 (UP), Paul Vecsei/Engbretson Underwater Photography; 282 (LO), Brandon Cole; 283 (UP), Peter Scoones/Science Source; 283 (LO), Dwight R. Kuhn; 284 (UP & LO), Andrew J. Martinez/Science Source; 285 (UP), Brandelet/Shutterstock; 285 (LO), Roger Munns, Scubazoo/Science Source; 286 (UP), Richard Herrmann/Minden Pictures; 286 (LO), Linda Pitkin/Science Source; 287 (UP), davidpstephens/Shutterstock; 287 (LO), Tom McHugh/Science Source; 288 (UP), Willyam Bradberry/Shutterstock; 288 (LO), Joyce Photographics/Science Source; 289 (UP), Andrew J. Martinez/Science Source; 289 (LO), Tom McHugh/Science Source; 290 (UP), holbox/Shutterstock; 290 (LO), Masa Ushioda/www.coolwaterphoto.com; 291 (UP), Tom McHugh/Science Source; 291 (LO), Fred Bavendam/Minden Pictures; 292 (UP), Naturfoto Frank Hecker; 292 (LO), Tosh Brown; 293 (UP), Norbert Wu/Minden Pictures; 293 (LO), Ingo Arndt/Minden Pictures; 294 (UP), D.P. Wilson/FLPA/Science Source; 294 (LO), Florian Graner/NPL/Minden Pictures; 295 (UP), Mark Boulton/Science Source; 295 (LO), Andrew J. Martinez/Science Source; 296 (UP), Juan Manuel Borrero/NPL/Minden Pictures; 296 (LO), Ken Lucas/Visuals Unlimited, Inc.; 297 (UP), Michael Patrick O'Neill/Science Source; 297 (LO), Cigdem Sean Cooper/Shutterstock; 298 (UP), Michael Patrick O'Neill/Science Source; 298 (LO), Lawrence Naylor/Science Source; 299 (UP), Jason Arnold/SeaPics.com; 299 (LO), Masa Ushioda/SeaPics.com; 300 (UP), Richard Herrmann/Minden Pictures; 300 (LO), Jones/Shimlock-Secret Sea Visions/Getty Images; 301 (UP), Steve Drogin/SeaPics.com; 301 (LO), Gilbert S. Grant/Science Source; 302 (UP), Mark Conlin/SeaPics.com; 302 (LO), Mike Howell; 303 (UP), Adrian E. Gray/sportfishimages.com; 303 (LO), Angelo Giampiccolo/NPL/Minden Pictures; 304 (UP), Fred Bavendam/Minden Pictures; 304 (LO), Ken Lucas/Visuals Unlimited, Inc.; 305 (UP), John Maraventano/Science Source; 305 (LO), Doug Perrine/NPL/Minden Pictures; 306 (UP), Alistair McGlashan/SeaPics.com; 306 (LO), Richard Herrmann/Minden Pictures; 307 (UP), D P Wilson/FLPA/Minden Pictures; 307 (LO), Jeff Rotman/NPL/Minden Pictures; 308 (UP), James D. Watt/SeaPics.com; 308 (LO), Andrew J. Martinez/Science Source; 309 (UP), Johnny Jensen; 309 (LO), Flip Nicklin/Minden Pictures; 310 (UP), Paul Nicklen/National Geographic Creative; 310 (LO), Flip Nicklin/Minden Pictures; 311 (UP), Masa Ushioda/SeaPics.com; 311 (LO), Brian Skerry/National Geographic Creative; 312 (LO), Jurgen Freund/Minden Pictures; 312 (UP), Randy Morse/SeaPics.com; 313 (UP), Brandon Cole; 313 (LO), Brian J. Skerry/National Geographic Creative; 314 (LO), Doug Perrine/SeaPics.com; 314 (UP), Paul Nicklen/National Geographic Creative; 315 (LO), Konrad Wothe/Minden Pictures; 315 (UP), Flip Nicklin/Minden Pictures; 316 (UP), Robin Chittenden/Minden Pictures; 316 (LO), Flip Nicklin/Minden Pictures; 317 (UP), Florian Graner/SeaPics.com; 317 (LO), Atsushi Sakurai/Nature Production/Minden Pictures; 318 (UP), Alex Mustard/NPL/Minden Pictures; 318 (LO), Charles V Angelo/Getty Images; 319 (UP), Jonathan Bird/SeaPics.com; 319 (LO), Georgette Douwma/NPL/Minden Pictures; 320 (UP), Ralf Kiefner/SeaPics.com; 320 (LO), Graham Eaton/NPL/Minden Pictures; 321 (UP), David Hall; 321 (LO), Michael Durham/Minden Pictures; 322 (UP), NaturePics/Alamy; 322 (LO), Kim Taylor/NPL/Minden Pictures; 323 (UP), Jim Young; 323 (LO), Andrew J. Martinez/Science Source; 324 (UP), Glenn Oliver/Visuals Unlimited, Inc.; 324 (LO), Jasper Doest/Minden Pictures; 325 (UP), Paul Bratescu; 325 (LO), Willard R. Culver/National Geographic/Getty Images; 326 (UP), Philippe Clement/NPL/Minden Pictures; 326 (LO), James Carmichael Jr./NHPA/Photoshot; 327 (UP), Francois Gohier/Science Source; 327 (LO), D P Wilson/FLPA/Minden Pictures; 328 (UP), D P Wilson/FLPA/Minden Pictures; 328 (LO), Doug Sokell/Getty Images; 329 (UP), Andrew J. Martinez/SeaPics.com; 329 (LO), Keoki Stender; 330 (UP), Les Wilk; 330 (LO), Andrew J. Martinez/Science Source; 331 (UP), WILDLIFE GmbH/Alamy; 331 (LO), Robert Zottoli; 332 (UP), Millard H. Sharp/Science Source; 332 (LO), Alex Mustard/NPL/Minden Pictures; 333 (UP), Andrew J. Martinez/SeaPics.com; 333 (LO), Vittorio Bruno/Shutterstock; 334 (UP), Stuart Wilson/Science Source; 334 (LO), Mauricio Handler/Getty Images; 335 (UP), Masa Ushioda/SeaPics.com; 335 (LO), Scott Leslie/Minden Pictures; 336 (UP), Alex Mustard/NPL/Minden Pictures; 336 (LO), Steven Kazlowski/SeaPics.com; 337 (UP), Ron Wolf; 337 (LO), Bob Evans/Photolibrary/Getty Images; 338 (UP), Ted Kinsman/Science Source; 338 (LO), Jon Corcoran; 339 (UP), Luciano Candisani/Minden Pictures; 339 (LO), D. R. Schrichte/SeaPics.com; 340 (UP), Gustav Paulay; 340 (LO), Gregory G. Dimijian, M.D./Science Source; 341 (UP), David Shale/NPL/Minden Pictures; 341 (LO), Andrew J. Martinez/Science Source; 342 (UP), Bill Coster/FLPA/Minden Pictures; 342 (LO), Brandon Cole; 343 (UP & LO), Fred Bavendam/Minden Pictures; 344 (UP), Alan Cressler; 344 (LO), Georgette Douwma/NPL/Minden Pictures; 345 (UP & LO), Andrew J. Martinez/Science Source; 346, papkin/Shutterstock; 349, Jim Richardson/National Geographic Creative; 352 (UP), Point-of-view/Alamy; 352 (LO), Lynette Schimming; 353 (UP), Marvin Dembinsky Photo Associates/Alamy; 353 (LO), J. L. Levy/Shutterstock; 354 (UP), George Grall/National Geographic Creative; 354 (LO), Danita Delimont/Alamy; 355 (UP), Nigel Cattlin/Science Source; 355 (LO), Neil Hardwick/Alamy; 356 (UP), Nigel Cattlin/Visuals Unlimited/Corbis; 356 (LO), Stuart Wilson/Science Source; 357 (UP), Scott Camazine/Science Source; 357 (LO), Judy Whitton/Shutterstock; 358 (UP), Simon Colmer/Getty Images; 358 (LO), Robert and Jean Pollock/Science Source; 359 (UP), Grant Heilman Photography/Alamy; 359 (LO), Rod Planck/Science Source; 360 (UP), Grant Heilman Photography/Alamy; 360 (LO), Premaphotos/Alamy; 361 (UP), Patrick Lynch/Alamy; 361 (LO), Ray Coleman/Science Source; 362 (UP), Darlyne A. Murawski/National Geographic Creative; 362 (LO), John T. Fowler/Alamy; 363 (UP), Arthur V. Evans; 363 (LO), Kallista Images/Getty Images; 364 (UP), Stephen P. Parker/Science Source; 364 (LO), lauriek/iStockphoto; 365 (UP), Lynette Schimming; 365 (LO), Kenneth H. Thomas/Science Source; 366 (UP), smuay/Shutterstock; 366 (LO), Matt Antonino/Shutterstock; 367 (UP), Charles W. Melton; 367 (LO), drsuth48/Shutterstock; 368 (UP), Arthur V. Evans; 368 (LO), Joseph Berger, Bugwood.org; 369 (UP), George Grall/National Geographic Creative; 369 (LO), Rolf Nussbaumer Photography/Alamy; 370 (UP), Jim Zipp/Science Source; 370 (LO), Arthur V. Evans; 371 (UP), Arto Hakola/Shutterstock; 371 (LO), Ryan Kaldari; 372 (UP), Phil Degginger/Alamy; 372 (LO), Dr. Morley Read/Shutterstock; 373 (UP), Premaphotos/Alamy; 373 (LO), Iurochkin Alexandr/Shutterstock; 374 (UP), Solodov Alexey/Shutterstock; 374 (LO), Premaphotos/Alamy; 375 (UP), Tyler Fox/Shutterstock; 375 (LO), Arthur V. Evans; 376 (UP), John M. Coffman/Science Source; 376 (LO), Joseph Berger, Bugwood.org; 377 (UP), Mark Wetmore; 377 (LO), Premaphotos/Alamy; 378 (UP), Jorge Casais/Shutterstock; 378 (LO), Scott Camazine/Science Source; 379 (UP), George Grall/National Geographic Creative; 379 (LO), Arthur V. Evans; 380 (UP), Guy Bruyea; 380 (LO), Susan Andrews/Alamy; 381 (UP), Visuals Unlimited, Inc./John Abbott via Getty Images; 381 (LO), Arthur V. Evans; 382 (UP), Gerry Bishop/Visuals Unlimited/Getty Images; 382 (LO), Kenneth M. Highfill/Science Source; 383 (UP), Arthur V. Evans; 383 (LO), Brian Lasenby/Shutterstock; 384 (UP), Danita Delimont/Alamy; 384 (LO), Gary McDonald; 385 (UP), Scott Camazine/Science Source; 385 (LO), Rod Planck/Science Source; 386 (UP), Daniel L. Geiger/SNAP/Alamy; 386 (LO), Robert Shantz/Alamy; 387 (UP), Arthur V. Evans; 387 (LO), Stuart Wilson/Science Source; 388 (UP), Clemson University-USDA Cooperative Extension Slide Series, Bugwood.org; 388 (LO), Dygiclick/Alamy; 389 (UP), Betsy Betros; 389 (LO), WebSubstance/iStockphoto; 390 (UP), Tom Murray; 390 (LO), anderm/Shutterstock; 391 (UP), Ron Rowan Photography/Shutterstock; 391 (LO), Bob Beatson; 392 (UP), Stephen Dalton/Science Source; 392 (LO), Bo Zaremba; 393 (UP), WebSubstance/iStockphoto; 393 (LO), Lynette Schimming; 394 (UP), Whitney Cranshaw, Colorado State University, Bugwood.org; 394 (LO), Barry Blackburn/Shutterstock; 395 (UP), Kallista Images/Getty Images; 395 (LO), Nigel Cattlin/Science Source; 396 (UP), nrpphoto/Shutterstock; 396 (LO), Tim Gainey/Alamy; 397 (UP), Kurt Hennige; 397 (LO), Sari ONeal/Shutterstock; 398 (UP), Bruce Raynor/Shutterstock; 398 (LO), Doug Lemke/Shutterstock; 399 (UP), Doug Lemke/Shutterstock; 399 (LO), Sari ONeal/Shutterstock; 400 (UP), John Serrao/Science Source; 400 (LO), Sari ONeal/Shutterstock; 401 (UP), Sari ONeal/Shutterstock; 401 (LO), Rasmus Holmboe Dahl/Shutterstock; 402 (UP), AtWaG/iStockphoto; 402 (LO), Leena Robinson/Shutterstock; 403 (UP), Doug Lemke/Shutterstock; 403 (LO), KellyNelson/Shutterstock; 404 (UP), Robert Shantz/Alamy; 404 (LO), Sari ONeal/Shutterstock; 405 (UP), J. L. Levy/Shutterstock; 405 (LO), Malcolm Schuyl/Alamy; 406 (UP), Marc Parsons/Shutterstock; 406 (LO), Millard H. Sharp/Science Source; 407 (UP), Dean Pennala/Shutterstock; 407 (LO), James Laurie/Shutterstock; 408 (UP), Leroy Simon/Visuals Unlimited/Corbis; 408 (LO), Nature's Images, Inc./Science Source; 409 (UP), Whitney Cranshaw, Colorado State University, Bugwood.org; 409 (LO), Sari ONeal/Shutterstock; 410 (UP), Jeremy Martin; 410 (LO), Elliotte Rusty Harold/Shutterstock; 411 (UP), Rolf Nussbaumer/imagebroker/Corbis; 411 (LO), Arto Hakola/Shutterstock; 412 (UP), Elliotte Rusty Harold/Shutterstock; 412 (LO), John Kaprielian/Science Source; 413 (UP), Scott Camazine/Alamy; 413 (LO), Gerald & Buff Corsi/Visuals Unlimited/Corbis; 414 (UP), Cristian Gusa/Shutterstock; 414 (LO), Whitney Cranshaw, Colorado State University, Bugwood.org; 415 (UP), Elliotte Rusty Harold/Shutterstock; 415 (LO), Bob Beatson; 416 (UP), Alex Wild; 416 (LO), Patrick Lynch/Alamy; 417 (UP), Alex Wild/Visuals Unlimited, Inc.; 417 (LO), DLILLC/Corbis; 418 (UP), Suzanne L. Collins/Science Source; 418 (LO), Birute Vijeikiene/Shutterstock; 419 (UP), Bo Zaremba; 419 (LO), Snowleopard1/iStockphoto; 420 (UP), Jason Patrick Ross/Shutterstock; 420 (LO), Ron Rowan Photography/Shutterstock; 421 (UP), Bill Frische/Shutterstock; 421 (LO), Gregory G. Dimijian/Science Source; 422 (UP), Jason Patrick Ross/Shutterstock; 422 (LO), Kathy Clark/Shutterstock; 423 (UP), AtWaG/iStockphoto; 423 (LO), Steven Russell Smith Photos/Shutterstock; 424 (UP), Derek R. Audette/Shutterstock; 424 (LO), Audrey Snider-Bell/Shutterstock; 425 (UP), Daniel L. Geiger/Alamy; 425 (LO), Robert Pickett/Visuals Unlimited/Corbis; 426 (UP), David Scharf/Getty Images; 426 (LO), Cristina Lichti/Alamy; 427 (UP), Science Source; 427 (LO), South12th Photography/Shutterstock; 428 (UP), Science Source; 428 (LO), Rene Krekels/FN/Minden Pictures/National Geographic Creative; 429 (UP), George Grall/National Geographic Creative; 429 (LO), Scott Camazine/Science Source; 430 (UP), Jason Patrick Ross/Shutterstock; 430 (LO), Art Directors & TRIP/Alamy; 431 (UP), Mitsuhiko Imamori/Minden Pictures/National Geographic Creative; 431 (LO), Hugh Lansdown/Shutterstock.

Index

Boldface indicates illustrations.